KU-330-554

SOLAR-TERRESTRIAL INFLUENCES ON WEATHER AND CLIMATE

SOLAR-TERRESTRIAL INFLUENCES ON WEATHER AND CLIMATE

Proceedings of a Symposium/Workshop held at the Fawcett Center for Tomorrow, The Ohio State University, Columbus, Ohio, 24-28 August, 1978

Edited by

BILLY M. McCORMAC
Lockheed Palo Alto Research Laboratory

THOMAS A. SELIGA
Atmospheric Sciences Program, The Ohio State University

D. REIDEL PUBLISHING COMPANY

DORDRECHT : HOLLAND / BOSTON : U.S.A.
LONDON : ENGLAND

Library of Congress Cataloging in Publication Data

Symposium/Workshop on Solar-Terrestrial Influences on Weather and Climate, Ohio State University, Columbus, 1978.
Solar-terrestrial influences.

Includes bibliographical references and index.
1. Solar activity–Congresses. 2. Atmosphere, Upper–Congresses. 3. Weather–Congresses. 4. Climatology–Congresses. I. McCormac, Billy Murray. II. Seliga, T. A. III. Title.
QC883.2.S6S94 1978 551.6′9 79–11425
ISBN 90–277–0978–5

Published by D. Reidel Publishing Company,
P.O. Box 17, Dordrecht, Holland

Sold and distributed in the U.S.A., Canada and Mexico
by D. Reidel Publishing Company, Inc.
Lincoln Building, 160 Old Derby Street, Hingham,
Mass. 02043, U.S.A.

Printed in The Netherlands

PARTICIPANTS

Row 1 (from L to R): S. K. Bringi, H. W. Kroehl, M. Briskin, J. J. Singh, J. W. King, S. T. Wu, B. M. McCormac, T. A. Seliga, and R. A. Goldberg

Row 2: B. E. Quate, R. N. Singh, S. S. Prasad, H. W. Ellsaesser, J. T. Prohaska, R. K. Seals, Jr., J. G. McKinley, D. G. McFarland, R. S. Davis, J. H. Shirley, and H. T. Morth

Row 3: H. P. Sleeper, N. J. F. Chang, R. J. Hung, C. G. Maclennan, E. W. Hones, Jr., M. Walt, R. Reiter, J. B. Smith, Jr., A. Lachapelle, H. H. Sargent, and S. Woronko

Row 4: J. W. MacArthur, S. Ramakrishna, L. K. Acheson, N. C. Maynard, R. S. Stolarski, C. A. Lundquist, N. R. Sheeley, Jr., J. C. Neill, E. G. Bowen, R. D. Hill, M. D. Lethbridge, and B. Jamison

Row 5: D. D. Bramch, J. Namias, M. Mendillo, D. F. Heath, S. Avery, R. D. Hake, D. Rote, R. A. Koehler, G. D. Nastrom, A. J. Weinheimer, S. Gathman, M. A. Bilello, M. M. Mitchell, and J. A. Eddy

Row 6: J. A. Carlson, N. W. Spencer, R. E. Hartle, C. M. Brown, W. E. Mitchell, Jr., A. B. Pittock, T. J. Sullivan, J. H. Binsack, W. C. Livingston, D. Hoyt, J. Keller, P. Haines, K. K. Mani, and R. Kakar.

Row 7: P. Waite, H. T. Suzuka, F. L. Bartman, S. Fritz, B. C. Parker, G. W. Brier, C. F. Sechrist, C. J. E. Schuurmans, D. N. Yarger, J. P. Basart, G. F. Shaw, J. T. Lutz, J. C. Jafolla, F. B. House, and W. C. Ryan

Row 8: R. Markson, M. D. Rodriguez, J. N. Lax, E. G. Kindle, P. M. Kelly, H. F. Diaz, M. F. Larsen, and C. A. Levis

TABLE OF CONTENTS

PREFACE

This book contains most of the invited papers and contributions presented at the Symposium/Workshop on Solar-Terrestrial Influences on Weather and Climate which was held at The Ohio State University on 24-28 July 1978.

The authors and publisher have made a special effort for rapid publication. The length of the individual papers in this book were deliberately limited by the editors.

Direct financial support for the Symposium/Workshop was provided by NASA.

Palo Alto — Billy M. McCormac

Columbus — Thomas A. Seliga

January 1979

SYMPOSIUM/WORKSHOP CONCLUSIONS

Billy M. McCormac
Department 52-10/B202
Lockheed Palo Alto Research Laboratory
3251 Hanover Street
Palo Alto, CA 94304 USA

Thomas A. Seliga
Atmospheric Sciences Program
The Ohio State University
2015 Neil Avenue
Columbus, OH 43210 USA

A. INTRODUCTION

The Symposium/Workshop on Solar-Terrestrial Influences on Weather and Climate was held at The Ohio State University on 24-28 July 1978. Its purpose was to provide an international forum for the presentation and discussion of recent research results and ideas regarding the question whether variations in solar outputs affect terrestrial weather and climate and, if so, to what extent and through what mechanisms. The Symposium focused on the results of previous studies and consisted of both invited and contributed papers. The Workshop, on the other hand, built upon these deliberations to develop ideas and directions for future research.

Over one hundred persons from eight countries attended the Symposium/Workshop. They represented research institutes, universities, private industry and federal agencies, and had professional interests ranging from practicing meteorologists to solar physicists. Considerable time was devoted to discussion; this produced important interdisciplinary interactions and greatly enhanced the scientific benefits for all participants.

1. MAJOR SCIENTIFIC ISSUES

Whether solar variability influences weather and climate is a fundamental scientific question, answers to which may have important practical implications for long-term weather and climate prediction. Unfortunately, we are not yet in a position to answer this question although evidence for significant relationships between solar variability and tropospheric

B.M. McCormac and T.A. Seliga (Eds.): Solar-Terrestrial Influences on Weather and Climate. 1-27.

phenomena has increased significantly over the last decade. Clearly, the energetics of solar variability are too small to directly affect tropospheric or stratospheric processes significantly. Therefore, it appears that, if the correlations between solar variability and lower atmospheric effects are indicative of actual physical linkages, the causative mechanism(s) would most likely be subtle, for example, trigger effects associated with changes in O_3 concentrations in the stratosphere and modifications of the atmosphere's electrical characteristics.

Sun-weather relationships may extend from days to thousands of years. The Milankovitch theory of climatic variability due to changes in solar insolation resulting from variations in the Earth's orbit over extended time periods is one generally accepted example of how long period climatic changes occur. Other much shorter, but socially significant, cycles effects seem to involve drought, thunderstorm activity, atmospheric vorticity, stratospheric-tropospheric exchange processes, global pressure pattern variations and atmospheric kinetic energy. In particular, 22-yr periodicity in areal drought over the western United States is very well correlated with the Hale sunspot cycle which governs the orientation of the interplanetary magnetic field. It is very interesting that the other most striking evidence for solar influences, albeit at time scales of days, involves the correlation between solar magnetic sector crossings at the Earth's orbit with atmospheric vorticity changes as measured by a vorticity area index (VAI). Perhaps these discoveries are telling us something about the role of the interplanetary magnetic field?

Other potential solar influences include the 11-yr sunspot cycle, the 27-day solar rotation and special solar events such as flares and coronal holes. There is a growing body of evidence that these events may influence weather either cyclically or, in the case of special events, on time scales of the order of days.

Practical use of solar variability as a tool for weather and climatic forecasting, other than through empirical approaches, remains to be developed. Nevertheless, some evidence suggests that the weather forecasts may depend upon solar events. Should this relationship prove to be correct, then indeed the study of solar-weather and climate relationships would have proven its worth.

2. WORKSHOP ACTIVITIES

The above issues dominated the activities of the Symposium and led to the Workshop deliberations which concluded the meeting. These sessions were organized to provide future directions for research into solar-weather and climate relationships. To do this, the attendees were invited to select one of six different groups to participate in the Thursday afternoon session. They were concerned with the following topics: 1. correlation studies; 2. solar influences on global circulation and climate models; 3. lower and upper atmospheric coupling, including electricity; 4. planetary motions and other indirect factors;

5. experimental approaches to sun-weather relationships; and 6. role of minor atmospheric constituents. The last session of the Symposium on Friday morning was devoted to presentations of the Workshop deliberations by the chairman of each working group followed by a general discussion. The major findings of each group are outlined in following sections. In addition, the Workshop deliberations prompted the assembly to introduce two related resolutions for dissemination to appropriate scientific organizations throughout the world. The complete text of these resolutions follows the Workshop findings.

B. WORKSHOP REPORTS

1. CORRELATION STUDIES

Chairman: J. Murray Mitchell, Jr., NOAA Environmental Data Service, Silver Spring, Maryland

Two guiding principles form the basis for consideration of correlation studies:

(1) The focus should be on how to approach problems, how to elucidate phenomena and cause-effect mechanisms, and how to test hypotheses, rather than on what needs to be done in the area of meteorological and climatological correlation studies.

(2) There appears to be essentially three categories of phenomena and time-scales that deserve intensive study:

(a) Solar-weather effects on the scale of days and weeks, such as those related to the Vorticity Area Index (VAI);
(b) Solar-climate effects on the scale of years and decades (e.g., the sunspot and Hale magnetic solar cycles), such as those involving drought, blocking, stratospheric warming events, and other large-scale atmospheric circulation effects; and
(c) Long-term climatic phenomena (e.g., the Little Ice Age) in relation to the "Maunder Minimum' class of anomalous solar behavior, on the scale of centuries and millennia.

Although long-term phenomena of the kind referred to in (c) were not explicitly dealt with in this conference, they are recognized as a major research area and should be included as a major topic in future conferences.

There are three main areas of concern regarding correlation studies: selection of data, considerations bearing on the conduct of statistical data analysis, and special considerations. Problems needing emphasis and perceptions on opportunities for the future are also examined.

1.1 Selection of Data

The choice of data to be applied in statistical correlation studies should be based on the following considerations:

(1) *Source* of data. First-hand, authoritative sources are usually the most reliable and the most likely to include all the relevant annotations to indicate the credibility and limitations of the data.

(2) *Quality* and *homogeneity* of data, especially in the case of data extending over many years, continuity of observing procedure and uniformity of observing conditions should not be taken for granted and should be verified whenever possible. Many long-term meteorological series suffer from the effects of station relocations, observing procedure changes and/or gradually changing biases due to nearby urbanization or other local environmental effects. Adequate annotations to data in such respects are not always provided along with the data themselves.

(3) *Representativeness* of data in both space and time. An example is precipitation data for an isolated station, especially in a semi-arid region. This may be an extremely "noisy", and often biased, estimator of precipitation over a region surrounding the station. A multi-station average of precipitation in a region might be a better choice of data to use, depending on the nature of the problem being addressed. Another example is thunderstorm frequency data based on cooperative station reports in the United States. Instructions to observers on the reporting of thunderstorms are sufficiently vague to leave much latitude for subjective judgment as to whether a storm sighted in the distance (for example) is to be reported as a thunderstorm in the same way as a thunderstorm directly overhead. The total incidence rate of thunderstorms, therefore, can (and does) depend on an individual observer's interpretation of his instructions for reporting thunderstorms.

(4) *Physical relevance* of the data. Many forms of data, whether pertaining to the Sun or to atmospheric events, are measures of symptoms (or correlates) of physical processes and not of the physical processes themselves. Questions arise as to the efficacy of such data in the elucidation of physical links between the Sun and weather or climate, in lieu of other, more fundamental data.

In all of the above respects, the selection of data for solar-weather or solar-climate correlation studies calls for resident competence and expertise in climatology and solar physics to assure the wisest possible selection and evaluation of the data, and to avoid later difficulties in interpreting the results of such studies at the beginning of the analysis.

1.2 Conduct of Statistical Data Analysis

A number of considerations bearing on the conduct of statistical studies of solar-weather and climate relationships are important for future research.

(1) It is important to know and to understand the statistical characteristics of the data, in both space and time, before undertaking any analysis involving the data.

(2) It is important not to regard statistical techniques as a magical "black box", drawn uncritically from a textbook without an adequate understanding of the conditions, if any, to be met by the data or by the design of the analysis to assure the validity of the techniques.

(3) It is important to tailor the choice of statistical analysis techniques to the peculiar nature of the data in relation to the objective of the analysis. Special problems arise when:

(a) a continuous process in one medium is to be correlated with impulsive events in another medium;
(b) multiple events occur in one medium that is being correlated with another, or events recur in rapid sequence; and
(c) confounding may occur, as, for example, confusion between a 27-day solar rotation effect and lunar tidal disturbance effects with periods in the 27 to 29-day range.

(4) In cases where either peculiarities in the statistical properties of the data or complexities in the experimental design raise doubts concerning the applicability of standard tests of the statistical significance of the results, one should not hesitate to determine statistical significance by employing empirical "nonsense" experiments based on monte carlo techniques to simulate conditions consistent with the null hypothesis. Modern computer capabilities make this approach to significance testing not only feasible but often economical.

(5) It is important to be aware of the hazards of *post hoc* hypothesis testing. In cases where the results of a correlation analysis are the sole basis of their own interpretation, with no *a priori* hypothesis in mind against which to judge them, there is always a risk that the results may be prejudiced by the behavior of the sample of data and lack any real meaning. When *post hoc* hypotheses are embellished in an effort to account for a number of details of the analytical results along with the basic result originally sought, one can soon exhaust all the available degrees of freedom in the data and end up with a perfect fit between hypothesis and data that means absolutely nothing.

(6) A singularly powerful measure of the validity of the results of a correlation analysis is the reproducibility of those results in an independent set of data. By "independent set of data" one usually means data collected in a different period of time, not in the same period of time at a different location. In most cases, an atmospheric parameter measured at one location tends to be highly correlated with the same parameter, at the same time, at another location even at intercontinental distances away. In a situation where additional data in another period of time are not available, then reproducibility of results should be investigated in different subsets of the original data set.

In all of the above respects, the proper design, conduct, and interpretation of any statistical correlation analysis calls for *resident* competence and expertise in mathematical statistics to assure the highest possible discriminative power of the experimental design and the most realistic assessment of the statistical significance of the results.

1.3 Special Considerations

A number of questions are highly relevant to correlation studies. These special questions include:

(1) *What aspects of weather and climate are most promising or most meaningful to examine?* Clearly, the problem of solar-weather and solar-climate relationships cannot be approached by considering the relationships of everything with everything else, for all time. In general, however, attention to two- and three-dimensional atmospheric descriptors deserves special emphasis, such as:

(a) dynamic structure of the low wave number (long wavelength) components of atmospheric circulation;
(b) energetics of atmospheric processes and their rates of change with time; and
(c) large-scale climatic patterns and features, such as blocking phenomena and stratospheric warming events.

Regarding the extension of various studies already in progress, the following should be made available:

(a) daily trough-ridge positions and global-scale indices of circulation, particularly at the 500 and 300 mb levels;
(b) daily ionospheric potential measurements;
(c) daily values of satellite-derived data on terrestrial IR flux, cloud cover, lightning discharges, ozone distribution, etc.; and
(d) daily verification scores for weather forecasts and upper-level prognostic charts from NOAA/NMC and elsewhere.

(2) *What is the transferrability of short-term solar-weather relationships to long-term solar-climate relationships?* With due consideration of their character, are short-term relationships likely to aggregate into significant long-term relationships, or not? How do short-term relationships depend on the "seasons" of longer-term solar cycles? Might the sun be capable of doing things fundamentally differently on long time scales than what it does on short time scales? Answers to such questions would have much to say as to whether solar-terrestrial effects on various time scales, from days to years to centuries, should be explored from a single unified conceptual framework or from different and essentially independent frameworks.

(3) *How does one explain the enigma of the apparent efficacy of 22-yr magnetic solar cycle effects, in contrast to the apparent inefficacy of 11-yr solar cycle effects, on the lower atmosphere and on climate?* Historically, most searches for 11-yr solar-climate relationships have yielded confusing and often contradictory results. On the other hand,

some new studies of 22-yr solar-climate relationships appear to have yielded strong and rather consistent results. Are there important 11-yr effects on climate, paralleling the strong 11-yr solar control on conditions in the very high atmospheric levels? Is it really solar magnetics that affect climate, or is the Sun capable of changing its irradiance on the time scale of 22 yr in a manner that has no parallel on the time scale of 11 yr? These are very difficult questions and it is suggested that the situation calls for designing such studies so as to stratify climatic data by phase of the 22-yr magnetic cycle, but with sufficient flexibility to recognize an 11-yr overtone in the data if one exists. By this procedure, one can readily collapse the results of an analysis into an 11-yr framework if that is called for, and one would not have to re-initiate a search for 22-yr effects.

(4) *What about the distinction between statistical significance and practical significance of the results of correlation studies?* If a result is statistically significant, it may or may not be practically significant as a weather or climate forecast aid. Many apparent solar-weather or solar-climate relationships appear to achieve credible levels of statistical significance but lack practical signifiance. In such cases, it is worth stressing that statistical significance may be sufficient to illuminate physical processes and cause-effect mechanisms, whether or not they can be used as a forecast aid. In any event, the two kinds of significance deserve to be carefully distinguished.

1.4 Problems Needing Emphasis and Opportunities for the Future

Four areas of problems and opportunities were identified as follows:

(1) *The need for continuing availability of basic solar and climate-related data on a timely basis.*

(a) There is an urgent need that certain solar indices which are widely used in studies of solar-weather and solar-climate relationships be continued (See Resolution II - On Solar-Geophysical Data Sets).

(b) Basic satellite data and atmospheric data (including operational NMC history tapes) should be preserved as a valuable resource for solar-weather and solar-climate investigations.

(c) NOAA and other agencies providing routine solar and atmospheric data should institute procedures to permit more rapid availability of such data, where possible, for real-time applications to solar-weather and solar-climate studies.

(2) *Opportunities to apply new satellite-based data and other novel data sets.* The present or pending availability of global-scale information on a number of solar and atmospheric parameters of potentially great value in solar-weather and solar-climate research is apparent and includes:

(a) incidence of lightning discharges;
(b) ionospheric potential;

(c) global-scale indices of cloudiness; and
(d) radiation fluxes (solar irradiance and global-scale terrestrial).

It is important that communications channels between scientists and NOAA and other sources of solar and atmospheric data remain open and timely in order to ensure that data processing and archiving be done to facilitate future solar-weather and solar-climate research activities.

(3) *Opportunities for expansion of solar and climatic data base into earlier centuries and millennia.* Burgeoning interest and capabilities for reconstructing past solar and climatic behavior by means of "proxy" techniques open up unprecedented opportunities for gaining a broader perspective on present-day behavior and for detecting possible long-term solar-climate relationships. These developments are important and full exploitation of such opportunities is encouraged.

(4) *The importance of case studies.* Because of the great complexity of both solar and atmospheric phenomena, elucidation of solar-weather linkages requires very careful and thorough study of these phenomena and their temporal relationships, in selected intervals of time immediately preceding (if possible) and following key solar events. Ideally, these selected intervals should be designated on a real-time basis, with provisions for making special observations (by means of radiosondes, rockets, etc.) to obtain detailed information on:

(a) tropopause motions and fine structure;
(b) vertical distribution of ozone;
(c) vertical distribution of stratospheric particles (using LIDAR);
(d) high resolution vertical electric potential gradient;
(e) vertical conductivity profile; and
(f) ionization rate parallel with gas reactions in the stratosphere.

With the accumulation of detailed measurements in a series of selected intervals, and their evaluation in the form of a series of case studies, it should be possible to sort out those phenomena on both the solar side and the atmospheric side that are possibly involved in cause-effect chains, and to distinguish these from all other phenomena on the basis of inappropriate timing relationships.

2. SOLAR INFLUENCES ON GLOBAL CIRCULATION AND CLIMATE MODELS

Chairman: C. J. E. Schuurmans, Royal Netherlands Meteorological Institute

Atmospheric modeling offers one approach to the study of prospective physical mechanisms linking solar variability with weather and climate. Furthermore, a hierarchy of such models exist or are under development. Therefore, it is essential that certain considerations be accounted for if the potential advantages of modeling are to accrue to research on solar-weather and climate relationships.

2.1 Potential Physical Mechanisms

The following is a list of physical mechanisms which are potential candidates for explaining solar effects on weather and climate; they are also candidates for inclusion in models:

(1) solar induced changes in O_3 content and distribution;
(2) solar constant variations, either regular or irregular;
(3) solar UV variations;
(4) changes in the thermal or dynamic structure of the upper atmosphere; and
(5) changes in the atmospheric electric field.

2.2 Choice of Models

There is a kind of hierarchy of atmospheric models ranging from global circulation models (GCM) and climate models on down to components of the system such as hurricane models, convective cloud models, etc. It appears that GCMs are not very useful for solar-weather and climate studies at present. They are too expensive to run and when the mechanism is only poorly known, it is usually difficult to insert it into such complicated models and expect to obtain any meaningful results. More often, one should start with pilot studies using very simple energy balance models. For instance, either the density or the vertical distribution of O_3 can be investigated, or sensitivity to the solar constant can be studied with models that are averaged over height and longitude.

2.3 Potential Experiments

There are four specific suggestions for possible modeling experiments. These are sensitivity experiments in which the amplitude of an input parameter is changed and the output is examined for effects. Although the relative values of the amplitude changes may be in error in such computer experiments, the importance of the changes may be discerned.

(1) Convective Storms. The first experiment suggested relates to changes in the background electric field in clouds. A convective cloud model could be used to see if and how clouds are influenced by changes in the atmospheric electric field. Existing cloud models should be modified or new ones developed to take into account this electric field.

(2) O_3 Distribution at High Latitudes. The second experiment is to change the amount and/or vertical distribution of O_3 in the auroral zone only. This could be done in any three-dimensional model which includes O_3 and its radiation and chemistry.

(3) UV Flux. The third experiment is to change the ultraviolet (UV) flux. Such an experiment could be done in a simple energy balance model which computes the radiative equilibrium states and includes photochemistry of the stratosphere. Such a model could calculate a realistic

equilibrium in the troposphere by inclusion of some convective adjustment schemes. Any promising results might then lead to experiments that change the UV flux in more complicated models employing a three-dimensional GCM.

(4) Weather Prediction. The fourth experiment is to attempt an indirect assessment of solar effects by comparing numerical weather predictions of the flow field for key days with the observed fields.

Such studies should compare the accuracy or deviations for various wave-number components as well as providing correlation indices between experimental outputs and observations.

3. PLANETARY MOTION-SUNSPOT-CLIMATE

Chairman: H. Prescott Sleeper, Kentron International, Huntsville, Alabama

There has been speculation for many years about the possible effects of planetary motion on sunspot activity and the Earth's weather and climate. In particular, it is essential to explore possible physical relationships between planetary motions, sunspot activity, and the Earth's weather and climate. The stage of development of these ideas is perhaps comparable with that of solar-weather relationships 10 to 15 yr ago. Two considerations make it timely to study this area in more detail: the current status of the study of solar-weather and solar-climate relationships; and the potential for developing deterministic models for changes in weather and climate.

3.1 Principal Areas of Study

The subject matter appropriate to this field of study falls generally into these categories:

(1) Mutual interactions among the principal astronomical bodies in the solar system, as it has evolved into a nonlinear phase locked system.

(2) The effects of changes in position or motion of the planets on the sun's dynamics, and the resultant consequences for sunspot production and flare activity.

(3) The effects of changes in solar activity on the solar wind and the planetary atmospheres.

(4) The effects of changes in position or motion of all the planets on the dynamics of any single planet.

3.2 Recommendations

If weather and climate are affected by solar activity which in turn is affected by the position and motion of the planets, then models to understand these synergistic processes should be developed. A complete understanding of these processes may permit better prediction of weather and climate and ultimately to new methods of climate control. There needs to be improved information transfer between the various workers on this subject and with the scientific community in general.

It is important to develop diagnostic and prognostic models for changes in solar system parameters which could affect solar activity or planetary weather and climate. These models cover the range from the effects of solar system evolution on the orbital and axial parameters of the Earth and their effects on climate (i.e., the Milankovitch model) or the Earth's weather. In particular, they cover the interaction of planets with sunspot and flare activity and the interaction of the moon on the Earth's weather.

Several models do exist. The Mörth-Schlamminger model postulates exchange of planetary angular momentum with the rotation of the Sun due to interaction between planets. This model should be simplified to a 3-body problem, and the dynamics worked out in more detail.

The Brier Physical-Statistical model is an empirical model emphasizing rigorous statistical tests. In the Sleeper-Jose planetary resonance model it was suggested that both the sunspot number time series and the climate have non-stationary characteristics, and that the change epochs tend to occur at certain planetary phase relationships. Due to the current coincidence of a number of planetary resonances, a change in sunspot and climate statistics may be due. Somewhat similar changes took place about 1800 and 1880. Important climate regime changes also occurred in 1923 and 1961. This particular model also suggests that sunspot cycle 21 will have a low magnitude in contrast with other current sunspot models. Because some of the models suggest that changes of climate regime are occurring or will occur in the very near future, there is a sense of urgency in attempting to develop and improve these models. Comparative tests of predictions based on this resonance model and other models with actual solar activity and the Earth's climate should be performed and scrutinized carefully.

4. LOWER AND UPPER ATMOSPHERE COUPLING AND ELECTRICAL COUPLING PHENOMENA

Chairman: R. N. Singh, Institute of Technoogy, Baranas Hindu University, Varanasi, India

There are several potential lower and upper atmospheric coupling mechanisms, including electrical coupling which may be important to Sun-weather and climate relationships. In this regard it is convenient to divide the atmosphere into two regions, namely, the weather region from 0 to 20 km and the upper atmospheric region from 20 km and above, including the magnetosphere. There is an obvious need for detailed study of these two regions with special reference to long-term and short-term phenoemena.

4.1 The Solar Constant and Solar Emissivity

The literature contains repeated statements by various reporters that the solar constant is almost constant. While this may be a fact on extended time scale and integrated emission basis, the need for data on the variation of the solar emissivity with frequency is essential. On a short-term scale the sun is known to emit enhanced EUV, impulsive hard X-rays and energetic particle streams. The emission of these enhanced radiations from specific areas of the solar disc would produce short-term effects on the Earth's atmosphere. Furthermore, the frequency and intensity of radiation would selectively affect the stratified Earth's atmosphere.

Coronal holes have been recognized to be a source of high velocity solar wind particles. These solar wind particles are known to interact with the geomagnetic field and compress the plasmasphere. The plasma stream particles enter the dayside of the atmosphere through the polar region and the nightside of the atmosphere after being accelerated through the neutral point in the plasma sheet region. These particles carry sufficient energy to initiate magnetic substorms having large energy. The key energy particle flux gives rise to aurora and changes the overall conductivity of the high latitude layer. The higher energy particles penetrate and give rise to atmospheric bremsstrahlung which penetrates to much lower heights and changes the conductivity of the region.

The solar EUV, X-rays, plasma streams, and interplanetary magnetic field play very important roles in disturbing the Earth's atmosphere, otherwise conforming to a steady state solution of hydrodynamic equation. The disturbed upper atmosphere is known to generate various wave modes, production of auroras, Joule heating, and electromagnetic induction in the Earth's surface. Gravity waves are one of the important wave modes and they transport energy from the lower region in the upward direction. It was realized that the reflection of these waves from the upper atmospheric layer and their subsequent attenuation in the lower atmosphere may play an important role.

The role of the O_3 layer and its response to solar UV radiations and cosmic ray flux is thought to play a very important role in controlling the energy balance of the weather region from 0 to 20 km and may provide a necessary triggering mechanism. A laboratory scale experiment on O_3 layer response to EUV, X-ray, energetic particles, and selected impurities should provide valuable information to establish O_3's role.

Mechanisms coupling the weather region with the upper atmosphere and magnetosphere are still in a primitive and indecisive state of development and are oftentimes ignored in the light of the near constancy of the solar constant. However, thunderstorm activity and atmospheric electricity are areas where the "needed" energy to control Sun-weather relationships may be adequate, since they do not critically depend on a change in the solar constant. It is interesting to note that thunderstorm energy in the lower region is of the same order of magnitude as that of substorm energy in the ionospheric region, i.e., 10^{10} to 10^{11} W. Probably the most important question is how thunderstorm as a controlling energy source for short-time effects is governed by the modification of electrical conductivity of the entire ionospheric region. The state of knowledge in this area, although limited, is much greater than in other competing coupling mechanisms, and stands on a much better scientific foundation. Nevertheless, detailed investigations are needed to establish the role of short-term solar ionizing electromagnetic radiations, energetic solar charged particle streams, and galactic radiation. In addition, the conductivity of the lower and upper atmosphere may play a very important role and should be monitored using extended ground-based and airborne sensors.

Another candidate for coupling the energy from the upper atmosphere to the lower atmosphere is the potential role of certain condensation nuclei. In particular, cosmic ray nucleation of supercooled water in cloud tops may play a very important role in coupling the energy from cosmic rays to the lower ionosphere.

4.2 Recommendations

In light of the above, a number of areas for future research have been identified. These resulted in the following recommendations:

(1) Long-term and short-term solar effects should be identified and their energy content and spectral features made available to workers in Sun-weather research.

(2) Solar induced changes in the upper atmospheric and magnetospheric conductivity by electromagnetic radiation and energetic particles should be fully investigated as they may relate to changes in the electrical generators in the weather region between 0 to 20 km.

(3) Longitudinal and meridional chains of ground-based stations should operate in conjunction with satellite sensors to measure the electrical characteristics with high spatial and temporal resolution.

(4) Cosmic ray modulation by interplanetary magnetic fields should be studied to determine their influence on the temporal and spatial variations of the ionospheric conductivity.

(5) Synoptic studies of electrical conductivity, ion production rate, and ionospheric electrical potential in the region of 0 to 25 km with a minimum of 1 measurement per day should be performed with a provision for measuring more frequently as necessary for specific cases.

(6) Possible cosmic ray nucleation of supercooled water in cloud tops should be investigated both theoretically and experimentally.

5. EXPERIMENTAL APPROACHES TO SUN/WEATHER RELATIONSHIPS

Chairman: Richard A. Goldberg, Laboratory for Atmospheric Sciences, National Aeronautics and Space Administration, Goddard Space Flight Center

There is a strong need to develop experimental techniques to identify, test, and evaluate causal mechanisms in the Sun/weather field. There is also a need to conduct appropriate and specific measurements for statistical studies which more closely relate cause and effect than the currently available data sets and correlations. The following considerations and recommendations are based on these premises.

5.1 Solar Outputs

In studying the atmospheric response to solar active radiations and effects, direct or induced, one must be concerned with three general categories of solar output:

(1) corpuscular radiations;
(2) electromagnetic radiations; and
(3) magnetic field fluctuations.

These can all perturb the atmosphere with respect to its chemical, dynamical, electrical, and/or thermal properties.

5.2 Specific Topics

Within the framework of Section 5.1, the following topics are considered important for immediate (next five years) study, particularly as they apply to the atmospheric response to solar active phenomena. They are:

(1) solar output;
(2) upper atmospheric-lower atmospheric coupling;
(3) atmospheric composition and phase changes;
(4) atmospheric response to high latitude energetic radiations;
(5) atmospheric circulation changes;
(6) atmospheric electrical structure and effects; and
(7) atmospheric wave structure and effects.

5.3 Recommendations

The following recommendations are provided as guidelines for any program designed to investigate the above topics.

(1) Conduct coordinated experiments involving techniques which combine solar output, interactive processes, and atmospheric response. These should include:

(a) The organization of multidisciplinary scientific teams for experimental planning within the next year;

(b) The selection of specific instrument groupings. These will arise from a study of the critical processes to be tested in any proposed Sun/weather mechanism. Identification of needs for new technology development should also be identified at that time.

(2) Whenever possible, coordinate such studies with experiments designed to evaluate anthropogenic-induced modifications of the atmosphere, e.g., with programs concerned with the implementation of solar power satellites.

(3) Create an accessible data base for widespread use by researchers.

(4) Define and perform relevant synoptic measurements.

(5) Utilize both remote sensing and in-situ techniques. The latter are particularly important for studying many parameters within the stratospheric-mesospheric domain.

(6) Conduct intercomparisons and calibrations of experimental techniques.

(7) Begin immediately and develop a strong level of effort within three years. This is essential to take advantage of the upcoming solar maximum and to assess anthropogenic-induced atmospheric modifications by future planned experiments.

6. ROLE OF MINOR ATMOSPHERIC CONSTITUENTS

Chairman: Chalmers F. Sechrist, Jr., Electrical Engineering Department, University of Illinois, Urbana

There are at least two aspects of the role of minor constituents in Sun-weather/climate relationships, viz., (1) how are the minor constituents affected by solar EUV and corpuscular variability and (2) how would the changes in the minor constituents affect the tropospheric weather/climate?

Minor constituents are an important part of the atmospheric radiative-chemical-dynamical system. Ozone is among the most important because it absorbs the UV radiation which drives the stratospheric dynamics and determines stratospheric temperature. NO_2 is another stratospheric constituent important for its role in radiative balance through its absorption of visible and near UV radiation. Minor constituents such as CO_2, H_2O, O_3, and others, are important in IR absorption and emission leading, for instance, to the greenhouse effect which warms the surface.

Several ways in which changes in atmospheric minor constituents may affect tropospheric weather or climate have been suggested. These can be grouped either into those which affect the global radiation balance or those which affect dynamical properties. In the first category are changes in either the total amount or altitude distribution of O_3, both of which have been calculated to change the global surface temperature. Changes in NO_2 may also affect the radiation balance as may changes in the stratospheric sink of minor radiative constituents such as N_2O. Dynamical changes may be induced via changes in the wind structure of the stratosphere which in turn affect the propagation properties of tropospheric planetary waves. Changes in troposphere-stratosphere exchange processes have also been suggested.

In the following sections, several sharply defined questions will be posed, several plausible links in the chain of physical processes between the sun and tropospheric weather and climate will be illustrated, and various types of mechanistic model sensitivity studies and some critical experiments to test the models will be recommended.

6.1 Major Questions

Assuming that solar-terrestrial influences on weather and climate are genuine, it is necessary to design strategies for the identification and exploration of plausible physical mechanisms that connect the sun to weather and climate in the troposphere. One strategy involves the preparation of several sharply defined queries. Some examples are:

(1) Does high-latitude particle precipitation affect the thermal structure and circulation of the stratosphere and troposphere?

(2) What is the relative importance of minor neutral gases, ions, and aerosols in the middle atmosphere, in the context of Sun-weather/climate relationships?

(3) Do solar EUV variations produce significant changes in the O_3 concentration in the middle atmosphere? If so, then how is the O_3 variability connected to changes in tropospheric weather/climate?

(4) Is there a close relationship between the IMF (interplanetary magnetic field) and high-latitude particle precipitation?

(5) What are the relative roles of solar emanations (radiations and particles), galactic cosmic rays, and precipitating energetic particles, in the context of Sun-weather connections?

(6) What is the relationship between auroras, magnetic substorms, high-latitude particle precipitation, and Sun-weather connections?

(7) Is energy dissipation (heating) through energetic particle precipitation at high latitudes (near the auroral oval) related to the high correlation between sector boundary passage and vorticity area index at high latitudes in winter?

(8) Do atmospheric waves (gravity, planetary, etc.) play a role in Sun-weather relationships?

(9) How does the middle atmosphere thermal structure affect planetary-wave propagation?

(10) What are the roles of planetary waves and the polar night jet (jet stream) in Sun-weather relationships?

(11) How do planetary waves influence tropospheric circulation and weather?

(12) What are the effects of solar activity (flares, particle events, etc.) on the middle atmosphere composition, thermal structure, and dynamics?

(13) What are the effects of geomagnetic activity on the middle atmosphere composition, thermal structure, and dynamics?

(14) What is the tropospheric response to changes in the middle atmosphere composition, thermal structure, and dynamics?

(15) What are the effects, if any, of solar activity, geomagnetic storms, and aurora on the intensity and location of the polar night jet?

(16) How sensitive are the middle atmosphere composition, thermal structure, and dynamcis to changes in solar radiation, energetic particle precipitation, auroras, and geomagnetic activity?

(17) How sensitive is tropospheric weather to the position and intensity of the polar night jet?

(18) What is the role of bremsstrahlung X-rays, associated with energetic electron precipitation, in Sun-weather relationships?

The questions posed above reveal the need for an array of scenarios, mechanistic model sensitivity studies, and crucial experiments to test the postulated physical mechanisms. However, while pursuing these, one question needs to be kept in mind: What do all of the correlative analyses reveal about the physical mechanisms of Sun/weather/climate relationships?

6.2 Plausible Mechanisms

A number of plausible physical mechanisms for solar-terrestrial connections have been proposed. Underlying these is a major concern that the energetically large response is much larger than the energetically small input. Possible mechanisms include:

(1) Effect of direct heating due to absorption of a variable component of the solar spectrum. Enhancement of the simple energetics may

arise due to effects on energy propagation in the upper stratosphere; this probably involves the 0.3 μm O_3 band as the plausible source.

(2) Modulation of the stationary waves produced by orography and heating anomalies through changing the reflection and absorption character of the upper stratosphere (probably through variations in O_3 heating). Since stationary waves are in a sort of resonance condition, a large surface response might be expected.

(3) Indirect physical processes of which least unlikely appears to be modification of cirrus cloud cover due to a flux of solar particles and ionization. Radiative repercussions may be large.

(4) Modulation of the initial perturbations that trigger baroclinic instability and lead to the initial development of the traveling waves associated with weather systems. While natural fluctuations appear to be large compared to solar, this may not be true at the large wavelength scale of solar perturbations.

6.3 Solar Variability and Minor Constituents

Figure 1 illustrates schematically several possible mechanisms, involving minor atmospheric constituents, through which solar and other extraterrestrial influences might affect tropospheric weather and climate. Given the hypothesis that atmospheric minor constituents could affect tropospheric weather or climate, in what ways might changes in the solar output affect minor constituents? Several possibilities are given below.

(1) *Protons.* The large solar proton event of August 1972 has been observed to reduce the O_3 column above 4 mb in general agreement with the theory using catalytic destruction by NO_x produced by the high energy protons (∿100 MeV) which penetrate to stratospheric levels. These large proton events do not directly correlate with general measures of solar activity but seem to prefer the declining phase of the activity cycle with some also occurring during the ascending phase. Smaller proton events occur much more frequently, are correlated generally with solar activity, and usually only penetrate to mesospheric altitudes where the sign of their effect on O_3 is uncertain.

(2) *Galactic cosmic rays.* Variations of galactic cosmic rays are out of phase with solar activity. They penetrate to the lower stratosphere and upper troposphere, producing ionization which leads to, among other things, NO_x. Current O_3 theories predict an increase in column O_3 due to NO_x injections below about 20 km thus implying an out-of-phase relationship with solar activity. Calculations of the magnitude indicate the effects to be very small.

(3) *Ultraviolet radiation.* The data base for UV variations with solar cycle and solar rotation is still small and somewhat controversial. The extreme UV below 0.1 μm certainly varies by as much as a factor of 2

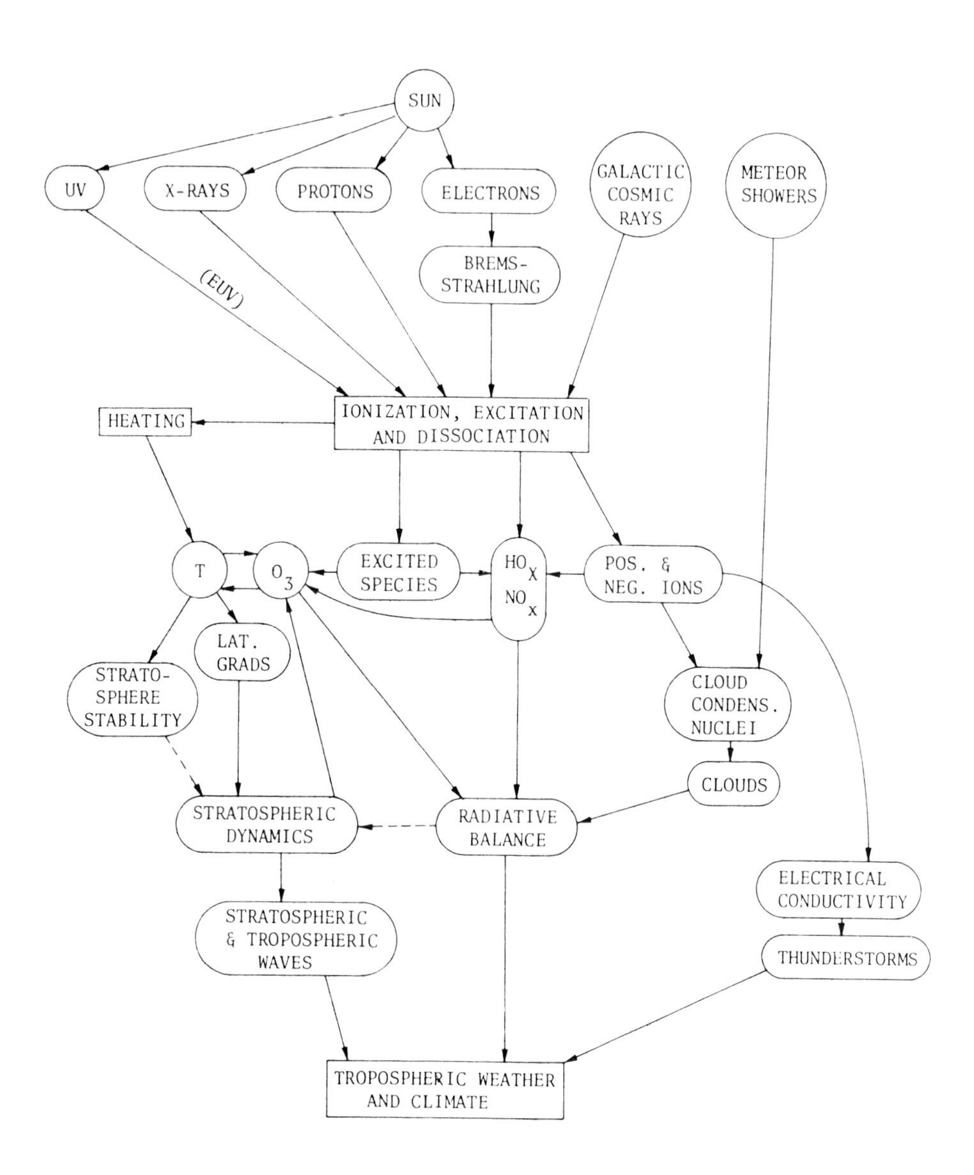

Figure 1. Diagram illustrating some possible mechanisms.

over a solar cycle. This affects mainly the thermosphere. The region from just below 0.2 μm to 0.3 μm varies less, perhaps by as much as 10 to 20%. These variations appear to be in phase with the solar cycle and, according to calculations, would cause a few percent in-phase variations of the O_3 column and larger in-phase variations of the upper stratosphereic O_3. Calculations also show a concomitant decrease in temperature around the stratopause.

(4) *X-Rays, relativistic electrons, and other flare-produced radiations.* Little work has been done on the flare effects on minor constituents. Since NO_x production will be one of the important minor constituent effects of their ionization, the sign of the overall effect on ozone concentration cannot be readily determined without a model calculation. High stratosphere NO_x injection is thought to lead to O_3 decreases while lower stratosphere NO_x injection is though to lead to O_3 increases. Therefore, the overall effect is a combination of opposite sign effects.

6.4 Examples of Scenarios

6.4.1 Scenario A: Solar UV

Recently, Callis and Nealy (1978) described a plausible and potentially significant physical mechanism which may link the 11-yr solar cycle and the chemical, and possible dynamic, state of the stratosphere. Essentially, their study revealed that solar variability in the 0.17 - 0.3 μm region might represent a physical mechanism which could link the solar cycle and the general circulation of the atmosphere. Basically, Callis and Nealy showed that the concentrations of minor neutral constituents and the thermal structure may be altered significantly in the 20 to 50 km altitude interval. In particular, they discovered that stratospheric concentrations of O_3, $O(^3P)$, $O(^1D)$, N_2O, NO, and OH are significantly sensitive to the solar UV flux variations for wavelengths less than 0.3 μm. Furthermore, temperatures above 30 km were found to be sensitive to the solar flux variations primarily for wavelengths below 0.21 μm.

If there are significant O_3 and temperature variations induced by solar UV flux changes, it implies that changes in the stratospheric wind profile and static stability would occur. As a consequence, one would expect corresponding changes in the meridional heat transport by planetary waves in the troposphere (Bates, 1977). Because these waves account for a large fraction of the total poleward heat flux in northern middle latitudes in winter, it seems plausible that a dynamical mechanism might be involved, through which variations in solar UV radiation and/or changes in the middle atmosphere O_3 abundance could lead to fluctuations in tropospheric weather/climate.

The above scenario is strengthened by the model investigations of Avery and Geller (1978) who determined the sensitivities of planetary waves to changes in zonal winds and static stability. They calculated that wave structure in the troposphere and stratosphere was found to be sensitive to changes in the strength of the polar night jet. Because the polar jet is a direct result of the meridional temperature gradient in the stratosphere, it is possible that changes in the O_3 abundance can influence the intensity and location of the jet. Consequently, O_3 changes may influence the structure of the stratospheric winds which, in turn, would modify the planetary-wave structure in the stratosphere and troposphere. Thus, different wave structures could modify the wave momentum and heat transports in the troposphere, as suggested by Bates (1977).

6.4.2 Scenario B: Energetic Electrons

The drizzle of precipitating energetic electrons from the outer radiation belt of the magnetosphere may cause a depletion of the O_3 concentration in the 50 to 70 km altitude region.

A considerable (5 yr) body of data giving electron precipitation from the outer belt now exists and more will be acquired from scheduled missions of low altitude polar satellites. From these data, one can compute the ionization in the 50 to 70 km level and infer O_3 depletion on about a daily basis. Fluctuations in both the intensity and latitude of this ionization source occur due to geomagnetic/solar influences. One would, therefore, expect variations in the mesospheric O_3 concentration in the same latitude region. Whether stratospheric O_3 or temperature varies from this mechanism has yet to be determined, but associated O_3 measurements should be of value in assessing whether this process can couple solar variability into the lower atmosphere.

6.4.3 Scenario C: Energetic Particles

Studies by Schwentek (1971) and Fritz and Angell (1976) suggest that energetic particles, and not solar UV radiations, may be responsible for the observed solar cycle variations of temperature at 35 km altitude over Berlin during the winter months. A suggested chain of physical mechanisms to explain this relationship follows:

- Energetic particles precipitate at high latitudes, creating auroras and heating the ionosphere by $\sim$200°K.

- The auroras occur in a narrow latitudinal zone (60° to 65°) and may emit 4.3 μm radiation from NO^+ ions. This wavelength lies within a CO_2 absorption band, with the result that this radiation may heat the 30 to 70 km region by about 2°K.

- This 2°K temperature change in a narrow latitudinal zone could reduce the west wind speed substantially near 70 km altitude and 60°N latitude.

- Planetary-wave theory indicates that a reduced westerly wind in the middle atmosphere is required in order to enhance the upward flow of wave energy from the troposphere into the stratosphere. This may produce a polar stratospheric disturbance in winter.

- Tropospheric temperature and circulation are affected through changes in planetary-wave amplitude and phase.

6.5 Recommendations

Because there are several specific physical-chemical processes which are unique to the problem of the influence of solar variability on minor constituents, there is an urgent need for "new" research relevant to the special problem of the effects of solar variability on the stratosphere. For example, more attention must be directed to investigations of: the chemistry of excited neutral constituents other than $O(^1D)$; ion composition and chemistry; and the formation of condensation nuclei in the middle atmosphere. These topics could be of considerable importance in studies of the effects of solar variability on the upper atmosphere. Several specific recommendations for future work are given below:

(1) *Dynamics and/or radiative mechanisms.* The block diagram shown in Figure 1 reveals very little about the connecting link between the stratospheric changes and tropospheric weather/climate. Much improvement is needed. Detailed investigations of the dynamic interactions between the troposphere and stratosphere are required. It is generally agreed that if the minor atmospheric constituents play a role in creating the Sun-weather effect that a dynamic or radiative coupling to the troposphere will be needed. Dealing specifically with a dynamic coupling it becomes very important to know the magnitude and spatial and temporal extent of these perturbations in order to understand how the propagation or general circulation characteristics of the middle atmosphere would be affected. In turn, dynamicists should determine the sensitivity of the middle atmosphere to changes in thermodynamic variables. If the radiative coupling is included as a possible mechanism, there is a need to know the source of the radiative changes, the time scales over which they occur, and what aspects of the energetics of the troposphere that are directly affected.

(2) *Model studies.* Most current atmospheric dynamics models are inadequate for assessing the impact of chemical and thermal changes (pulse and long-term cycle). Model deficiencies which might plausibly affect results (e.g., lacking mesospheric chemistry, etc.) should be corrected to the extent current knowledge and theory will permit. Some simplifications may be possible as the model scale moves from 1 to 3D, i.e., all calculated 1-D effects might be couched in terms of O_3 changes, stratosphere temperatures, and tropopause stability changes.

(3) *Tests of Scenarios.*

(a) 1-D(and possible 2-D) models should be employed to assess the impact of plausible solar effects on the distribution of constituents and temperature through chemistry and coupled/uncoupled thermal effects.

The response amplitude and lifetimes of the various constituents and temperatures should be the goal of such studies. This modeling effort could then provide guidance for 2- and 3-D calculations as well as identification of sensitive constituents and thermal responses versus altitude for development of future measurement programs.

(b) 2-D calculations might be used to assess the latitudinal extent of some of the chemical/thermal effects and thus provide crude guidance for spatial measurement programs designed to confirm or deny hypothesized phenomena.

(c) 3-D model calculations should be made based on the most plausible stratospheric perturbations which theory and 1- and 2-D model calculations suggest. Since these calculations would be very expensive, at least for the type of model which has reasonable interactions, experiments should be judiciously chosen. Calculations should be designed to determine the short-term dynamic response of the general circulation to stratospheric O_3, temperature, and tropopause static stability changes (taken singly and in various combinations). Recent and future satellite data could also provide the basis for realistic initial conditions. Model results should be compared with observations of VAI, surface pressure changes, etc.

(4) *Global perspective.* The investigation of causal mechanisms which could produce the many correlations reported in the area of Sun-weather phenomena would benefit enormously from global coverage available from satellite observations. In addition, little is known of the efficiency of coupling solar activity with tropospheric phenomena. To this end it is felt that satellite observations of the middle atmosphere over extended time periods might contribute significantly to our understanding of causal mechanisms in Sun-weather phenomena.

For the middle atmosphere, one should consider the use of middle atmospheric data which have been acquired by Nimbus 4-6 and the Nimbus G satellite which is scheduled for launch in September 1978. In the future, the Upper Atmosphere Research Satellite (UARS) and Spacelab could carry some experiments which are especially suited for the investigation of these causal mechanisms. Measurements of minor constituents in the middle atmosphere can be especially valuable because they can be used to infer solar-terrestrial interaction, through induced changes in minor constituent composition or heating. Minor constituents such as O_3 can also be used as Lagrangian tracers of atmospheric motions especially in response to solar activity. During the early 1980's the Tiros-N series operational satellites will be measuring important solar and middle atmospheric parameters such as O_3, temperature structure, and total and UV solar spectral irradiances.

These approaches should alleviate a major weakness of past investigations of Sun-weather effects which has been the inability to perform correlations on a global basis with sufficiently fine grid size.

REFERENCES

S. K. Avery and M. A. Geller, Effects of middle atmosphere winds on planetary waves in the troposphere, submitted to Geophys. Res. Lett., 1978.

J. R. Bates, Dynamics of stationary ultra-long waves in middle latitudes, Quart. J. Roy. Met. Soc., 103, 397, 1977.

L. B. Callis and J. E. Nealy, Solar UV variability and its effect on stratospheric thermal structure and trace constituents, Geophys. Res. Lett., 5, 249, 1978.

S. Fritz and J. K. Angell, Temperature and wind variation in the tropical upper stratosphere and lower mesosphere during a sunspot cycle, J. Geophys. Res., 81, 1051, 1976.

J. S. A. Green, An assessment of possible mechanisms for Sun-weather relationships, presented at the International Symposium on Solar-Terrestrial Physics, Innsbruck, Austria, June 1978.

H. Schwentek, The sunspot cycle 1958/70 in ionospheric absorption and stratospheric temperature, J. Atmos. Terr. Phys., 33, 1839, 1971.

RESOLUTIONS OF

THE SYMPOSIUM/WORKSHOP ON SOLAR-TERRESTRIAL INFLUENCES ON WEATHER AND CLIMATE, THE OHIO STATUS UNIVERSITY, 24-28 JULY 1978

(as transmitted to scientific organizations and other concerned agencies)

On Friday, 28 July 1978, participants at the Symposium/Workshop met in a general session to hear reports of Working Groups and to consider issues of greatest concern to the science of Sun-weather and climate relationships. As a result of this, the assembly unanimously approved two resolutions addressed to the International Council of Scientific Unions and other appropriate scientific organizations as follows:

I. On Solar-Weather and Climate Research

Background

The Symposium/Workshop on Solar-Terrestrial Influences on Weather and Climate was conducted in an open atmosphere of scientific inquiry. It consisted of invited and contributed papers, extensive discussion and participation in working groups which formulated plans and recommendations for future research. Major topics of the Symposium included (1) a review of solar-terrestrial influences and solar phenomena, (2) correlation studies, (3) experimental data, and (4) physical processes. The working groups dealt with (1) meteorological and climatic correlation studies, (2) solar influences on global circulation and climate models, (3) solar influences on minor atmospheric constituents, (4) solar system dynamics, (5) electricity and upper-lower atmospheric coupling, (6) experimental approaches to Sun-weather relationships, and (7) extra-terrestrial inputs.

Resolution

Whereas participants of the Symposium/Workshop on Solar-Terrestrial Influences on Weather and Climate have presented and discussed evidence and suggested plausible mechanisms for the effects of solar variability on weather and climate; the subject is relevant to many aspects of human activity and welfare; and

the proceedings of the Symposium/Workshop will be published early 1979.

Therefore, be it resolved that

the participants in the Symposium/Workshop wish to bring to the attention of international scientific bodies, particularly the International Council of Scientific Unions and the World Meteorological Organization, the proceedings of the conference; and, in view of the multidisciplinary and interdisciplinary character of the subject, the participants urge that further studies be conducted in an internationally coordinated mode, establishing appropriate data distribution and analysis channels and identifying types of measurements to be conducted in the future.

II. On Solar-Geophysical Data Sets

Background

Long synoptic solar geophysical data sets often bear the mark of a single individual who has struggled, through the years, to maintain the continuity, consistency, and quality of the data set. Examples of such data sets are:

1. The Zurich sunspot numbers created first by Wolf and, more recently, maintained by Waldmeier (since 1818);

2. The McMath calcium plage regions compiled by Prince (since 1942);

3. The Ottawa 10.7 cm solar radio flux measurements conducted by Covington (since 1947); and

4. The aa-index of geomagnetic activity created and maintained by Mayaud (since 1868).

These are examples of ground-based synoptic data sets which are of great value because of their continuity, consistency, and quality. These sets have been used extensively and are available to the research community because responsible individuals championed them. They were highly dedicated and determined to continue and maintain their data sets, often against great odds. All of these people have now retired or are about to retire. During the Symposium/Workshop, questions abounded about the fate of these particular synoptic data sets. Since many of the Symposium/Workshop participants are using one or more of these indices as fundamental data in their investigations, they are justifiably alarmed at the prospect of losing these vital tools for their research.

Resolution

Whereas the participants in the Symposium/Workshop on Solar-Terrestrial Influences on Weather and Climate have learned that certain vital synoptic solar-geophysical data sets listed above may be jeopardized by the retirement of their chief proponents;

the value of these data to scientific inquiry into Sun-weather and climate relationships is of obvious importance to the participants;

the continuity and quality of these data sets and indices seem to require an individual's special attention, rather than that of an institution or agency; the problem may be resolved by finding new champions of the threatened synoptic data sets and continued support for their work;

there is a need to focus attention on this problem at the highest level of international scientific cooperation and to encourage agencies and institutions to provide the necessary financial support for continuation of these data sets; and

the results of research associated with use of these data sets will benefit all of mankind, regardless of national and idealogical differences.

Therefore, be it resolved that

the International Council of Scientific Unions and its constituent bodies and national scientific organizations concern themselves with the fate of these synoptic data sets and indices which are vital to research in solar-weather and climate relationships and take steps to ensure their continuity and quality.

Respectfully submitted:

Thomas A. Seliga, Director, Atmospheric Sciences Program, The Ohio State University

Billy M. McCormac, Lockheed Palo Alto Research Laboratory

Co-Chairmen of the Symposium/Workshop on Solar-Terrestrial Influences on Weather and Climate

INTRODUCTORY REVIEW OF SOLAR-TERRESTRIAL WEATHER AND CLIMATE RELATIONSHIPS

Walter Orr Roberts
Aspen Institute for Humanistic Studies
1919 Fourteenth St., #811
Boulder, CO 80302 U.S.A.

ABSTRACT. I shall review some of the best established empirical Sun-weather relationships, and suggest how these relationships identify certain atmospheric parameters and phenomena that merit intensified observation and theoretical study. It is a matter of practical and theoretical importance to know whether there is indeed a significant Sun-weather interaction at short or long time scales. Some data suggest that short-term weather forecasting accuracy suffers at the same epochs that we find short-term responses to solar sector boundary passages and other Sun-weather effects.

1. HELIOGEOPHYSICAL ENTHUSIASTICS

It is extremely heartwarming for me to present this introductory review, after having spent a good part of my professional life working on the effects of variable solar activity on the weather. These proceedings contain the results by so many friends and colleagues who have made substantial contributions to the advancement in this field over these many years. Together we have watched this field grow from modest beginnings to a point where a good, solid scientific symposium like this can be held — with contributors from many parts of the world participating. We are also glad to have some skeptics here like Barrie Pittock to echo Andrei Monin's warning against "heliogeophysical enthusiastics."

It used to be thought by nearly everybody that we Sun-weather researchers were doing no more than successfully engaging in what Monin called autosuggestion and unconscious manipulation of our data. Now it is a bit different. Not everybody thinks that we are victims of our own autosuggestion. And I hope that the critics amongst us will work responsibly and hard to help us to avoid the pitfalls in our data analysis and interpretation, because there are many opportunities for error in our work, and none of us wants to spend time following false scents.

Nature has a way of being extraordinarily complex and nonlinear in its interactions. Nature could just be in a position to frustrate Andrei

B.M. McCormac and T.A. Seliga (Eds.): Solar-Terrestrial Influences on Weather and Climate. 29-40.

Monin's hope for simpler causal relationships. Monin felt that it would be a tragedy for meteorology if solar variability did play a significant role, because then we would have to forecast the Sun as well as the weather. But it may just turn out that this is true, and that the Sun's activity does play a significant role. Our task here, as I see it, is to evaluate the evidence and to design the strategies that will help us replace speculation with knowledge. After all, that is the goal of all good science.

Let me now turn to my very small role in this important task. I am going to speak first about the importance of the search for the influences, if any, of the Sun's variability on weather and climate. Then, rather than trying to make a comprehensive review of the field, I am going to try to pick out two or three salient features of the evidence before us and say a few words about the significance of these and indicate where, in my own opinion, we may find promising hunting grounds for future research strategies.

First of all it seems clear to me, as I am sure it is to you, that in all the world people are concerned about weather and climate. From the pampas of Argentina to the deepest reaches of the new lands of Siberia, from East Africa to the high plains of the United States, people are earnestly searching in the skies for advance signs of change of weather and climate.

There are very factors in the natural environment of man that play so large a role in the well-being of humanity as does the constancy of weather and climate. Yet climate and weather vary from the mean, as we all know, often by very large amounts. And they do so on all time scales and all spatial scales. The variations that occur in weather and climate are an essential part of the norm. Accommodating to these changes is, thus, an essential part of the process of living. But when the variations are too big, they are also a source, as Mayor Tom Moody of Columbus, Ohio, pointed out, of grave dislocations of the affairs of humanity.

2. THE IMPORTANCE OF WEATHER AND CLIMATE

There are very few factors in the socio-economic environment that exceed weather and climate in importance to the production of food, to the conduct of commerce, and to the pursuit of a better life. Crops in the heartland of Europe depend sensitively on having the right amounts of Sun and rain and warmth at the right times and with the right distribution. Irrigation water in the altiplano of Mexico is abundant at times when the hurricanes from the Gulf of Mexico assist in thrusting the moist air up over the mountains and into the altiplano. Irrigation water there is in short supply when the families of hurricanes, for some reason, avoid the Gulf.

Japan's very finely tuned railroad system degrades badly when deluge rains occur. Indonesian rice suffers, even in an abundantly rainy country, when the rain slacks off. Perhaps the most serious effects of all occur in the vast arid and semi-arid lands of the Earth. These lands are already heavily pressured by population growth and by the fragility of their soils and vegetation. In these arid lands large gains and losses result from the waxing and waning of rainfall. Desertification and accelerated loss of productive agricultural soil occur at times of heightened drought.

Forecasting the ebb and flow of weather systems has occupied scientists and laymen alike from the very threshold of history. Statesmen, warriors, merchants, brigands, and even mayors have adjusted their plans and strategies to fit the weather and the climate. Intricate networks of observational stations have been established at great cost in all parts of the world in order to improve the data base from which to forecast and from which to do fundamental research. Cooperative exchanges of research, like this symposium, are set up to improve our understanding of basic processes, and they are directed ultimately towards better foreknowledge of climate and weather. Meetings like this are one of the great means of international cooperation and exchange.

3. THE SLOW PACE OF RESEARCH PROGRESS

Yet for all the focus that we have developed over the years, progress in the field of long-range meteorological forecasting has been deliberate. The forecast skills that we have achieved, when you go beyond a few days, fall far short of the hopes and expectations of a keenly concerned local and world public. It is clear from official pronouncements, for example in the Congress of the United States, the Parliament of the United Kingdom, the Central Committee of the Communist Party of the USSR, that those bodies are all prepared to commit resources to the study of this problem if there is a realistic hope that better understanding will lead to progress in forecasting. Similar attitudes prevail in other countries throughout the developed and the developing world.

Yet, in spite of these high stakes in the field of climate and weather forecasting, it is not clear, at least to me, how much farther we can push our forecast skills. The most important forecast lead time, I suspect, for food and agriculture, is one year, or possibly two years.It seems to me quite uncertain that we will ever succeed, for specific regions, in developing the ability to predict departures from the mean of rainfall and temperature with any accuracy. The reason is that we are not sure that we understand all of the forces that are at work in the so-called "normal" weather fluctuations after the lead time exceeds something like five days or a week.

There are clearly some responsible researchers in the field who believe that prediction at this long lead time (beyond a week or two) is likely forever to frustrate us. They believe that the system is essentially unpredictable at these lead times. There are others, myself included, who think that there are discoverable causal mechanisms at work that are not yet adequately taken into account. We consider that these offer hope for improvement in our understanding of weather processes and from this in our ability to forecast. Of course there are people with positions in between these extremes.

4. THE SEARCH FOR SOLAR INFLUENCES AND THEIR CAUSAL MECHANISMS

It is precisely because of this possibility of finding new mechanisms that the field of Sun-weather research is so important. We now have, I venture, pretty solid evidence that there are some bona fide influences of variable solar activity on the troposphere that operate at the weather lead time of a few days and on up to climatic lead times of seasons, years, decades, and even longer.

The effects that we observe of solar activity on the weather are, one should admit, small and even esoteric. But they do apparently cause discernible responses in the weather. And they are responses that we observe even though prevailing meteorological theories suggest that we should see no such effects.

This fact is, I believe, of extremely great theoretical importance. The Sun-weather effects, if they are indeed real, as I believe they are, signal to us that something is going on that we do not yet include in our forecast models. It signals to us that "there is gold in them thar hills," and we had better get out and dig for it. For if we can unravel what is happening physically, it may have profound significance on the improvement of conventional forecasting. It may even indicate a new prospect for large-scale weather and climate modification, for all of the pros and cons that that will involve. Because, if solar activity can do it, with the extremely small amounts of energy that flow into the upper atmosphere, then there must be some kind of trigger mechanism or positive feedback involved. It may just be that the trigger can be pushed by human activity, either deliberately or inadvertently. For my part, I hope such powers are not in the cards.

The real payoff of Sun-weather research, as I see it, is the hope that it will lead us to understand some important, but now neglected, factors in atmospheric processes. From this may come new practical benefits, and I am also sure from it may come new problems, as Monin feared.

Even with the present crudity of knowledge, there is, as will be discussed later, some evidence that solar activity affects routine weather forecasting even at the 12 and 24 h lead time.

Now let me turn, for a moment, to a somewhat personalized account of some of the Sun-weather relationships that merit particular attention at this time. To me they represent unexplained but reasonably solid evidences of directions in which for us to look in the future. I would like, in respect to each of these, to highlight just a few of the key points regarding our present knowledge. In relation to all of these, the central point is that the amount of change in the energy fluxes to the Earth that result from the solar activity is trivial when compared to the energetics of the stratosphere and troposphere. Thus, we are clearly dealing with some kind of subtle feedback or trigger mechanism that releases energy stored in the atmosphere. It seems to me that the important thing now is for us to discover how these trigger mechanisms are operating.

The first example that I would like to speak about has to do with the so-called "signature" of the hemispheric vorticity area index (VAI) after solar sector boundaries in the interplanetary magnetic field have passed the Earth. Figure 1 was prepared by Hines and Halevy [1] a couple of

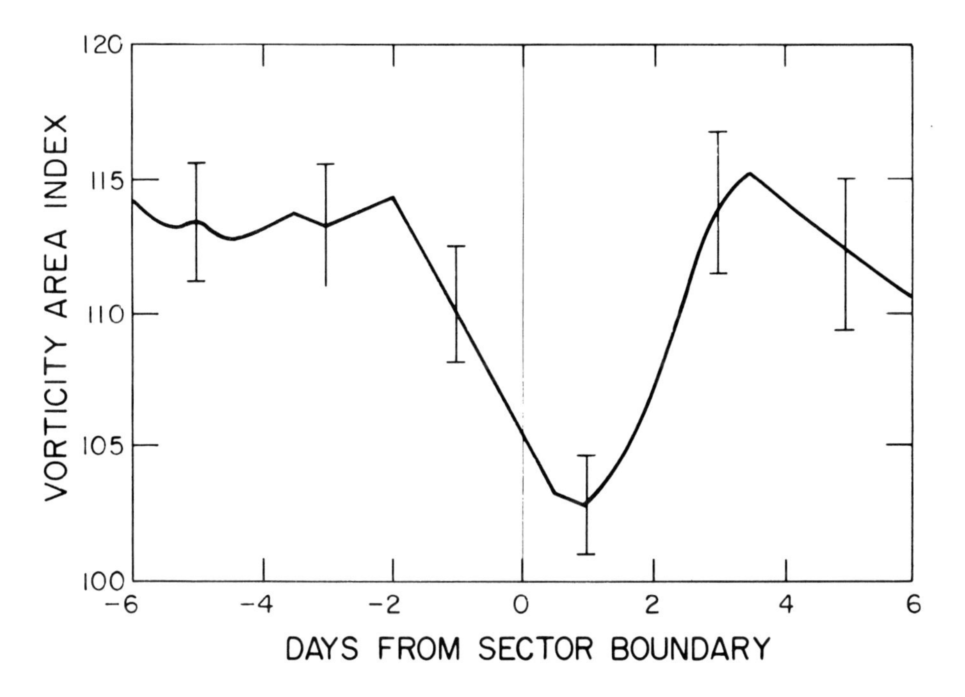

Figure 1. Superposed epoch analysis of vorticity area index at 300 mb for Northern Hemisphere vs. solar sector boundary dates.

years back, and it shows the total area of the Northern Hemisphere over which strong positive cyclonic vorticity was measured. The index, called the hemispheric VAI, was developed by Roger Olson and me. It is based on the area of the hemisphere north of 20° latitude where the absolute vorticity exceeded $20 \times 10^{-5} s^{-1}$, plus the area where the vorticity exceeded $24 \times 10^{-5} s^{-1}$. So the vorticity area index measures the total area where the cyclonic absolute vorticity is very strong. This Figure displays the result of a superposed epoch analysis. The ordinate displays the average value of the hemispheric VAI on the days when a solar sector boundary swept past the Earth — and on the days before and after such sector passages. The data are taken from something like 20 yr of vorticity analysis. They are all from the winter half year.

Olson and I developed this particular index not to elucidate mechanisms of Sun-weather relationships, but to get around the troublesome and plaguing subjectivity of the previous indices that we had used to search for a Sun-weather effect. We believed it to be an insensitive index to solar influence, but felt that if it did reveal statistically significant results, they would be the "proof of the pudding," after which better mechanism-finding studies could be devised. The VAI is totally objective, it covers large-scale features of the Northern Hemisphere of the Earth, and it comes out of the computer untouched by human hand. Its purpose, as I said, was simply to establish whether there is a Sun-weather effect.

Olson and I used the index originally to search for a relationship between large abrupt geomagnetic disturbances and the vorticity over the Northern Hemisphere. We published affirmative results of this some years ago, and these appeared to verify the earlier researches that we had done with more subjective indices.

John Wilcox then suggested, after our earlier results, that it would be worthwhile for us to take a look at the VAI for days preceding and following the passage of solar sector boundaries. The results that we found gave a signature of change showing a dip about a day after the sector boundary passage that very closely resembles Figure 1. However, the result was greeted with interest and some skepticism. Perhaps we were victims of our own autosuggestion!

I am happy now to be able to say that it was extensively tested by others. In particular Hines and Halevy [1] tested our finding. Their effort to verify the result was clearly the most careful and the most elegant, and the value in Figure 1 can scarcely be distinguished from the value that we developed ourselves, and their tests of significance were far more sophisticated than ours. The value in Figure 1 is for the winter half year. The effect, other work shows, is exhibited most clearly at 300 mb level, but it also shows up well at the 500 mb level and can be detected at most other levels that we studied. It has a large latitudinal extent, it is not just confined to a small high latitude region. It works, apparently, equally well with the sectors of both kinds: the toward-to-away, and the away-to-toward. The results indicate that space magnetic fields are probably involved in this meteorological response, rather than electromagnetic radiation of the Sun.

This finding shows up in nearly all winter season time spans that we have studied. It establishes a relationship between solar activity and the weather that is very hard to disbelieve, as implausible as it may seem. But the VAI is an extraordinarily crude tool. There must be, if we can discover the mechanism that underlies the change that you see in this curve, far better atmospheric parameters for us to use to study what is causing this to occur. It seems to me that this legitimizes an intensified effort to find specific explanatory mechanisms.

I feel that most future work should be directed to the testing of specific mechanisms. I strongly recommend that we "graduate" from looking at this averaged hemispheric VAI to looking at more significant meteorological parameters, such as Gareth Williams, Stephen Businger, and others have started to do. It is now worthwhile to look more sharply at such things as zonal kinetic energy, eddy kinetic energy, and so forth. Also, that we should turn to specific geographical regions for study. It seems quite likely that effects over major land masses and oceans will differ; hemispheric averaging may "wash out" important clues.

Figure 2 shows a study carried out by P. B. Duffy [2], working with Wilcox and colleagues, about a year ago. This shows the VAI in a slightly different form from Figure 1 where the vorticity index was averaged over the entire Northern Hemisphere. In this instance, my colleague Roger Olson identified the VAI associated with each specific low pressure trough as it moved into the Gulf of Alaska from west of longitude 180°. He tied the vorticity index to each low pressure cell, and followed the cell, calculating twice-daily values of its vorticity index, as it migrated eastward in succeeding days. The zero day in this graph is the first day after recognition of the low pressure trough system in the Gulf of Alaska area. On subsequent days shown on the abscissa, it migrated eastward. The figure shows that the troughs had a higher VAI, on the average, when they were first detected in the Gulf of Alaska during the time when there was a solar sector pointed away from the Sun (a plus sector as shown by the upper curve). The ones with the oppositely directed sector had, on the average, a smaller VAI. Again, this is a winter half year phenomenon.

The result is on a much less solid statistical base than the results shown in Figure 1. The significance of this, however, is that it indicates that there is a specific geographical region, the Gulf of Alaska, where the Sun-weather effect is especially strong — probably due to ocean temperature and land temperature contrasts in the winter season. Perhaps it is significant that the region is also at a latitude close to the auroral zone during the times of strong auroral activity. The region, the Gulf of Alaska area, is moreover a location of long-known, extremely strong cyclogenesis in winter. Studies of specific areas like this, by specific means like this, it seems to me, give us opportunity to get clues to mechanisms. I will not say anything more about this because Wilcox is going to present, later in this volume, newer results that focus the effect even more sharply to the Gulf of Alaska area. I present this, however, in the hope that it will serve as an example of the kind of thing to which we need now to pay additional attention.

The next thing that I want to discuss briefly has to do with thunderstorms and atmospheric electricity. I am not going to use a figure on this because Reinhold Reiter and Ralph Markson, who are pioneers in this field, have presented a number of results later in this volume. I want to speak of it principally to acknowledge that to my mind this explora-

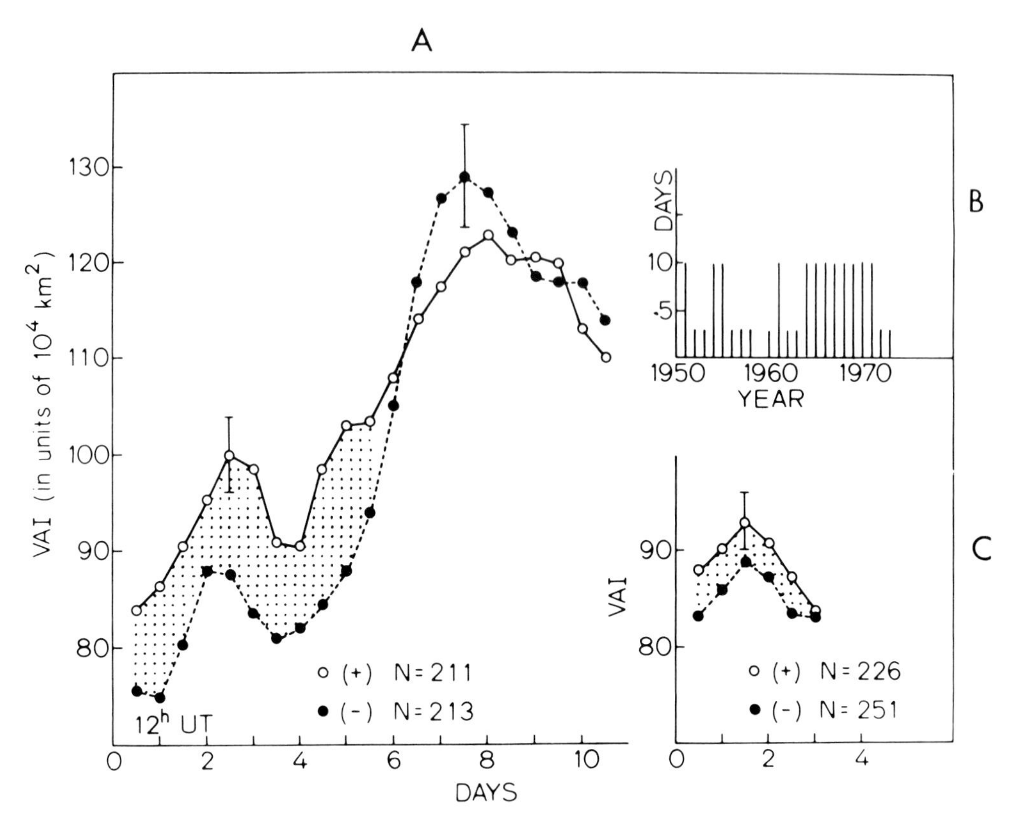

Figure 2. A. Open circles show the average area of low pressure troughs during the 10 days after the trough passed 180°. The troughs were classified as "away" troughs if the direction of the interplanetary magnetic field was away from the sun on the day that the troughs were observed east of 180°. The filled circles are the same for "toward" troughs. During the first 5 days after the trough passed 180° the area of the "away" troughs is significantly larger than the area of the "toward" troughs. The error bars are ± the standard error of the mean.
B. An indication of the winters for which trough area data was available for only 3 days. C. The same as the first 3 days but computed for the years in which trough area data was available for only 3 days. The area of the "away" troughs is again larger than the area of the "toward" troughs.

tion of thunderstorms and atmospheric electricity is the most promising direction in which to turn at the present time in the search for mechanisms. Ralph Markson was the first, I believe, to suggest that sector boundaries in their passage of Earth might show an effect on thunderstorm activity, but the effect also exists apparently with flares and other solar parameters.

Studies in this field point to the possibility of a viable physical mechanism to explain Sun-weather relationships. The general value of explaining these may far transcend Sun-weather field, and may find use in ordinary meteorology and climate research, unrelated to solar activity. Such studies may even lead to knowledge of large-scale weather modification possibilities. But for the present, as far as I am concerned, the atmospheric electricity consideration opens up a whole new family of testable ideas about Sun-weather mechanisms. Many things that are discussed later in this volume have to do with possibilities of measuring systematically the ionospheric potential relative to the Earth potential. I cannot overemphasize the importance of observing this parameter well, and studying its relation to thunderstorm frequency, to solar activity, and to cosmic ray activity. Years ago Ney suggested explaining sunspot cycle relationships to climate via the sunspot cycle modulation of cosmic rays. It will also be important to relate it to the Forbush decreases in cosmic rays that occur after certain large solar flares.

The next matter that I would like to mention is the western United States droughts and the double sunspot cycle. Figure 3 was produced by Mitchell et al. [3]. It shows the spectral density of a drought area index that they developed based on tree ring data. It shows, as you can see at the left, strong and presumably significant energy at about 22 yr. The drought index is a "surrogate" Palmer drought index based on western zone tree rings. The index correlates highly with the meteorologically derived Palmer index for the period during which weather records and tree ring data overlap. The data from which this spectrum was derived go back to about 1700.

The significance of this result, if it is verified, is that it shows a solar forcing function operative over a long time scale. It also indicates the special importance of a specific geographical region — namely the western part of the United States, probably affected by the large orographic barrier represented by the Rockies. The result of this study challenges us to relate the climatic time scale to the day-to-day time scale in our Sun-weather efforts.

In my view, the time has come for us to dedicate far more effort to seeing how far we can go in interrelating the various different kinds of Sun-weather correlations that we appear to have observed. Are they manifestations of a single cause-effect relationship? Or are we going to be forced to our dismay to have to confront the evidence of multiple causation? I hope it is not the latter!

5. THE EFFECT ON FORECASTING

Finally, I present Figure 4, from a paper by Larsen and Kelley [4]. This figure shows a superposed epoch analysis with the solar sector boundary passage as the zero date, and the days before and after sector boundary passage as the abscissa. The ordinate is the average correlation coefficient of a forecast of a meteorological parameter vs. the actually observed meteorological parameter. The forecast is based on the National Weather Service 12 h and 24 h forecasts. The forecasted

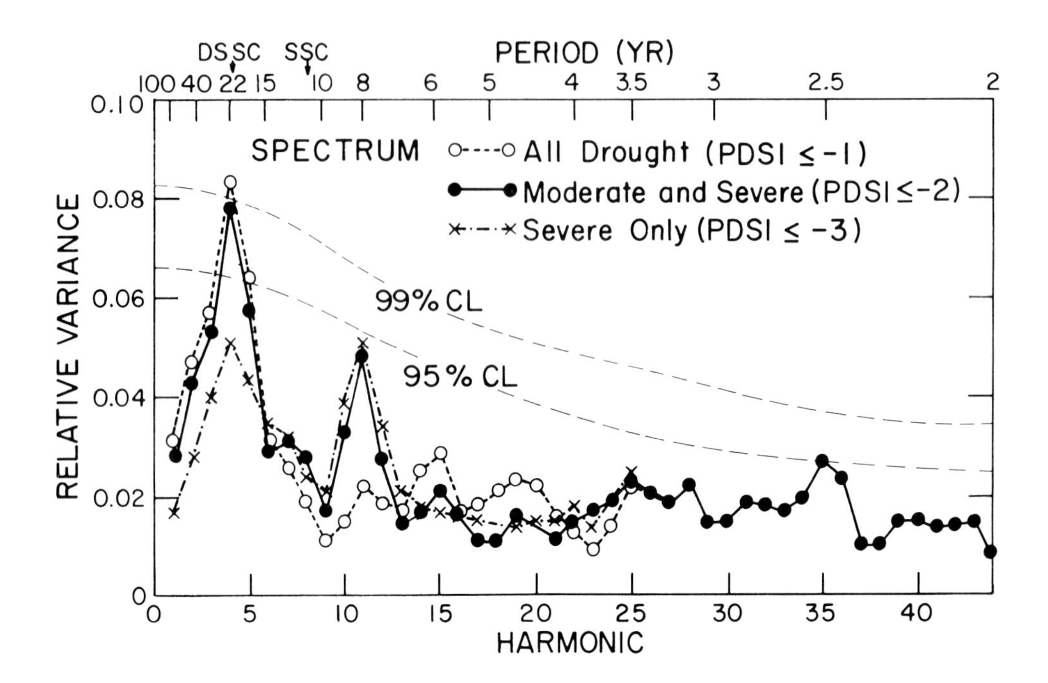

Figure 3. Spectral density functions for drought area index series. Confidence limits based on assumption of first order Markson process generating function.

parameter, in this study, is the hemispheric VAI derived from these maps. The vorticity index, of course, has previously been shown to have a statistical relationship to solar sector boundary passages. There is a dip, as you can see, very close to the same position, a day or so after the sector boundary, that showed the dip in our previous work. The significance of this is that it shows that there is some energy source related to the sector boundary that affects the troposphere but is ignored in the forecast model. This seemingly causes the forecast to depart from its usual accuracy after sector passage. It results in a dip in accuracy from about 85% to about 68%, as you can see, on both the 12 h and 24 h forecast lead times, even though these are lead times for which our forecast skill is best. I point out that this figure contains a very small time sample. It cries out for extension to a larger time period, and for a greater effort to find in what parts of the map the principal errors come about. It shows that the Sun's activity can have practically significant effects.

6. CONCLUDING REMARKS

Let me now make a very few concluding remarks. I want, first, to thank Billy McCormac and Tom Seliga and the Committee and others for bringing this workshop and symposium to pass. I am confident that this will be, in future perspective, a significant step towards realizing

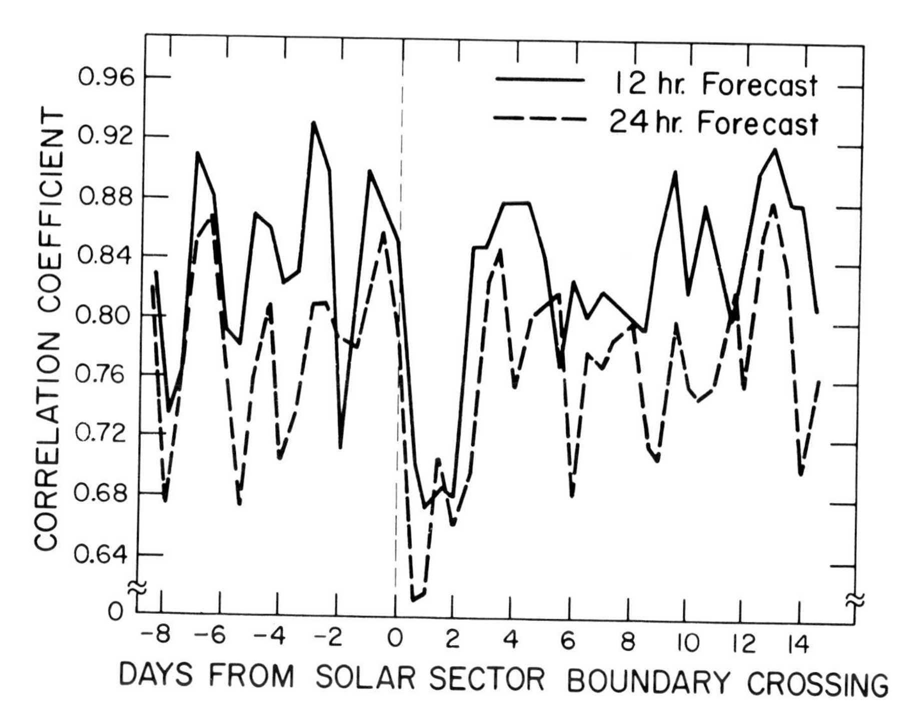

Figure 4. Correlation coefficient between forecast and observed vorticity area index (500 mb).

practical benefits through study of these now small and frustratingly fleeting clues to mechanisms of Sun-weather connections. In 5 to 10 yr from now, I suspect, understanding of these factors will result in practical and routine benefits to everyday forecasting. If I am wrong I hope, at the very least, that we will forever have laid some ghosts of autosuggestion and unconscious data manipulation in coffins, hopefully not ever to be revived to mislead us again.

In some 25 to 30 yr we will inevitably have a nearly doubled global population. We will then live in a world that needs three times more food than we now produce. This food will probably have to be produced on only 50% more arable land than we now have under cultivation. Then, as now, climate and weather fluctuations will almost certainly be the source of the greatest uncertainty in crop forecasting and crop production. In the light of these considerations nothing could be more challenging than the effort to understand new mechanisms affecting weather and climate.

REFERENCES

1. C. O. Hines and I. Halevy, J. Atmos. Sci. 34, 382, 1977.

2. P. B. Duffy and J. M. Wilcox, EOS. 58, 1220, 1977.
3. J. M. Mitchell, C. W. Stockton, and M. Meko, in press, 1978.
4. M. F. Larsen and M. C. Kelley, Geophys. Res. Lets. 4, 337, 1977.

"THE ENIGMA OF DROUGHT-A CHALLENGE FOR TERRESTRIAL AND EXTRA TERRESTRIAL RESEARCH"

Jerome Namias
Scripps Institution of Oceanography
University of California, San Diego
La Jolla, California 92093

The term drought has never been given a quantitative definition acceptable to everyone. In loose terms it is an extended period characterized by lack of precipitation, the consequence of which is generally reflected in some important phase of the changing economy. Under this definition dry spells in areas at times when little or no rain normally falls are not considered drought periods, and the anomalous character of precipitation is an important part of the definition. It is this phase of the subject which shall be discussed in this article with particular emphasis on the departures of precipitation from long-term normals (the climatic anomalies of precipitation) as they relate to the atmosphere's broad-scale or general circulation.

Climatic fluctuations are continually taking place over large areas of the world. These fluctuations may be of the order of a week, a month, a season, a year, a decade, a century, and upward to millennia. Meteorologists recognize these fluctuations as integral parts of the total weather picture. If there were no substantial deviations from normal of weather conditions in some portions of the Nation during a month or season the meteorologist would indeed be astonished. In short, abnormality of cumulative weather is in fact a "normal" condition.

The meteorological time series which express these phenomena are statistically coherent. The coherence exists not only between successive days but, equally important, exists also in a persistent recurrence of similar weather regimes during a specified period. These climatic fluctuations, particularly those which last from a season to several years, have always attracted the attention of the layman as well as the meteorologist, because they are of immense importance to various segments of the American economy.

The most recent climatic fluctuation with which the general public is now familiar is the drought which plagued the West Coast during 1976 and 1977, although a large segment of the world's population is also aware of the drought in the Sahel (the sub-Saharan area) which held sway during the first half of the 1970's.

B.M. McCormac and T.A. Seliga (Eds.): Solar-Terrestrial Influences on Weather and Climate. 41-43.

At the start it should be emphasized that a complete understanding of drought is not at hand, although its diagnosis has progressed substantially so that it is frequently possible to predict for a season in advance its incipience and its demise during or at the end of that season. Prediction of droughts for series of years is well beyond the state of the art. Therefore, no stone should be left unturned in the quest for understanding and prediction. This includes of course, solar weather influences. This report, by virtue of ignorance of the author and fuzziness of the topic, will only describe some of the physical features associated with droughts which occur in different regions and at different times of the year.

At the start, the local signature of drought is usually characterized in the free air by stable stratification--that is, temperatures which are relatively high aloft and therefore associated with sinking motions of the order of several hundred meters per day. This sinking motion provides for adiabatic heating and drop in relative humidity--factors which inhibit the formation of cloud. The subsidence in turn is usually associated with horizontal divergence of air, and thus with anticyclonic curvature of the isobars particularly at midtropospheric levels. Thus drought is a manifestation of upper level high pressure and anticyclonic curvature. This seems to be almost universally true in the temperate latitudes. It held for the West Coast Drought during 1976 and 1977, the Russian Drought of 1972, the British Drought of 1976, the Southwest Drought of 1952-54, and the great droughts of the Dust Bowl of the 1930's.

Into these anticyclones there is usually a spiralling inward flow of dry sinking air which is flung southward from the prevailing westerlies, while moist air currents are forced around the periphery of the drought. During summer over the continents, dry air is almost always associated with excessive heat caused in part by the subsidence and more directly by the high insolation associated with the lack of cloud.

The central problem is how the high pressure areas are forced to recur persistently over the drought prone area. Recent studies show that for the Plains of the United States, for example, an anomalously strong upper level high pressure area needs for its sustenance strong high pressure areas in the Atlantic and in the Pacific as well. Thus, a series of standing waves becomes established with areas of sinking motion separated by troughs with rising motion. These are the well known Rossby waves. More recent evidence suggests that these anomalous upper level wind configurations are closely associated with abnormalities of the sea-surface temperature over the oceans. The anomalous heating or cooling of the water can frequently be explained as a result of several synergistic processes e.g., insolational heating, Ekman Drift (horizontal advection) upwelling or downwelling of water, and exchange of latent and sensible heat with the overlying air. There is some indication that the oceanic abnormal thermal structure feeds back to influence the atmosphere, and, therefore, provides "a memory" for it. This sea-air feedback may indeed play a role in forcing drought because the sea has a time constant about an order of magnitude higher than the air.

In this paper several examples of past droughts are shown to bolster the above hypothesis, and to cite areas of our ignorance regarding physical causes.

In addition, there is statistical evidence indicating that soil moisture conditions antecedent to and during droughts may contribute to their maintenance and even expansion. Perhaps this relationship results from the modulation of the radiation budget or through the cloud physics associated with high cloud droplet concentration associated with dust from the dry soil usually found during periods of drought.

Since droughts frequently occur over spells lasting years, it is entirely possible that some other external force besides ocean temperatures or soil moisture operates. Irregular variations of solar activity is one candidate. However, it seems to the present author that if this is so, the initial state of the atmosphere and its lower boundary must modulate the solar influence. By way of suggestion, in such solar-drought relationships it might be advisable to stratify droughts in accordance with region, season, and with companion large-scale cells external to the drought area. In this way, some statistics might emerge which offer clues to how solar influences can play a role in production of drought.

SOLAR INPUT TO THE TERRESTRIAL SYSTEM

W.C. Livingston
Kitt Peak National Observatory*
P.O. Box 26732
Tucson, AZ 85726 USA

ABSTRACT. We review the physical origin, spectral nature, energetics and known variability of all solar emanations: photons, particles and fields. The potential role of line-blanketing as a cause of irradiance change is given.

1. INTRODUCTION

The Sun's input to the Earth environment is apportioned as shown in Table I (a revision of White, 1977). Thanks to the proximity of the Sun and the zeal of observers the distribution in photon flux is described over 14 decades in wavelength and 27 decades in flux (Figure 1). Our search for the solar connection to the weather/climate demands that, initially at least, we take a look at all forms of solar emanations. This will be attempted here. Fortunately some immediate priorities of effort seem obvious. The bulk of the energy is in the visible and near IR photon flux, and here the task is to look for small changes, a fraction of 1%, in a large quantity. We must also be aware that alteration of irradiance, or spectral energy distribution, can occur without affecting S, the total energy output. Energy escape routes such as solar faculae, or energy roadblocks such as sunspots or spectrum line-blanketing, can in principle change the "Planckian" energy distribution, but at other wavelengths (in order to appreciably affect the solar constant) we must look for larger fractional variations in smaller quantities. A known region of change is the EUV. Particles and changing magnetic fields carried by the solar wind are another area of importance. Fortunately the discovery and recent elucidation of coronal holes are an asset to understanding solar wind fluctuations (Zirker, 1977). How the Sun evolves magnetically at the photospheric level is illustrated in Figure 2.

---

* Operated by the Association of Universities for Research in Astronomy, Inc., under contract with the National Science Foundation.

B.M. McCormac and T.A. Seliga (Eds.): Solar-Terrestrial Influences on Weather and Climate. 45-57.

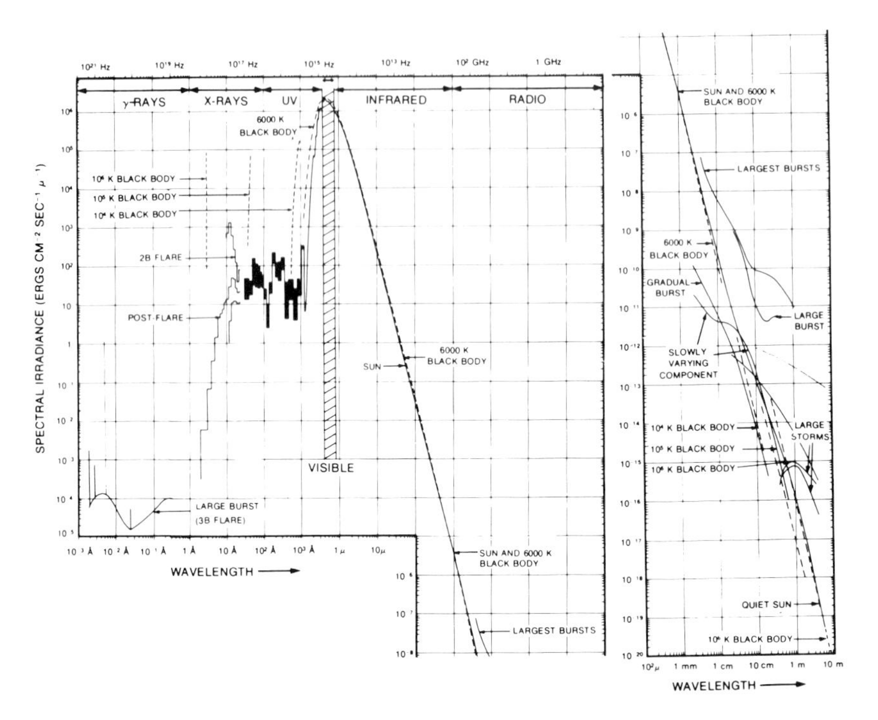

Figure 1. Spectral energy distribution of the quiet/active Sun (Malitson after White, 1977).

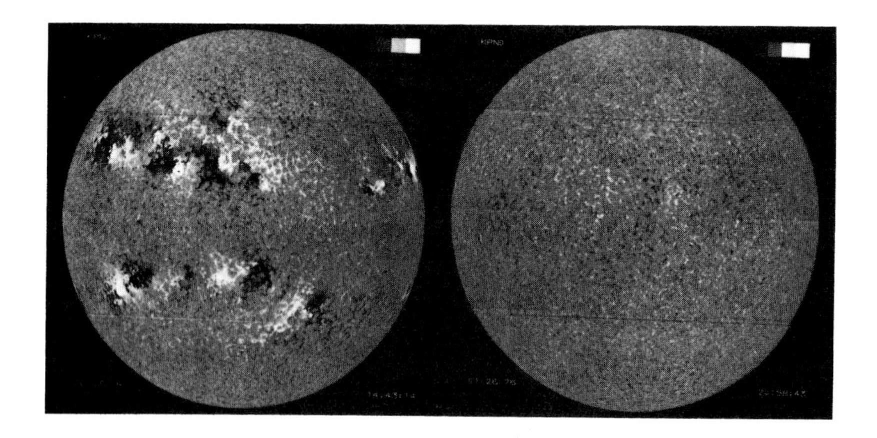

Figure 2. Photospheric magnetographs showing solar cycle changes. Right, quiet Sun 26 January 1976; Left, active Sun, 23 June 1978.

TABLE I

Total Energy
(erg cm^{-2} s^{-1} at 1 AU)

radiation	particles and fields	
solar constant $S = 1.4 \times 10^6$	(neutrinos	5×10^4)
	solar wind	1
	magnetic field	10^{-2}

Distribution of Energy

radiation		particles and fields	
4 μm to ∞	700	(neutrinos	5×10^4)
3000 Å to 4 μm	98% of total		
(excluding neutrinos)			
1200 to 3000 Å	1.6×10^4	energetic particles	
Lyman α	3-6		0.05
300 to 1200 Å	2	neutrons	
30 to 300 Å	1	galactic	
10 to 30 Å	0.01	cosmic rays	0.0006
0.1 to 10 Å	$10^{-3}10^{-5}$		
0 to 0.1 Å	$0\text{-}10^{-6}$		

The far IR and radio emission is ineffectual in the present context, being swamped by the local radiation field of the Earth. Also it seems to us that flare-related phenomena must take a back seat simply because of their temporal rarity. Or perhaps flares are not so rare. At any given instant, quiet Sun or active, several X-ray bright points may exhibit flare characteristics (Golub et al.,1975). No doubt others at this conference will more thoroughly discuss the role of flares (cf. Schuurmans, 1979).
Besides the changing energetics of the solar output we must bear in mind the modulation that arises simply from geometric causes: solar rotation, ellipticity of the Earth's orbit, and the inclination of that orbit to the Sun's equatorial plane. Table II summarizes these factors.

2. RADIATION

2.1 Visible-near IR, 0.3 μm < λ < 4 μm, 98% of(solar constant)S. The solar continuum in the visible arises from free-bound and free-free transitions of the negative hydrogen ion in the photosphere.

TABLE II

Geometric Effects

EFFECT	period	modulation amplitude
solar rotation	27^d	λ-dependent: exponential decrease with λ in UV 25% λ1200 Å 1% λ3000 Å (ref. Heath)
	∿ $27^d/2$	0 -0.4% faculae
Earth-Sun distance	365^d	± 3.4% (Max on 2 Jan; precesses through year with 26,000 yr period)
inclination of sun's equator to the ecliptic	365^d	(B_o = 0 5 Jan., 6 Jun. = -7°.2 7 Mar. = +7°.2 8 Sept.)
		particles only
Archimedes spiral	-	(optimum solar long: 50°W of central meridian sun-earth delay : 1-4^d)

Why should S change? In the absence of nuclear sources the present luminosity could be maintained by a gravitational contraction of only 10^{-4} arc s yr^{-1}. If the Sun were living on its stored heat and gravitational capital, the change would have been imperceptible over historical times, even to modern instruments (Eddington, 1926). What then is the basis for concern that S may fluctuate on time scales of a year, or a few years? To give but one example, a big unknown in the theory of stellar interiors is the role of convection (Parker, 1976). Convection is a stochastic process in which ℓ/H, the ratio of the mixing length to the pressure scale height, defines the efficiency. Dearborn and Newman (1978) show that if ℓ/H were to vary by 0.02, $\Delta S/S$ would respond by 1%.

<u>What is the evidence for a temporal variation in S</u>? Solar constant values were determined simultaneously at Montezuma, Chile and Table Mountain, California from 1923 to 1942 and continued to 1955, by the Smithsonian program (Abbot et al. 1942). Sterne and Dieter (1958) determined the co-variance between 10 day means of the two stations. The co-variance of independent measurements of a changing quantity equals the variance of that quantity independent of the errors of observation. Let Z be the simultaneous value of S present at the two stations. At one station they measured $X = Z + \epsilon$, where ϵ is an error that is statistically independent of Z. At the other station $Y = Z + \eta$, where η is also independent of Z. Assuming η and ϵ are independent of each other, then the variance in Z is given by $\sigma_Z^2 = \overline{XY} - \overline{X} \cdot \overline{Y}$. They found $\sigma_Z (=\Delta S/S) = 0.2\%$, which represents an upper limit of change for the entire 30 years of records.

This rather stringent limit of 0.2% to variability in S must be qualified in the sense that it applies to the full 30 years of data set. It does not rule out the possibility that sub-sets would display a larger variance. Unfortunately we cannot profitably inspect sub-sets because then the magnitude of ε and η is too high for the number of available measurements. For a comprehensive review see Hoyt (1979).

Foukal et al. (1977) have analyzed both the Mariner 6 and 7 solar radiometry monitor returns and the Smithsonian S-values in a search for correlations with sunspots or faculae. The Mariner radiometer sampled S for 5 months at solar maximum in 1969, a time when many spots and faculae transited the disk. Yet no modulation in S was detected: $\sigma = 0.05\%$ for spots, 0.07% for faculae. This places a strong upper limit on variability, though again the time span is short.

To suppress errors and accentuate any relation between solar features and the Smithsonian S-values, correlations were formed between pairs of days of maximum and minimum activity within month intervals. No significant correlation was found for sunspots but for faculae a slightly less than 2σ positive correlation emerges. Taking the position that the spacecraft results negate a direct S dependence on faculae, Foukal et al. (1977) suggest that atmospheric O_3 may account for this ground-based dependence (cf. Angione et al., 1976). But again a short term S-faculae relation is not ruled out.

What is the expected modulation of S due to sunspots and faculae? A succinct answer is - we don't know. Smith and Gottlieb (1974), from geometric arguments, equate a Zurich sunspot number $Z_S = 200$ to a $\leq 0.1\%$ reduction in S. This is an upper limit because the missing umbral flux may surface undetected around the spot.

The effective contrast of faculae is uncertain and there is always a question about the contribution of surrounding unresolved dark elements, i.e., pores or micropores. Chapman (1975) measured an excess brightness near the limb due to faculae of $\sim 1\%$ over 3% of the disk. Livingston (1978) reported a 0.4% modulation due to faculae in the full disk at the onset of cycle 21, but the effect has become less clear with increasing activity (See also report by Hoyt, this volume.)

Can Solar irradiance vary if S is held fixed? If the Sun were a blackbody, according to the Planck function the answer is no. However, the Sun is far from a blackbody. True, its spectral energy distribution in the visible and IR, $6000\ \text{Å} < \lambda < 1$ mm, mimics a 6000 K Planckian curve (Figure 1, and Pierce and Allen, 1977) but this is somewhat misleading. Solar irradiance is determined by the H^- absorption coefficient in the visible-IR and by the source function as modified by line-blanketing for $1600\ \text{Å} < \lambda < 6000\ \text{Å}$. As we scan the spectrum in wavelength we see to different depths in an atmosphere having a steep temperature gradient (cf. Gingerich et al. 1971).

Line absorption blocks an increasing fraction of the continuum flux as we go to the blue $\lambda < 4000$ Å and this line-blanketing causes a backwarming of ∿225 K in the low photosphere (Carbon, 1972). Certain strong Fraunhofer lines vary in strength with the solar cycle. For example, the central intensity of Ca II H and K increased ∿5.7% in 1977 compared to the quiet sun value (White and Livingston, 1978). The strength of other lines depends on a poorly understood broadening mechanism called microturbulence which could reflect the density of subsurface wave phenomena. Backwarming gives rise to a characteristic temperature T_c which differs from the solar constant temperature T_{eff} depending on the total line blanketing coefficient (Carbon and Gingerich, 1969)

$$\frac{T_c}{T_{eff}} = \frac{1}{1 - \eta}^{1/4} .$$

For the Sun $\eta = 0.138$, i.e., ∿14% of the total flux is intercepted by spectrum lines. If η increases by 1% the IR irradiance will rise at the expense of the UV by ∿0.5%.

We know of but one extended time base study of solar irradiance, that of Pettit (1932), who monitored 3200 vs. 5000 Å for 7 yr. He found a strong correlation of the violet component with sunspot number which has never been explained.

What is the rate of change of S and T_{eff} due to main sequence evolution?
Negligible; $\Delta S/S$ ∿$2(10^{-10})$ yr^{-1}; ΔT_{eff} ∿$4(10^{-8})$ K yr^{-1}.

2.2 IR - far IR, 4 μm < λ < 1 mm, $\Delta S/S$ ∿0.06%

This radiation arises from free-free transitions of the negative hydrogen ions. The level of origin passes from the high photosphere (at λ = 4 μm) to the low chromosphere and a temperature minimum (at 1 mm). Since the amount of energy radiated in this wavelength interval is small, the overall percentage change in the solar constant is negligible, particularly considering that the Earth's own radiation peaks at ∿10 μm.

2.3 Radio, 1 mm < λ < ∞

Free-free electron transitions in the chromosphere (at λ = 1 mm) to the corona (λ > 1 m) show great variability as indicated in Figure 1, but again the percentage energy change is negligible.

2.4 Ultraviolet, 1220 < λ < 3000 Å), $\Delta S/S$ ∿1%

This is a complex and important region where the spectrum goes from a bound-free continuum (3000 Å), to bound-bound emission lines λ < 1750 Å, and no absorption lines λ < 1500 Å. The proportion of UV flux from active regions grows with decreasing wavelength and the 27^d rotation modulation becomes more noticeable (Heath and Thekaekara, 1977). Recent summaries by Simon (1978) and Schmidtke (1978) stress the uncertainty of our knowledge in this UV window. According to the latter, "There is a long-term variability of the EUV flux with solar cycle. However, the data available do not suffice to derive quantitatively the changes during solar cycle 20." This has been because of instrumental difficulties. The Solar Ultraviolet Spectral Irradiance Monitor (SUSIM), a part of the Space Shuttle program, may clarify the role of UV variability providing it can be flown repeatedly. SUSIM scans the region from 1200 to 4000 Å and compares irradiance against an internal calibrated source. (Nimbus G, to be launched in Fall 1978, will also measure the UV irradiance at 1600 to 4000 Å, but lacks a calibration reference.) Rocket and spacecraft integrated flux measurements will continue to be essential to fix the value of the solar constant and determine the limit to variability (Willson and Hickey, 1977).

2.5 Lyman α, 1216 Å, $\Delta S/S \sim 2\text{-}4 \times 10^{-4}\%$

The flux in this line equals that of all integrated photon emission at shorter wavelengths. The 27^d rotation modulation $\leq$ 30%. Two components are seen: the quiet Sun part originating in the network, and active regions. From solar maximum to minimum the network flux varies x 2. Solar flares produce a full disk enhancement as follows (Smith and Gottlieb, 1974).

importance	relative frequency	Ly α enhancement
1	0.89	2%
2	0.10	8%
3	0.01	20%

2.6 EUV, $300 < \lambda < 1200$ Å, $\Delta S/S \sim 2 \times 10^{-6}\%$

This spectral region consists of emission lines from the transition zone (which still outline chromospheric network features), and of coronal lines which are confined to active regions and coronal condensations. The transition component shows a $\pm$ 15% modulation with 27^d rotation while coronal emission varies 200 to 300% from this cause. No solar cycle variability statement seems firm but Timothy (1975) has not ruled a factor of two change.

2.7 Soft X-ray-EUV, $30 < \lambda < 300$ Å, $\Delta S/S = 1 \times 10^{-4}\%$

Besides coronal active region manifestations the so-called "X-ray bright points" are seen at these wavelengths. The X-ray bright points are the shortest wavelength signature of the quiet Sun, being identified with short-lived miniature bipolar magnetic features. Approximately 10% of all points appear to flare (Golub *et al.* (1975)) while ordinary flares produce up to a x 20 enhancement.

2.8 Soft X-ray, $10 < \lambda < 30Å$, $\Delta S/S = 1 \times 10^{-6}\%$

Free-free thermal continua and emission lines from 10^6K ions, such as O VIII and Fe XVII, make up this region. Solar rotation induces up to X 5 and the activity cycle X 100 variability (Donnelly, 1977).

2.9 Hard X-ray, $0.1 < \lambda < 10Å$, $\Delta S/S = 1 \times 10^{-7} - 10^{-9}\%$

Primarily confined to flare events where the spectra display a power-law continuum. Several orders of magnitude variation in individual events are seen (Kreplin et al., 1977).

2.10 Gamma-rays, $\lambda < 0.1Å$

Nuclear reactions in solar flares produce detectable gamma-ray lines, but these are mainly used for flare diagnostics (Ramaty and Lingenfelter, 1975).

3. PARTICLES AND FIELDS

Plasma and fields are inseparable in the solar wind and must be considered together. Unlike the photon flux, substantial modifications to particles and fields can take place during transit between the Sun and Earth. Hence our interest shifts to 1 AU as a boundary condition. For example, flare neutrons never make it to the Earth because of the 11^m half-life. Low and high speed wind streams interact to amplify the accompanying magnetic field with geomagnetic consequences. In this section our attention expands to the heliosphere within the Earth's orbit although, when possible, the physical source at the Sun will be given.

3.1 Solar wind

The bulk properties of the solar wind at 1 AU are summarized by Feldman et al. (1977), Table III.

To a first approximation the data in the table are solar cycle invariant. Although a subtle modulation is produced by the preferential occurrence of high speed streams to precede solar minimum (Bame et al., 1976; Nolte et al., 1978). The solar wind consists of equal numbers of protons and electrons with a somewhat variable fraction of α-particles and heavier elements. Although $N_\alpha/N_p = 0.047 \pm 0.019$ is not constant, particularly in low speed flows (Feldman et al., 1977), this ratio is definitely in disagreement with the spectroscopic value from prominences $N\alpha/N_p = 0.10 \pm 0.25$ (Heasley and Milkey, 1978). Prominences are formed from coronal or chromospheric gas. This disagreement emphasizes the fact that our knowledge about the origin of the solar wind decreases as we approach the solar surface. What we have learned in recent years is that coronal holes are a source of the high speed component.

TABLE III

Bulk Properties of the Solar Wind

	low speed	high speed	average
particle no. density N (cm^{-3})	12	4	9
particle velocity V ($km\ s^{-1}$)	327	702	468
particle energy ($erg\ cm^{-2} s^{-1}$)	1.68	2.15	2.05
composition $N\alpha/Np$	0.038	0.048	0.047
magnetic field $B(\gamma)$	-	-	6.2 ± 2.9
magnetic energy density $B^2/8\Pi$ ($erg\ cm^{-3}$)	-	-	1.7×10^{-10}
magnetic energy flux $V(B_2^2/8\Pi$ ($erg\ cm^{-2} s^{-1}$)	0.0056	0.012	0.008

Coronal holes are regions in the corona of abnormally low density and temperature (Zirker, 1977). X-ray photographs ($\lambda < 60$ Å) show virtually no emission in holes while elsewhere over the disk there is everywhere some radiation. Most revealing is that holes are regions of open, diverging magnetic fields. Eclipse photographs have suggested this but today potential field extrapolations from photospheric magnetograms most clearly demonstrate the open field association (cf. Levine, 1977).

One cannot simply look at a full disk magnetogram and predict where the coronal holes will be. But if one compares a coronal hole photograph with a same day magnetogram most people will exclaim, "Of course, that's where the fields are predominantly open and that's the region (on the magnetic map) which outlines the hole." The problem is that at the photospheric level many bipolar closed magnetic features reside within hole boundaries. Only when the field is extrapolated upward to $\sim 1.5\ R_O$ does the open structure clarify. Coronal gas within this region escapes radially outward along the field lines, forming an identifiable stream and in the process cooling and reducing the density of the cavity left behind.

As mentioned, coronal holes are most clearly visible on X-ray pictures, but they are also evident in He 304Å spectroheliograms. Chromospheric He 10830 is in part excited by coronal He 304, and ground-based observations of the former display the holes nicely (Harvey and Sheeley, 1977). From 5 years of synoptic He 10830 pictures Bohlin and Sheeley (1978), Sheeley and Harvey (1978), and others (cf. Hundhausen, 1977) conclude:

a. Coronal holes are a source (possibly the source) of high speed streams in the solar wind. Streams from polar holes can at times reach down to the ecliptic near 1 AU.

b. Holes, particularly the smaller ones, are most effective when they lie at a heliographic latitude coincident with the plane of the ecliptic (B_0 in Table II). Amplitude of the wind stream is proportional to the area of the hole.

c. While the declining phase of cycle 20 produced long-lived holes, several extending from the pole to the equator, presently (onset of cycle 21) the holes tend to be weak, short-lived and at the high latitudes coincident with the sunspots.

Why are we interested in high speed streams? Because when a low speed stream is followed by a high speed the latter tends to catch up, compress, and thus amplify the accompanying magnetic field. The south directed component of B_z is the most effective in inducing geomagnetic storms (cf. Burlaga and Lepping, 1977). The amplitude of the southern component will undergo a 22 yr periodicity owing to its partial origin with the polar holes.

This brings us to the fact that it is the interplanetary magnetic field and not the particles which is crucial to many solar-terrestrial relationships. Perhaps more fundamental than coronal holes is the magnetic sector structure (cf. Wilcox, 1968). Just as holes map into unipolar regions at 1.5 R_0 the entire sun maps into 4 or 2 zones of polarity when viewed at low resolution. These zones extend to 1 AU as the familiar magnetic sector structure. Cross correlation of the interplanetary field at 1 AU and photospheric mean fields indicates a transit time of $\sim 4^d$. The relation of sector structure to the pattern of coronal holes is described by Hundhausen (1977).

3.2 Flare particles

Great flares produce cosmic rays which reach the Earth in 20^m to 1^h. Only 26 relativistic particle events have occurred since 1942 (Lanzerotti, 1977) and we can assume that this rarity precludes a Sun-weather effect.

Far more common are particle clouds and interplanetary shock waves. A "Catalog of Solar Particle Events 1955-69" has been prepared by Svestka and Simon (1975) which records 732 events. The optimum root point for efficient particle propagation to the Earth is 50° west of the central meridian of the Sun, but there is considerable leeway (Van Hollebeke, *et al.*, 1975). Besides protons and electrons, heavier nuclei are present. Some events seem to produce electrons only. The interplanetary shocks can modulate galactic cosmic rays and it follows there is a solar cycle dependence on this source.

3.3 Particle events of unknown origin

Frequently geomagnetic disturbances cannot be identified with a coronal hole, flare, or other solar source. It may be that certain active regions spew out particles continuously but this is unproven according to Svestka (1975). Over the limb flares and slow-moving coronal mass ejections are other candidates (Gosling *et al*., 1977). In any case the relative energy is small.

3.4 Neutrino emission

According to theory, neutrinos from the p-p reaction account for a non-negligible 3.5% of the Sun's total output (Bahcall and Sears, 1972). The failure of Davis' experiment to detect this flux could arise from an error in the ν - ^8B cross section (unlikely). Neutrinos could decay before reaching the solar surface (Bahcall and Sears, 1972) in which case the interior would energetically couple to the surface (unlikely). Hence it seems that neutrinos are only of academic interest in the present context.

4. SUMMARY

With a minimum of priority assignments we have tried to assess the bewildering array of solar inputs to the Earth. Although not yet proven, it seems probable the solar constant will turn out to be time invariant. In this case irradiance is the prime candidate for the sun-climate connection with line-blanketing as the underlying mechanism, transferring UV flux to the visible-IR. A secondary candidate is the solar wind - both the particles and fields - mainly because of its known efficiency in producing geomagnetic effects. We recall that only the interplanetary magnetic field carries the 22 yr signature that certain drought cycle studies seem to require.

Finally it seems that observational plans to improve our knowledge of irradiance variation are inadequate. If SUSIM works, it needs to fly regularly and indefinitely into the future. Extension beyond 4000Å into the near IR may be desirable.

ACKNOWLEDGMENT

In preparing this review I have benefited immeasurably from conversations with D. Carbon, T. Duvall, J. Harvey, K. Pierce, N. Sheeley, and O. R. White.

REFERENCES

Abbot, C.G., Aldrich, L.B., and Hoover, W.H., Annals. Astrophys. Obs. Smithsonian Inst. 6, 85, 1942.
Angione, R.J., Medeiros, E.J., and Roosen, R.G., Nature, 261, 289, 1976.
Bahcall, J.N., and Sears, R.L., Ann. Rev. Astron. and Astrophys., 10, 25, 1972.
Bame, S.J., Asbridge, J.R., Feldman, W.C., and Gosling, J.T., Astrophys. J. 207, 977, 197.
Bohlin, J.D. and Sheeley, Jr., N.R., Solar Phys. 56, 125, 1978
Burlaga, L.F. and Lepping, R.P., Planet Space Sci. 25, 1151, 1977.
Carbon, D.F., thesis, Harvard Univ. 1972.
Carbon, D.F. and Gingerich, O., The Grid of Model Stellar Atmospheres from 4000° to 10,000°, in: Theory and Observation of Normal Stellar Atmospheres, ed. by O. Gingerich, MIT Press, Cambridge, P. 377, 1969.
Chapman, G.A., Ann. N. Y. Acad. Sci., 262, 481, 1975.
Dearborn, D.S.P. and Newman, M.J., preprint, (Cal. Tech.) February,1978.
Donnelly, R., Extreme Ultraviolet and X-rays (> 10Å), in: The Solar Output and its Variation, ed. by O. R. White, Colorado Assoc. Univ. Press, p. 33, 1977.
Eddington, A.S., The Source of Stellar Energy, in: The Internal Constitution of the Stars, Cambridge Univ. Press, London, p. 289, 1926.
Feldman, W.C., Asbridge, J.R., Bame, S.J. and Gosling, J.T., Plasma and Magnetic Fields from the Sun, in: The Solar Output and its Variation, ed. by O. R. White,Colorado Assoc. Univ. Press, p. 351, 1977.
Foukal, P.V., Mack, P.E., and Vernazza, J.E., Astrophys. J. 215, 952, 1977.
Gingerich, O., Noyes, R.W., Kalkofen, W., and Cuny, Y., Solar Phys. 18, 347, 1971.
Golub, L., Krieger, A.S., Silk, J.K., Timothy, A.F., and Viaiana, G.S., Time Variations of Solar X-ray Bright Points, in: Solar Gamma-, X-, and EUV Radiation, D. Reidel Pub. Co., Dordrecht, p. 23, 1975.
Gosling, J.T., Hildner, E., Asbridge, J.R., Bame, S.J., and Feldman, W.C., J. Geophys. Res. 82, 5005, 1977.
Harvey, J.W. and Sheeley, Jr., N.R., Solar Phys. 54, 343, 1977.
Heasley, J.N. and Milkey, R.W. Astrophys. J. 221, 677, 1978.
Heath, D.F. and Thekaekara, M.P., The Solar Spectrum between 1200 and 3000Å, in: The Solar Output and its Variation, ed. by O.R. White, Colorado Assoc. Univ. Press, Boulder, p. 193, 1977.
Hoyt, D.V., The Smithsonian Astrophysical Observatory Solar Constant Program (submitted) Rev. Geophys. and Space Physics.
Hundhausen, A.J., An Interplanetary View of Coronal Holes, in: Coronal Holes and High Speed Wind Streams, (ed. by J.B. Zirker) Colorado Assoc. Univ. Press, Boulder, p. 225, 1977.
Kreplin, R.W., Dere, K.P., Horan, D.M., and Meekins, J.F., The Solar Spectrum Below 10Å, in: The Solar Output and its Variation, ed. by O.R. White, Colorado Assoc. Univ. Press, Boulder, p. 287, 1977.
Lanzerotti, L.J., Measures of Energetic Particles from the Sun, in: The Solar Output and its Variation, ed. by O.R. White, Colorado Assoc. Univ. Press, Boulder, p. 383, 1977.
Levine, R.H., Large Scale Solar Magnetic Fields and Coronal Holes, in: Coronal Holes and High Speed Wind Streams, ed. by J.B. Zirker, Colorado Assoc. Univ. Press, Boulder, p. 103, 1977.

Livingston, W.C., Nature, 272, 341, 1978.
Malitson, H.H., GSFC Report based on The Solar Output and its Variation, ed. by O.R. White, Colorado Assoc. Univ. Press, Boulder, 1977.
Nolte, J.T., Davis, J.M., and Sullivan, J.D., Solar Wind Stream Structure During the Early Phase of Solar Cycles 20 and 21, submitted to Solar Physics, 1978.
Parker, E.N., The Enigma of Solar Activity, in: Basic Mechanisms of Solar Activity, ed. by V. Bumba and J. Kleczek, D. Reidel Pub. Co., Dordrecht, Holland 1976.
Pettit, E., Astrophys. J. 75, 185, 1932.
Pierce, A.K. and Allen, R.G., The Solar Spectrum between .3 and 10 μm, in: The Solar Output and its Variation, ed. by O.R. White, Colorado Assoc. Univ. Press, Boulder, p. 169, 1977.
Ramaty, R. and Lingenfelter, R.E., Gamma-ray Lines from Solar Flares, in: Solar Gamma-, X-, and EUV Radiation, ed. by S.R. Kane, D. Reidel Pub. Co., Dordrecht, Holland p. 363, 1975.
Schmidtke, G., Planetary and Space Sci. 26, 347, 1978.
Schuurmans, C., this volume, 1979.
Sheeley, N.R., Jr. and Harvey, J.W., Solar Physics, (in press), 1978.
Simon, P.C., Planet.Space Sci. 26, 355, 1978.
Smith, E.V.P. and Gottlieb, D.M., Space Sci. Rev. 16, 771, 1974.
Sterne, T.E. and Dieter, N., Smithsonian Contrib. to Astrophys. 3, 9, 1958.
Svestka, Z., Solar Flares, D. Reidel Pub. Co., Dordrecht, Holland, 1975.
Svestka, Z. and Simon, P., Catalog of Solar Particle Events, 1955-1969, D. Reidel Pub. Co., Dordrecht, Holland, 1975.
Timothy, J.G., B.A.A.S. 7, 407, 1975.
Van Hollebeke, M.A.I., Ma Sung, L.S., and McDonald, F.B., Solar Phys. 41, 189, 1975.
White, O.R., Activity and Variability in Other Stars, in: Solar Output and Its Variation, p. 44, 1977.
White, O.R. and Livingston, W., Astrophys. J. (in press), Dec. 1978.
Wilcox, J.M., Space Sci. Rev. 8, 258, 1968.
Willson, R.C., Hickey, J.R., in: The Solar Output and its Variation, ed. by O.R. White, Colorado Assoc. Univ. Press, p. 111, 1977.
Zirker, J.B., Coronal Holes - an Overview, in: Coronal Holes and High Speed Wind Streams, ed. by J.B. Zirker, Colorado Assoc. Univ. Press, Boulder, p. 1, 1977.

ANALYSIS OF REGRESSION METHODS FOR SOLAR ACTIVITY FORECASTING

Charles A. Lundquist and William W. Vaughan
Space Sciences Laboratory
George C. Marshall Space Flight Center
National Aeronautics and Space Administration
Marshall Space Flight Center, AL 35812 USA

ABSTRACT. A key input to the utility of any solar-weather-climate relationship is the ability to forecast solar activity. The regression method of McNish and Lincoln has received considerable application in forecasting the sunspot cycle behavior. This paper provides a sensitivity analysis of the behavior of such regression methods relative to aspects such as cycle minimum, time into cycle, composition of historical data base, and unnormalized versus normalized solar cycle data. Comparative solar cycle forecasts for several past cycles are given relative to these aspects of the input data. Implications for the current cycle, No. 21, are likewise given.

Assuming that a mode of solar influence on weather can be identified, advantageous use of that knowledge presumably depends on estimating future solar activity. Projections of weather trends several months or a few years in advance would require anticipating the course of an individual solar cycle, and perhaps the initial phase of the following cycle. The already observed solar activity within an ongoing cycle is certainly the most immediate information about the course of that cycle.

The subject of this paper is the potential use of the most recent solar data to project trends in the next few years. The work reported arose from the current problem of estimating future density values in the high atmosphere for spacecraft lifetime calculations (particularly Skylab), but similar considerations could eventually enter into weather projections based on expected solar activity. We will not discuss possible use of long-term periodicities in successive solar cycles, although that topic has received much attention (e.g., [1,2]). Nor will we address the very short-term problem of predicting specific solar events (for example, flares) hours or days before their occurrence, although that important topic also has been widely considered [3-5].

A frequently used technique for solar cycle predictions is a linear regression procedure along the lines formulated by McNish and Lincoln [6]. This is a procedure used by the Environmental Data Service of the

B.M. McCormac and T.A. Seliga (Eds.): Solar-Terrestrial Influences on Weather and Climate. 59-64.

National Oceanic and Atmospheric Administration in the "Solar-Geophysical Data Prompt Reports." Slight variations of the technique are employed also at the National Aeronautics and Space Administration's Marshall Space Flight Center.

Briefly, the general procedure begins with a historical data base, for example, the Zurich smoothed sunspot numbers $\overline{R}_{13}(i)$, defined later. These historical data are used statistically at a current time i months past the minimum of a cycle to predict values R_{13} (i+n) at a later month, i+n, using Equation (1).

$$\overline{R}_{13}(i+n) = K(i,i+n)\ \Delta\overline{R}_{13}(i) + \overline{\overline{R}}_{13}(i+n) \quad (1)$$

where

$R_{13}(i)$ = smoothed 13-month running mean

$$= \frac{1}{24}\left[\overline{R}_1(i-6) + \overline{R}_1(i+6) + 2\sum_{j=-5}^{+5}\overline{R}_1(i+j)\right]$$

$\overline{R}_1(k)$ = mean for month k of daily sunspot numbers R(d)

$$= \frac{1}{m}\sum_{d=1}^{m} R(d)$$

$\overline{\overline{R}}_{13}(i)$ = mean of N cycles

$$= \frac{1}{N}\sum_{q=1}^{N}\overline{R}_{13}(i,\ \text{cycle } q)$$

$\Delta\overline{R}_{13}(i)$ = departure from mean

$$= \overline{R}_{13}(i) - \overline{\overline{R}}_{13}(i)$$

$K(i,i+n)$ = linear regression coefficients.

Admittedly, this is an empirical procedure with no physical basis other than an implicit premise that there is some systematic similarity in the shape of solar activity cycles. It assumes that if the current cycle is running above (or below) the mean cycle, then in the future the current cycle is likely to follow the course of previous cycles than ran correspondingly above (or below) the mean.

One measure of the validity of this premise is the statistical uncertainty derived from checking Equation (1) against historical data. After the K(i,i+n) have been derived, Equation (1) can be employed for each past cycle to calculate a "predicted" value, $\overline{R}_{13}$ (i+n) to be compared with the actually observed value. The standard deviation between these values can then be calculated. Figure 1 illustrates this, using solar cycle 20 as an example and cycles 1 through 19 as the historical data. The central curves give the predicted value of $\bar{R}_{13}(i,55)$ as a function of the month i for which Equation (1) was evaluated. The actual observed value of $\bar{R}_{13}(i,55)$ is also indicated, and differences from the prediction are apparent. The $+1_\sigma$ curves give the standard deviation between predicted and actual values from cycles 1 through 19. Cycle 20 was a cycle with ideal behavior from the linear regression standpoint; unfortunately, not all cycles follow its pattern.

The $\pm 1_\sigma$ spread in Figure 1 (and in other cases like it) is substantial for appreciable values of n. In an effort to reduce the predicted spread, several authors have investigated variants of the linear regression method, with more or less success [1,7]. One simple modification reduces the minima at the start of all cycles to the value zero, then performs the linear regression calculation and, finally, adds the minimum value back into the prediction. This seems a natural modification inasmuch as a minimum higher (or lower) than the mean minimum does not imply that a cycle higher (or lower) than the mean will follow. This normalization of all minima to the same value (zero) seems particularly significant in the case of the current cycle, 21, which started from the highest minimum yet recorded.

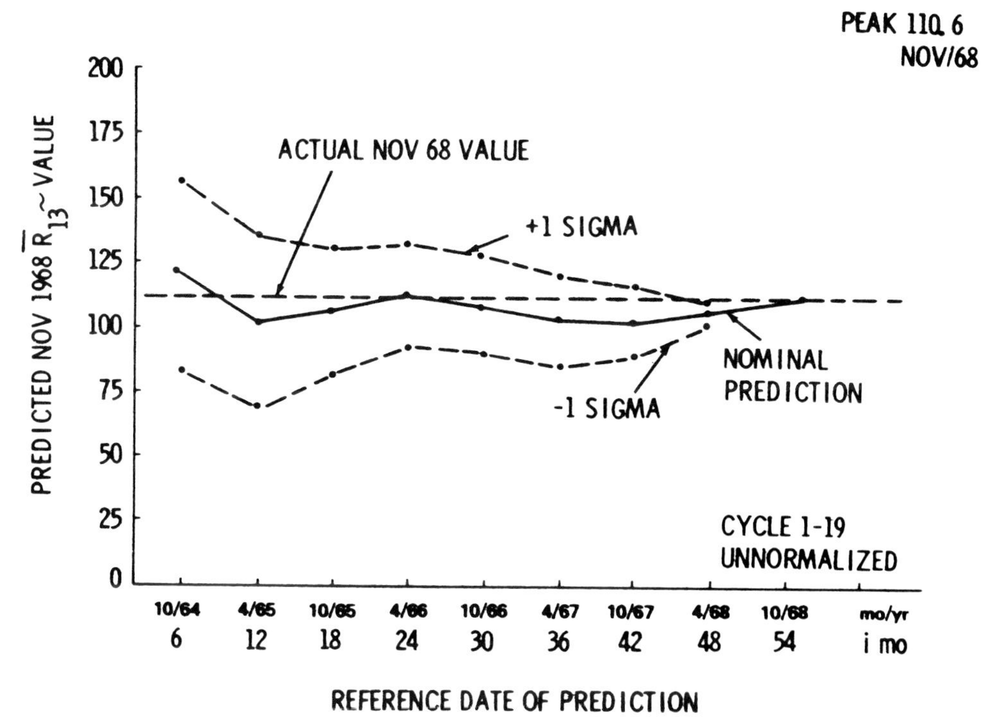

Figure 1. Predicting cycle 20 by linear regression.

Figure 2 illustrates the predicted maximum of cycle 18, as a function of i, for the normalized and unnormalized treatments. The predictions are quite different early in the cycle but differ very little later in the cycle. Incidentally, cycle 18 is an example in which the linear regression method is much less satisfying than the previous example.

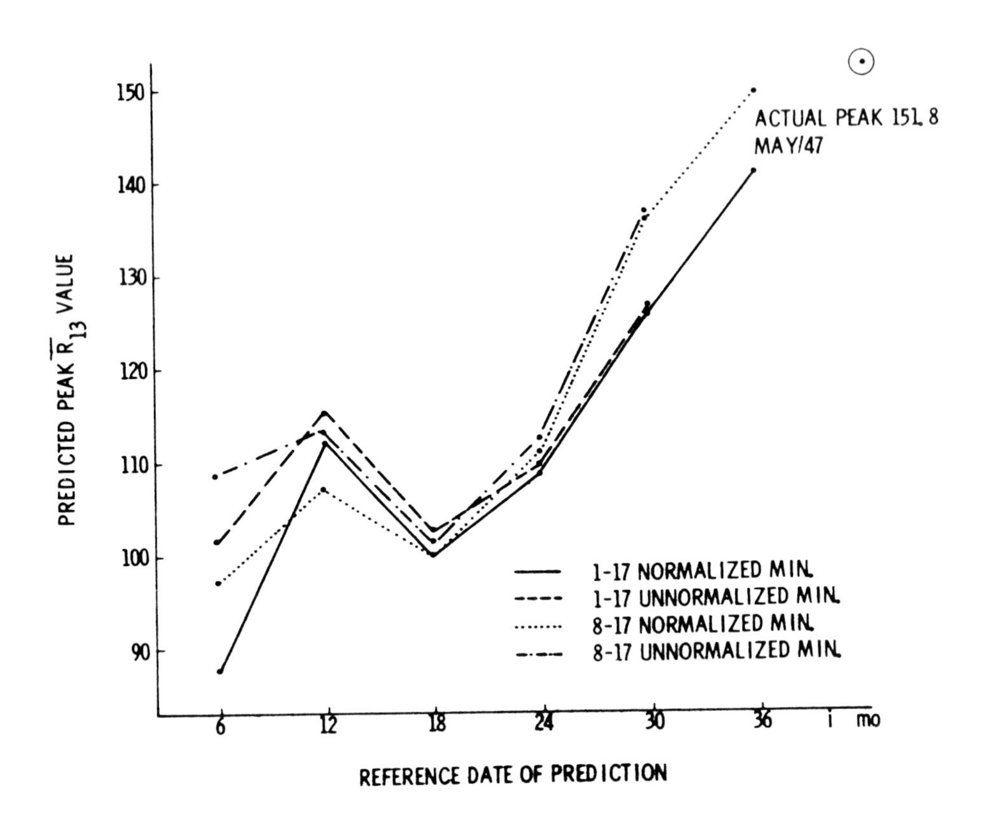

Figure 2. Predicting cycle 18 by linear regression.

Another way in which applications of the linear regression method can differ results from different choices of the historical data base. Values of $\overline{R}_{13}$ are tabulated for 20 cycles, with increasing reliability at more recent times [8]. Some groups, recognizing that the earlier cycles are less reliable, elect to delete cycles 1 through 7 [7]. The choice seems to be the added strength of a longer data base versus a shorter but more accurate base. Figure 2 also compares the results for cycle 18 using different historical bases. After the cycle has progressed for a year or so, the effects of this choice appear negligible.

The current cycle 21 presents an unusual problem in that 2 months, March 1976 and June 1976, had identical minimum values of $\overline{R}_{13}$; i.e., there was a double minimum. When one uses a linear regression procedure, the projected peak $\bar{R}_{13}$ value for the cycle depends critically upon the date adopted for the minimum. One way out of this dilemma is to use the shape of the ascending and descending curve to interpolate a minimum date; for example, using the times at which $\bar{R}_{13}$ (i) reached values such as 20, 30, or 40 before and after the 1976 minimum relative to corresponding times from the mean minimum curve. Such an interpolation gave April 1976 as an appropriate choice for the minimum.

Figure 3 summarizes the July 1978 state of affairs with application of the linear regression approach to the current solar cycle. The choice of a minimum date strongly influences the result. The data base and normalization choices are of less significance at this phase of the cycle. We have found a presentation such as Figure 3 to be useful in representing the range of rational linear regression projections of the coming maximum.

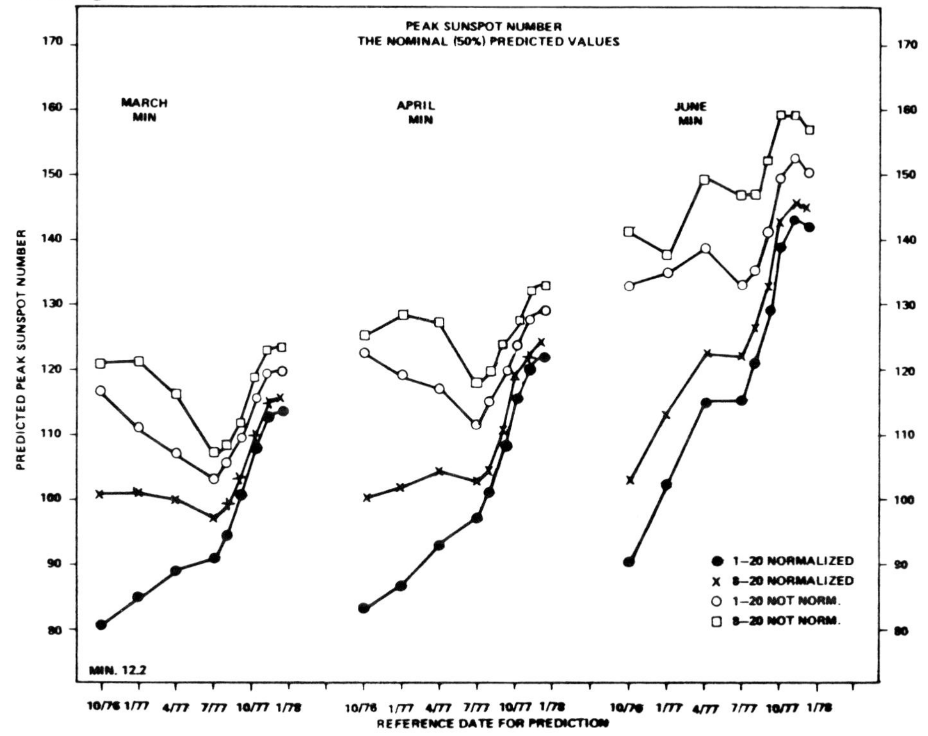

Figure 3. Predicting cycle 21 by linear regression.
Figure 3. Predicting cycle 21 by linear regression.

Clearly, the depicted situation leaves much to be desired. Perhaps something better than a linear regression technique can be found, but what that might be is not clear to us.

REFERENCES

1. Yu. I. Vitinskii, Solar-Activity Forecasting, translated from Russian, NASA TT F-289, Israel Program for Scientific Translations, Jerusalem, 1965.
2. J. R. Hill, Nature, 266, 151, 1977.
3. P. McIntosh and P. Simon, Solar Activity and Predictions (ed. P. S. McIntosh and M. Dryer), MIT Press, Cambridge, Massachusetts, 343, 1972.
4. J. B. Smith, Solar Activity and Predictions (ed. P. S. McIntosh and M. Dryer), MIT Press, Cambridge, Massachusetts, 429, 1972.
5. M. Pick and P. Simon, Proceedings of the IEEE, 61, 1303, 1973.
6. A. G. McNish and J. V. Lincoln, Transactions American Geophysical Union, 30, 673, 1949.
7. H. Slutz, B. Gray, L. West, G. Stewart, and M. Leftin, Solar Activity Prediction, Contractor Report prepared by National Oceanic and Atmospheric Administration, NASA CR-1939, November 1971.
8. M. Waldmeier, The Sunspot Activity in the Years 1610-1960, Schulthess and Co., Zurich, 1961.

VARIATIONS IN THE SOLAR CONSTANT CAUSED BY CHANGES IN THE ACTIVE FEATURES ON THE SUN

Douglas V. Hoyt
National Oceanic and Atmospheric Administration
Environmental Research Laboratory
Boulder, CO 80303 USA

ABSTRACT. The solar constant is calculated as a function of time using known contrast values for active features on the Sun. The photospheric effective temperature is assumed to be constant in this model and the change in radiant output is assumed to be caused solely by active features on the Sun which have different effective temperatures than the surrounding photosphere. The theoretical annual mean solar constant varies by no more than 0.102 mW cm^{-2} or 0.075% of its value. For the more active Sun the solar constant is lower. If the energy flux from the deep interior of the Sun is constant on the time scale of a century, the decreased radiant output at sunspot maximum may be balanced by energy storage in the active features or by increased convective energy. A compensating change in the photospheric effective temperature of 1.1^{o}K at sunspot maximum is also sufficient to give a constant solar radiant output.

The possibility that active features on the Sun, such as sunspots or photospheric faculae, can cause a change in the solar constant has long been a subject for speculation. Langley [1] was apparently the first to investigate this problem and concluded that the solar constant would change by less than 0.1% because of solar activity. Subsequently, isolated calculations of the effect of a large sunspot on the solar constant have been made by Mitchell [2], Smith and Gottlieb [3], and many others. All these authors find the effect is of the order of 0.1% and the motivation for the calculation generally is to show that the effect is small. The subject has not heretofore been systematically investigated to find the effect on the solar constant due to changes in solar activity over the past century. This paper investigates the possible changes in the solar constant caused by changes in the active features on the Sun under the assumption that the photospheric effective temperature remains constant.

Photospheric faculae, umbrae, and penumbrae all have a different brightness than the photosphere and therefore can affect the solar constant. The ratio of the intensity of radiation (I) in an active region to that of the quiet photosphere (I_{ph}) is known as its constrast. Photospheric

B.M. McCormac and T.A. Seliga (Eds.): Solar-Terrestrial Influences on Weather and Climate. 65-68.

faculae are bright regions on the surface of the Sun with a contrast of about 1.15 [4-6]. Umbral regions have a contrast of about 0.28 [7] and penumbral regions are about 0.75 [8]. The uncertainty in each of the three contrast figures is about 0.05.

The yearly mean projected areas of the faculae, umbrae, and penumbrae corrected for foreshortening are given by Jones [9] from 1874 to 1954 and by Strickland [10] from 1955 to 1970. Using these areas and the contrast figures above, the solar constant can be calculated for each year. The results are summarized in Figure 1. The maximum excursion in the solar constant is -0.102 mW cm^{-2} or 0.075% of its mean of 136.7 mW cm^{-2} [11]. Prior to 1940 the variations from year to year are less than 0.03%. If the solar energy output is a constant on these time scales, the decreased radiant output at sunspot maximum may be balanced by the generation of active features on the Sun such as the prominent coronas observed at sunspot maximum, or by an increase in convective activity or by increased magnetic field strengths and so forth. Alternatively, if the photospheric effective radiation temperature were not more than 1.1°K warmer during sunspot maximum than during sunspot minimum, the radiant output of the Sun would be a constant. The variations in the solar constant in Figure 1 should be interpreted as an upper limit on the real variations that may exist because possible compensating changes in the photospheric effective temperature are neglected.

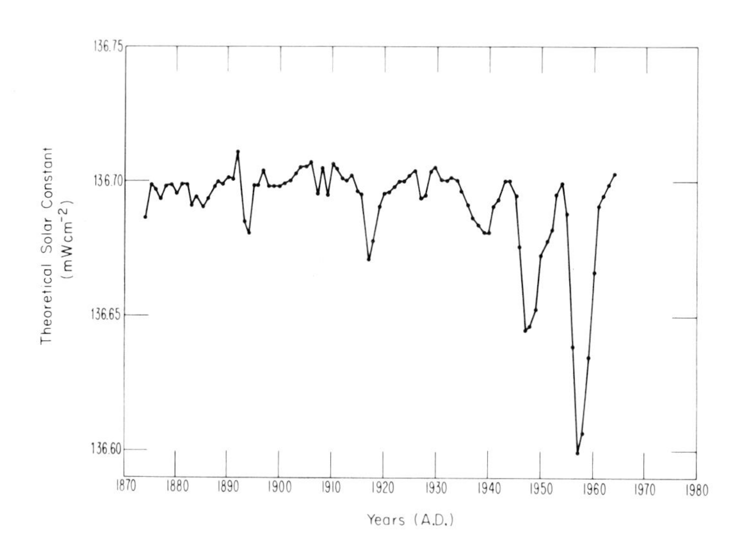

Figure 1. A reconstruction of the solar constant using known contrasts for active features on the Sun. The quiet Sun has a solar constant equal to 136.7 mW cm^{-2} [12].

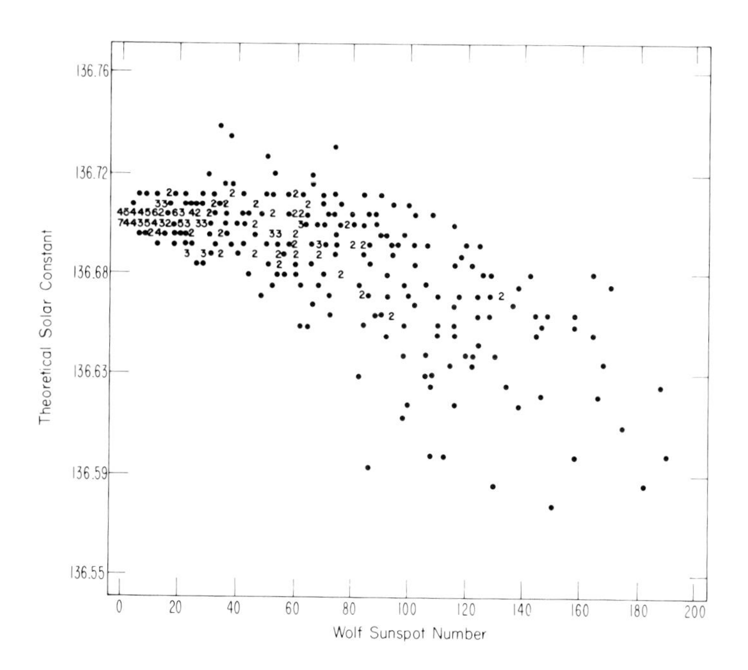

Figure 2. The dependence of the calculated solar constant on Wolf sunspot number between 1923, and 1954. In this figure the quiet Sun has a solar constant equal to 136.7 mW cm^{-2}.

The theoretical solar constant values are plotted against the Wolf sunspot number [9] in Figure 2 for the period 1923 to 1954. In general, the calculated solar constant decreased with an increase in sunspot number. A linear least squares fit through this plot gives:

$$S_o = (136.7 \pm 0.00014) - (3.9 \times 10^{-5} \pm 0.2 \times 10^{-5}) R \qquad (1)$$

where S_o is the solar constant in mW cm^{-2} and R is the Wolf sunspot number. The one standard deviation uncertainties are given in the equation. For a Wolf sunspot number of 200, the decrease in the solar constant from the quiet Sun conditions (R=0) is 0.11%. This variation is less than the observed variation of the solar constant with Wolf sunspot number reported by Abbot [12], Kondratyev and Nikolsky [13], or Hoyt [14] and in the opposite direction. Unless there is also a concurrent modulation of the photospheric or faculae temperature over long periods of time, the present ground-based and balloon measurements of the variation of the solar constant with Wolf sunspot number can best be explained by experimental error.

Meteorologists using climatic records to investigate for evidence of a solar signal [15-16] would probably have the best chance if they confined themselves to the period 1940 to 1963. The theoretical solar constant has its largest variations in this time and volcanic activity was minimal. The amplitudes of the Earth surface temeprature variations, however, are probably less than 0.1°C and therefore will be difficult to discern in the noisy climatic records available.

We strongly emphasize that the variations in the solar constant represented in this paper are the variations which arise from a known mechanism as the title itself indicates. If one or more of the assumptions used in this model, such as the assumed constant photospheric temperature, should prove to be false, then the actual behavior of the solar will be considerably different from that illustrated here. The validity of the assumptions that went into this model are now under investigation.

REFERENCES

1. S. P. Langley, Mon. Not. Roy. Astron. Soc., 37, 5, 1876.
2. J. M. Mitchell, Jr., in Proc. of the Seminar on Possible Responses of Weather Phenomena to Variable Extra-Terrestrial Influences, NCAR Tech. Note TN-8, National Center for Atmospheric Research, Boulder, Colorado, 1965.
3. E. v.P. Smith and D. M. Gottlief, in Possible Relationships Between Solar Activity and Meteorological Phenomena, Preprint X-901-74-156, NASA-Goddard Space Flight Center, Greenbelt, Maryland, 1974.
4. O. N. Mitropolskaya, Iz. Krim. Ap. Obs., 11, 152, 1954.
5. O. N. Mitropolskaya, Iz. Krim. Ap. Obs., 15, 130, 1955.
6. J. B. Rogerson, Ap. J., 134, 331, 1961.
7. R. J. Bray and R. E. Loughhead, Sunspots, Wiley, New York, 1964.
8. E. Tandberg-Hanssen, Solar Activity, Blaisdell Publishing Co., Waltham, Mass., 1967.
9. H. S. Jones (compiler), Sunspots and Geomagnetic-Storm Data derived from Greenwich Observations, 1874-1954, H. M. Stationery Office, London, 1955.
10. D. J. Strickland, Private communication, 1977.
11. R. C. Willson and J. R. Hickey, In the Solar Output and its Variation Colorado Associated University Press, Boulder, Colorado, 1977.
12. G. G. Abbot, The Sun and the Welfare of Man, Smithsonian Institution Series, Inc., New York, 1934.
13. K. Ya. Kondratyev and G. A. Nikolsky, Q. J. R. Meteor. Soc., 96, 509, 1970.
14. D. V. Hoyt, Revs. of Geophysics and Space Physics, in preparation, 1978.
15. E. J. Gerety, J. M. Wallace and C. S. Zerefos, J. Atmos. Sci., 34 673, 1977.
16. C. Mass and S. H. Schneider, J. Atmos. Sci., 34, 1995, 1977.

THE VARIABILITY AND ABSOLUTE MAGNITUDE OF SOLAR SPECTRAL IRRADIANCE

Glenn E. Shaw
Geophysical Institute, University of Alaska
Fairbanks, Alaska 89701 U.S.A.

Claus Frölich
World Radiation Centre, CH-7260
Davos-Dorf, Switzerland

ABSTRACT. It has been hypothesized that the sun's irradiance may fluctuate at near UV and blue wavelengths due to changes in solar line blanketing associated with solar activity. Preliminary determinations of the solar spectral irradiance in the visible spectrum, made over a 6 mo period at the Mauna Loa Observatory, indicated that the solar radiation was constant to at least 1%. There were indications, however, that the blue light of the sun ($\lambda \simeq 400$ nm) may have varied by a few tenths of one percent. It is recommended that solar spectral irradiance be measured in absolute (SI) units so that long-term (decades) changes, that may be related to terrestrial weather, can be quantized. It is now possible to determine spectral irradiance to $\pm$ 1% accuracy by incorporating a new dye laser-active cavity radiometer calibration method.

1. INTRODUCTION

Heath and Thekaekara [1] reported measuring changes in the solar UV irradiance of about 1% over a solar rotation and about 10% over an 11 yr cycle at $\lambda = 300$ nm; generally the variability decreased monotonically with increasing wavelength. By extrapolating Heath and Thekaekara's measurements into the visible spectrum, there are indications that spectral irradiance may vary over a solar cycle by as much as several percent at 400 nm wavelength. If true, this would be a potentially important solar link to photochemical and weather processes. In this paper we are reporting on preliminary results of an investigation of spectral solar irradiance made from the Mauna Loa Observatory, Hawaii (altitude 3380 m).

2. DESCRIPTION OF EXPERIMENT AND RADIOMETRIC CALIBRATION

The experimental work was directed at measuring the sun's irradiance integrated in wavelength over nine individual narrow passbands (10 nm

B.M. McCormac and T.A. Seliga (Eds.): Solar-Terrestrial Influences on Weather and Climate. 69-73.

wide) within the visible spectrum. Wavelength regions were not contiguous, but were placed between terrestrial gas absorption bands [2]. The radiometer incorporated dielectric interference filters to isolate individual passbands and a solid state PIN-doped silicon photodetector. Filter leakage outside the passbands was measured with a dye laser down to optical transmissions of 10^{-7}, and a detailed relative solar spectrum of Beckers was convolved with the passbands in order to be able to compare our results with other workers' tabulations. Two reference standards were used for calibration purposes: (1) a 1000 W halogen cycle standard lamp calibrated by the U. S. National Bureau of Standards and (2) a primary electrical substitution absolute cavity radiometer and dye laser at the World Radiation Centre, Davos, Switzerland. The accuracy of the absolute irradiance values depended on wavelength: it was estimated to be $\pm$ 3% for λ < 500 nm, $\pm$ 2% for λ > 500 nm for the lamp calibrations and 0.8% for one wavelength band (wavelength 615.9 nm), which was calibrated with an active cavity radiometer (3). A great deal of attention was devoted to identifying and reducing sources of systematic error in the experiment.

3. THE OBSERVING STATION AND EXTRAPOLATIONS THROUGH THE ATMOSPHERE

Mauna Loa Observatory, Hawaii (elevation 3380 m) was chosen for the measurements because of its isolation from interfering aerosol contamination. The aerosol optical thickness above the station varied from 0.014 at λ = 400 nm to 0.009 at λ = 600 nm. Extrapolations through the atmosphere were done with the Langley method; Figure 1 shows typical deviations of measured points about the regression-fitted Langley line. The constancy from day to day in slope of the Langley lines and their linearity indicated that one can routinely extrapolate through the atmosphere at Mauna Loa to determine extraterrestrial irradiance to 0.5% accuracy, provided that one chooses wavelengths that are free from contamination from terrestrial gas absorption bands.

4. RESULTS

The relative standard deviations of extrapolated extraterrestrial solar irradiance for the sun at one astronomical unit of distance and over the 6 mo period March to August, 1976 are listed in Table I. Though the standard deviations are small, <1%, the fluctuations in solar irradiance must have been even smaller, since the measurements themselves were uncertain by $\sim\pm 0.5\%$ due to accumulated random errors in the photometer calibration and in the atmospheric extrapolation process. The conclusion is that the sun's spectral irradiance was constant to better than a few tenths of one percent at wavelengths longer than $\sim$ 500 nm, but there may have been slight fluctuations of the order of 0.3% in spectral irradiance in the region λ < 420 nm. This finding of near-constancy of the sun's spectral radiation contrasts sharply with the deviations in values of solar spectral irradiance that have been

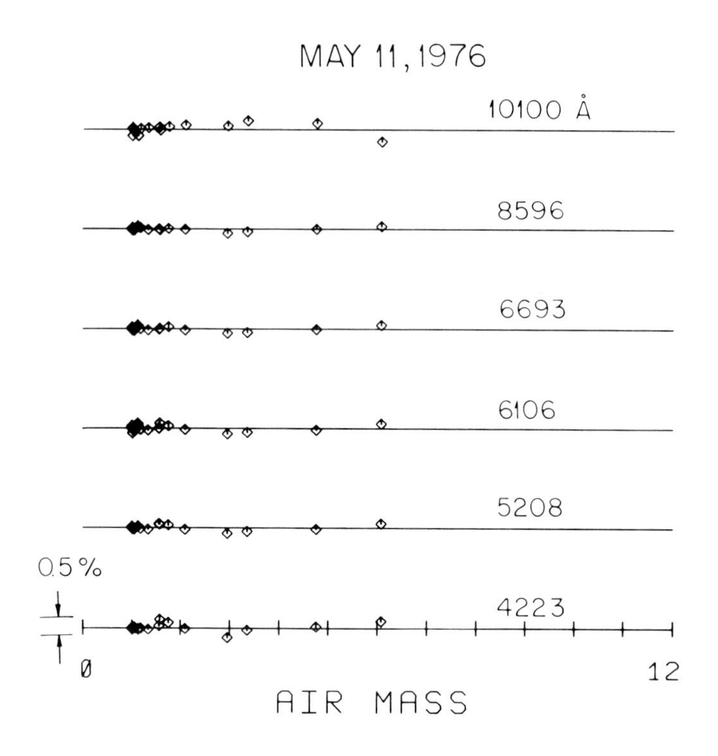

Figure 1. Deviations (in percent) of measured points about least-square fitted Langley Plot lines.

TABLE I

RELATIVE STANDARD DEVIATION (IN PERCENT) OF EXTRAPOLATED ZERO AIR MASS PHOTOMETER VOLTAGE INTERCEPTS ABOUT THE REGRESSION-FITTED PHOTOMETER DRIFT CURVE FOR THE PERIOD MARCH TO AUGUST, 1976

λ (nm)	400.7	422.3	486.2	520.8	537.4	610.6	659.5	669.3	750.3	859.6	1010.0
σ (%)	0.8	0.7	0.3	0.4	0.5	0.5	0.5	0.7	0.6	0.5	0.2

measured and tabulated in the literature by other workers. These tabulated values disagree with each other, especially at bluer wavelengths, and in some instances by as much as 10 or 20%. We believe that the reason for the disagreement is to be found in unrecognized systematic errors that were present in many of the past experimental measurements.

Figure 2 shows the absolute measured solar irradiance of the sun in terms of Labs and Neckel's [4] tabulated values. In deriving the data

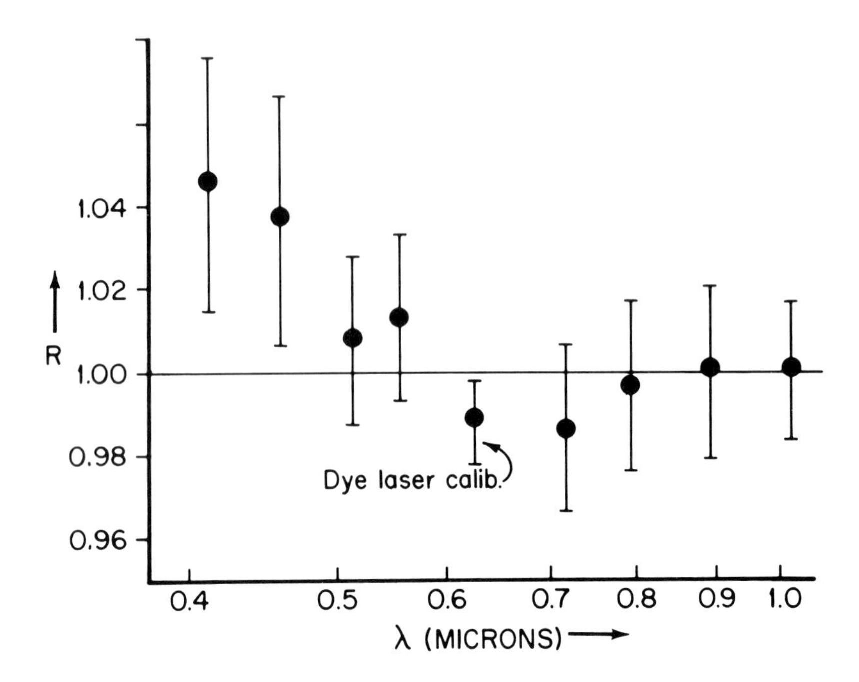

Figure 2. Ratio of absolute mean solar irradiance measured at Mauna Loa in February, 1978 to Labs and Neckel's values.

illustrated in Figure 2 we integrated a high resolution spectrum of the sun with each filter's optical transmission bandpass characteristics. The high resolution atlas had already been put on Labs and Neckel's irradiance scale by Beckers [5] and was available on magnetic tape.

The agreement between the measured absolute values of spectral irradiance with those reported by Labs and Neckel is quite satisfactory at wavelength $\lambda > 500$ nm, but the measured values at the two blue wavelength bands centered at $\lambda = 410$ and $\lambda = 460$ nm were about 4% larger than those reported by Labs and Neckel. The reason for the discrepancy at shorter wavelengths is not known; but it must be emphasized that the reference standard we used was a 1000 W quartz iodine lamp, however, our experience was that this type of lamp is not reliable enough to use for accurate determinations of irradiance on an absolute scale.

One wavelength passband ($\lambda = 615.9$ nm) was calibrated to an estimated 0.8% absolute accuracy with a dye laser and active cavity radiometer (3) at the World Radiation Centre in Davos. This calibration is believed to be at the best existing state-of-art accuracy in the science of radiometry. Figure 2 shows that the result of the determination compares very favorably with Labs and Neckel's values, being only 1% lower.

5. DISCUSSION

Our findings, although preliminary, indicate that the sun's irradiance in narrow wavelength bands (10 nm wide) was quite constant over 6 mo, though evidence was found that the near UV radiation may be somewhat changeable at a few tenths of one percent level. This result was not entirely unexpected, but obviously it is of the greatest importance to monitor changes in solar irradiance, especially at the near UV since (1) our data indicate that radiation in this wavelength band may vary, and (2) the near UV solar radiation contains significant amounts of energy that are deposited in the lower stratosphere by O_3 absorption, hence the variations could provide a solar link to terrestrial weather phenomena. It is strongly recommended that future determinations of solar spectral irradiance be referred to absolute SI system of units so long-term changes over at least one cycle of solar activity can be assessed. The dye laser-active cavity radiometer approach to calibration provides the means for the first time of making absolute measurements of irradiance to better than 1% accuracy; this new technique should be important for future research in Solar-Terrestrial physics.

ACKNOWLEDGMENT

This research was done under NOAA, Office of Geophysical Monitoring for Climatic Change, Contract No. 03-6-022-35154. Calibration facilities at World Radiation Centre, Davos, were used for this study. We thank Doctors J. Geist and J. Tech, U. S. National Bureau of Standards, for their help.

REFERENCES

1. D. F. Heath and M. P. Thekaekara, in The Solar Output and Its Variation, Ed. by O. R. White, Colorado Associated University Press, Boulder, 1977.
2. G. E. Shaw, Geophysical Institute Report, University of Alaska, Fairbanks, UAG-R-251, January, 1978.
3. J. Geist, B. Steiner, R. Schaefer, E. Zalewski, and A. Corrons, App. Phys. Lett., 26, 309, 1975.
4. D. Labs and H. Neckel, Solar Physics, 19, 3, 1971.
5. J. M. Beckers, C. A. Bridges, and L. B. Gilliam, Air Force Geophysics Laboratory Report AFGL-TR-76-0126, Vol. 1 and 2, 1976.

EMPIRICAL MODEL OF THE EARTH'S RESPONSE TO ASTRONOMICAL VARIATIONS OF INSOLATION

Frederick B. House
Dept. of Physics and Atmospheric Science
Drexel University
Philadelphia, PA 19104, U.S.A.

ABSTRACT. An empirical model of the Earth-atmosphere system is formulated from satellite observations of the response to monthly variations of insolation. Observations of reflectance at different latitude zones are portrayed as functions of insolation magnitude, its spatial gradient and variation in time during the year. Assuming heat transport and storage terms remain the same as today, conservation demands a radiant emittance (and corresponding surface temperature) which balances the energy equation.

Sensitivity studies are performed to determine mean surface temperature changes associated with variations in obliquity, eccentricity, and longitude of perihelion of the Earth's orbit. Time series of temperature changes between 40 and 70°N are presented for a period from 500,000 yr before present to 100,000 yr after present. Maximum oscillations of temperature occur for orbital conditions of larger eccentricity, and for longitudes of perihelion at 135° (warmest) and 270° (coldest) in the absence of positive feedback mechanisms. Periods of glacial build-up favor orbit conditions of larger eccentricity. In addition the range of the obliquity cycle may be an important factor in long-term temperature variations of the Earth.

1. INTRODUCTION

The Milankovitch theory [1] of climate control has recently gained general acceptance as a factor in the long-term variation of climate. Investigations by researchers in the CLIMAP project, especially the study by Hays et al.[2], have shown that some of the cyclic patterns in proxy records of climatic fluctuations are coupled to secular variations in the Earth's orbit. Their spectral estimates of proxy time series identify the 21,000-year precessional cycle and the 41,000-year obliquity cycle of the Earth's orbit. In addition they show that more than half of the variance seems to be coupled to the eccentricity of the Earth's orbit, having a period of about 100,000 yr. Milankovitch viewed the variation in eccentricity as only a modulator of the amplitude of the precessional cycle, affecting the winter-summer insolation at a latitude.

B.M. McCormac and T.A. Seliga (Eds.): Solar-Terrestrial Influences on Weather and Climate. 75-81.

Wigley [3] points out that the Earth-atmosphere system must have a non-linear response for the variations of eccentricity to be important since the annual insolation at a latitude is nearly constant for any eccentricity.

The fourth power dependence of radiant emittance W_e on emitting temperature T_e may be an important non-linear property of the Earth-atmosphere system. This may be approximated by Stefan-Boltzmann's Law where $W_e = \sigma T_e^4$. The derivative of the expression is of interest here where $dT_e = dW_e/4\ \sigma T_e^3$, showing that the temperature change dT_e varies inversely as the third power of temperature. To illustrate the consequences of this property, consider a thought experiment. Suppose a change in emittance is required to compensate for an increase of insolation at some time. The temperature increase depends on the absolute temperature of the system at that time, being larger for colder temperatures and smaller for warmer temperatures. Thus, variations of insolation during the year will propagate as temperature changes within the system in a non-linear manner, depending on the absolute temperature of the system (season of the year) at the time of change.

This paper considers only the change of the surface temperature climate from today's values for different distributions of insolation. Calculations employ a steady-state, empirical model, based on the observed response of the Earth system today, to predict its response at other times. Satellite observations of radiant reflectance are used to establish the functional relationship between system response and the monthly distribution of insolation. The investigation explores only the consequences of the non-linear dependence of radiant emittance on temperature and its coupling to surface temperature.

2. EMPIRICAL MODEL

The net radiation at the top of the atmosphere for some month and latitude zone is written as $Q_n = H_s - W_r - W_e$. Q_n is the net flux positive to the system, H_s the insolation, W_r the radiant reflectance and W_e the radiant emittance. Q_n also represents the net export and storage of heat (NE&S) with regard to a column bounded below at a subsurface level where the energy flux is only the geothermal component, and bounded above by the top of the atmosphere (nominally 30 km). If Q_n is positive, the excess energy within the column is stored as energy and/or exported to adjacent columns. Solving for the radiant emittance in the above equation,

$$W_e = H_s - W_r - \text{NE\&S} \tag{1}$$

W_e is the quantity computed in the model that meets the condition of energy conservation. H_s is determined from orbital considerations for a specific obliquity, eccentricity, and longitude of perihelion, assuming a solar constant of 1360 Wm^{-2}. NE&S is assumed to have a magnitude of Q_n that is observed today, based on satellite observations from data sources [4-7].

In Equation (1), W_r is the Earth response to a given distribution of insolation. Its value is determined from an empirical multiple regression equation of the form $W_r = a_1 + a_2(H_s) + a_3\ (-\nabla H_s) + a_4(\dot{H}_s)$. Here, W_r depends on the magnitude of insolation at a latitude, its meridional gradient $(-\nabla H_s)$ and its change with time $(\dot{H}_s)$. Regression coefficients were determined for 10° latitude zones from computed values of insolation for today's Earth orbit and from satellite observations [4-7].

The surface temperature T_s at a latitude is computed from the value of W_e in Equation (1) through a linear regression equation coupling surface emittance σT_s^4 and observed radiant emittance. That is, $T_s = ((c_o + C_1 W_e$ $T_s = ((c_o + c_1 W_e)/\sigma)^{1/4}$. Monthly average surface temperature data (Jenne et al.[8], Jenne [9]) were used in conjunction with satellite observations to determine the regression coefficients (c's). Mean surface temperatures $\overline{T}_s$ at a latitude are computed from monthly temperatures by time-weighted averaging. Of interest in this study is the annual mean temperature change $\Delta\overline{T}_s(t)$ at some some time t for different orbital configurations, $\Delta\overline{T}_s(t) = \overline{T}_s(t) - \overline{T}_s(\text{now})$. For brevity in the text, annual mean temperature changes are hereinafter referred to as ΔT_s.

3. RESULTS

Calculations of $\Delta\overline{T}_s$ were performed for ranges of obliquity (ε = 22.2° to 24.4°, eccentricity (e = 0 to 0.06) and longitude of perihelion (Π = 0° to 360°). These results establish sensitivities to orbital elements which allow predictions of ΔT_s for time-varying elements, calculated by Berger [10].

Figure 1 presents results of the variation of $\Delta\overline{T}_s$ with obliquity for a circular (e = 0) Earth orbit. For the latitudes shown, ΔT_s at 75°N spans a maximum range of 3.5 K for obliquity extremes. At lower latitudes, the temperature sensitivity to obliquity gradually decreases, becomes zero at 47°N as mentioned by Milankovitch [1] and then exhibits a negative slope. It is noted that the change of ΔT_s again becomes positive in tropical regions between 0 and 20°N (not shown in Figure 1).

Figure 2 shows the variation of $\Delta\overline{T}_s$ during the precessional cycle for 45 and 55°N at a high obliquity of 24.4°. In these results, e = 0.06 in order to emphasize maximum sensitivity to eccentricity. The range of variation at 55°N is about 1.7 K, and becomes a maxima of 1.9 K for the Northern Hemisphere at 45°N. The variation of $\Delta\overline{T}_s$ exhibits a sinusoidal character that is not quite symmetrical in angle Π. Warmer temperatures occur for Π = 135° while colder temperatures occur at Π = 270°. The shift from 90° to 135° is a result of the non-linear variation of Earth albedo with insolation during the year. For comparison purposes, the dashed curve shows the variation of ΔT_s at 55°N for a low obliquity of 22.2°. The entire variation is about 1.1 K colder due to the decrease of obliquity.

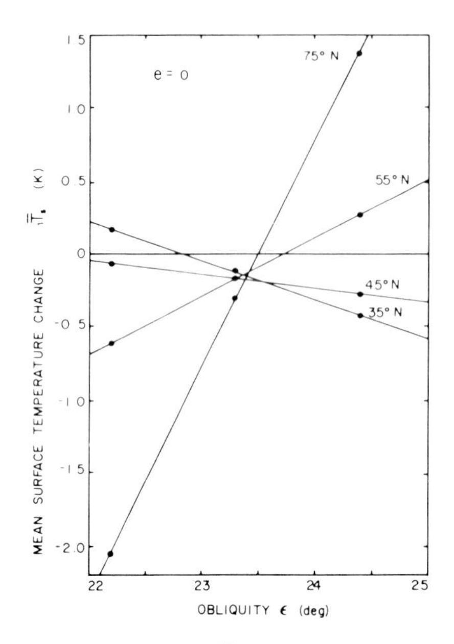

Figure 1. Variation of $\Delta\overline{T}_s$ with obliquity for different latitudes in the Northern Hemisphere (zero eccentricity).

Model calculations indicate that the coldest $\Delta\overline{T}_s$ poleward of 47°N occur for low obliquities, large eccentricities and perihelion occurring during the Northern Hemisphere summer. The preference is for a short, hot summer and a long, cool winter to take advantage of the non-linear dependence of radiant emittance on temperature. These results are similar to the combination of orbital parameters proposed by the Milankovitch hypothesis except that he advocates perihelion during winter instead of summer.

Figure 3 shows the variation with obliquity of the warmest and coldest temperatures at 55°N latitude. It is convenient to compare these values with those for a circular orbit. The warmest $\Delta\overline{T}_s$ at Π = 135° are about 0.4 K higher than the circular orbit reference, whereas the coldest $\Delta\overline{T}_s$ are lower by -0.9 K. The bounding lines also indicate the temperature range associated with possible excursions of orbital eccentricity, indicating a range of 1.3 K. The net cooling caused by maximum eccentricity, after averaging over the precessional cycle, is about -0.4 K to -0.5 K below today's values.

Figure 4 presents a time series of $\Delta\overline{T}_s$ during the period 500,000 yr before present (B.P.) to 100,000 yr after present (A.P.). These data consist of an area-weighted average $\Delta\overline{T}_s$ for the region bounded by latitudes 40 and 70°N, and should be interpreted only as temperature variations from today's values (with no albedo feedback mechanisms). Several features of the ensemble of variations should be discussed. The main oscil-

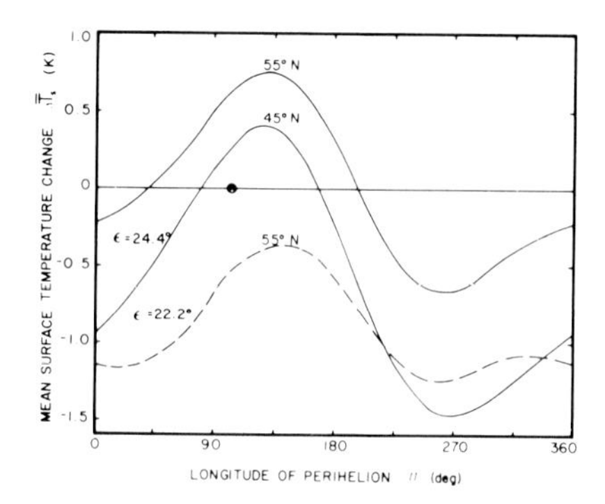

Figure 2. Variation of $\Delta\overline{T}_S$ during the precessional cycle for 45°N and 55°N with eccentricity = 0.06 in magnitude.

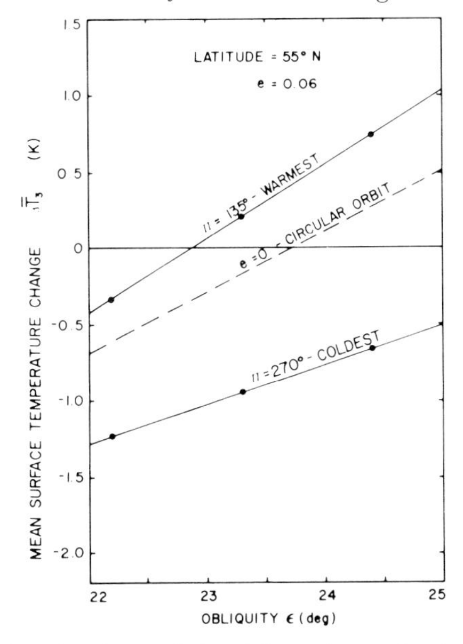

Figure 3. Variation with obliquity of the warmest and coldest temperature changes at 55°N.

lation is associated with the 41,000 yr obliquity cycle, caused primarily by latitudes between 55 and 70°N. The higher frequency variations which appear to increase and decrease in amplitude (see variations at 230,000 and 180,000 yr B.P.) correspond to the precessional cycle with amplitude modulation by the eccentricity. The warmest $\Delta\overline{T}_S$ value is 0.56 K at 207,000 yr B.P., and the coldest is -0.95 K at 198,000 yr B.P. The maximum range of variations is then about 1.5 K.

The calculations reveal an interesting temperature variation associated with the changing range of max-min angles during the obliquity cycle. This is seen as the varying max-min envelope of the data in Figure 4. For example, the maximum temperature range occurs between 250,000 and

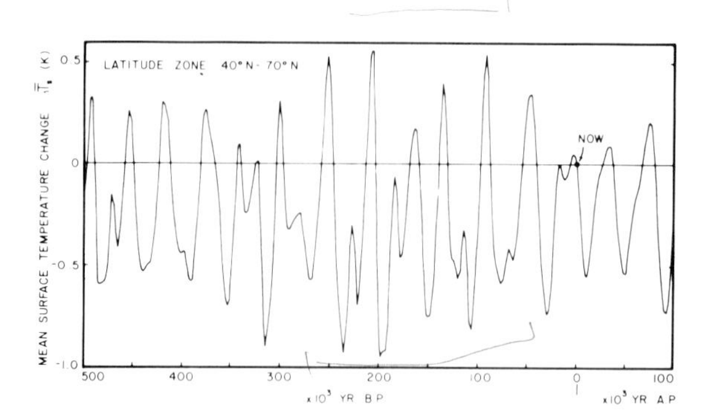

Figure 4. Time series of $\Delta\overline{T}_S$ during a 600,000-yr period.

200,000 yr B.P., and it is associated with a 2.3° angular variation of obliquity. In addition, the eccentricity during this period is 0.05 in magnitude, causing an enhancement of the cooling. It appears from these results that both the magnitude of eccentricity and the range of obliquity may be important contributors to long-term variations in the temperature climate of the Earth.

The magnitudes of $\Delta\overline{T}_S$, by themselves, do not explain temperature variations associated with glacial events. The results do indicate that periods of glacial build-up favor orbital conditions of larger eccentricity yielding a net cooling during each precessional cycle. Furthermore, it is noted that regions in the centers of continents would have ΔT_S variations about 75% larger than the results shown in Figure 4. It is emphasized that no albedo feedback mechanisms are incorporated in the model at this time. If they were added, there would tend to be a cumulative addition of successive temperature changes with time, and a corresponding phase shifting of maximum and minimum values, compared to those in Figure 4.

4. CONCLUSIONS

An empirical model is presented which employs satellite observations of radiant reflectance to define the Earth's response to varying distributions of insolation. The model explores the non-linear dependence of radiant emittance on temperature and its coupling to surface temperature. Sensitivity studies are presented for varying orbital conditions in terms of mean surface temperature changes from today's values ($\Delta\overline{T}_S$). The results indicate: (1) latitudes north of 50° have temperature variations ranging from 1 K to more than 3 K; (2) 55 and 45°N are more sensitive to variations of eccentricity than other latitudes, exhibiting 1.7 to 1.9 K ranges of variation during the precessional cycle; (3) maximum eccentricity of 0.06 causes a net change of about -0.5 K; (4) results support the Milankovitch hypothesis except that a preference is indicated for a perihelion occurring during the Northern Hemisphere summer rather than the winter; (5) $\Delta\overline{T}_S$ values range a maximum of 0.56 to -0.95 K during a 600,000 yr period, based on time-varying orbital elements, and

(6) both the range of obliquity variations as well as eccentricity may be important factors in long-term temperature changes of the Earth.

The magnitudes of predicted temperature changes demonstrate the importance of the non-linear character of the Earth's radiant emittance. Work is currently underway at Drexel to incorporate the effects of non-linear albedo feedback within the model to predict time-varying temperature trends.

ACKNOWLEDGMENTS

Special thanks are expressed to Karl Guenther, an undergraduate student at Drexel, for coding and running all the computer programs related to this research. This effort was supported by the National Environmental Satellite Service, NOAA, under Grant No. 04-6-158-44061.

REFERENCES

1. M. Milankovitch, Canon of Insolation and the Ice-age Problem, translated from German, U. S. Dept. of Commerce, CFSTI, Sprinfield, VA (-Belgrade) 1941.
2. J. Hays, J. Imbrie, and N. J. Shackleton, Science, 194, 4270, 1976.
3. T. M. L. Wigley, Nature, 264, 629, 1976.
4. T. H. Vonder Haar and V. E. Suomi, J. Atmos. Sci. (3), 28, 305, 1971.
5. E. T. Raschke, T. H. Vonder Haar, W. R. Bandeen, and M. Pasternak, J. Atmos. Sci., 30, 341, 1973.
6. F. B. House, unpublished compilation of ESSA 7 satellite earth radiation budget observations, 1978.
7. W. L. Smith, J. Hickey, H. B. Howell, H. Jacobowitz, D. T. Hilleary, and A. J. Drummond, Appl. Optics, 16, 306, 1977.
8. R. L. Jenne, H. van Loon, J. J. Taljaard, and H. L. Crutcher, NOTOS, 17, 35, 1968.
9. R. L. Jenne, Zonal Means of Climatological Analyses of the Northern Hemisphere, available at the NCAR data bank, 1978.
10. A. Berger, Astron. and Astrophy., 51, 127, 1976.

SOLAR WIND-MAGNETOSPHERE-IONOSPHERE COUPLING

Edward W. Hones, Jr.
University of California
Los Alamos Scientific Laboratory
Los Alamos, NM 87545 USA

ABSTRACT. The flowing solar wind imparts mass, momentum, and energy to the magnetosphere. It does so in the outer region of the magnetosphere, the boundary layer, both by direct entry of solar wind plasma into closed flux tubes of the boundary layer and by connection of interplanetary magnetic field (IMF) lines to the Earth's magnetic field. The latter mechanism seems to be dominant when the IMF has an appreciable southward component and several observations demonstrating its importance are discussed. The former may be responsible for the enduring presence of the magnetotail during even prolonged intervals of northward IMF. By either mechanism of interaction the basic process, transfer of momentum, is accompanied by electric polarization of the interaction region. The polarization charge leaks away as a depolarizing current that flows along magnetic field lines to the ionosphere where it imparts momentum to the ionospheric plasma. At the same time the momentum exchange in the interaction region stretches the field lines there downstream to form the magnetotail — a reservoir of stored magnetic energy that frequently releases stored energy to the nightside of the Earth — creating substorms. Recent evidence suggests that the field aligned currents from the interaction region create field aligned electric fields at altitudes of 1 to 2 R_E and that these are responsible for the acceleration of the particles that produce auroral arcs.

1. INTRODUCTION

The object of this paper is to describe the influence of the Sun's particulate emissions upon the Earth's magnetosphere and thence upon its ionosphere. The Sun's particulate emissions can be considered in two categories — the low energy electrons and ions that constitute the solar wind plasma and the high energy electrons and ions that are ejected from solar flares and active regions. The solar wind plasma particles move coherently and interact with the magnetosphere as a fluid. The high energy particles, having much lower spatial density than the plasma particles, move independently of each other and exert little influence on the electrical properties of the ambient medium. This paper is concerned only with the solar wind interaction.

B.M. McCormac and T.A. Seliga (Eds.): Solar-Terrestrial Influences on Weather and Climate. 83-100.

2. PLASMA, MOMENTUM, AND ENERGY TRANSFER FROM THE SOLAR WIND TO THE MAGNETOSPHERE

In this section we describe two conceptually different (but concurrently acting) mechanisms whereby transfer of plasma, momentum, and energy from the solar wind to the magnetosphere are thought to take place. One is by direct (possibly diffusive) entry of solar wind plasma into a "magnetospheric boundary layer" of closed magnetic flux tubes. The other is by connection of solar wind magnetic field lines to the Earth's field lines. Observational evidence available today does not permit us to deterime quantitatively the relative importances of these two mechanisms. The rate of transfer by the first mechanism may depend only upon the rate of momentum transport (i.e., the dynamic pressure) of the solar wind and thus may provide a lower limit below which the total transfer rate never falls. The transfer rate by the second mechanism, which depends on the solar wind magnetic field, is highly variable and is probably the dominant mechanism during geomagnetically disturbed times.

2.1 Transfer via a magnetospheric boundary layer

A particle moving with velocity $\overline{V}$ in a magnetic field $\overline{B}$ experiences a Lorentz force,

$$F_L = q(\overline{V} \times \overline{B}) \tag{1}$$

that is perpendicular to $\overline{B}$ and in opposite directions for opposite signs of the charge, q. This is illustrated in the upper left part of Figure 1 where $\overline{B}$ is directed out of the figure and electrons and protons are deflected to the left and right respectively. The application of this phenomenon for the generation of an electric current is illustrated at upper right in Figure 1. Plasma from a plasma generator is forced through the magnetic field, $\overline{B}$. Deflected electrons accumulate at the left edge of the $\overline{B}$ region and deflected protons accumulate at the right. In the absence of an external conducting circuit this "polarizing current" of deflected particles would quickly produce charge layers at the edges of the $\overline{B}$ region whose electric field would exactly compensate the Lorentz force on subsequent particles and they would then flow straight across the field, suffering no deflection. The external circuit shown in the figure provides a path for a "depolarizing current" that will tend to neutralize the accumulating charge. Note that the polarizing current, $\overline{J}_p$, flowing from left to right through $\overline{B}$ exerts a force $\overline{J}_p \times \overline{B}$ (downward in the figure) that opposes the plasma flow from the generator, i.e., the plasma's momentum is being reduced by the interaction. Note, also, that if the $\overline{B}$ field were extended to include the "electrical load" or the load were moved so as to lie within the shown $\overline{B}$ field (but out of the plasma flow region, i.e., in front of or in back of the plane of the figure) the depolarizing current J_D, flowing from right to left, through $\overline{B}$ would exert a force $\overline{J}_D \times \overline{B}$ (upward in the figure) that is in the direction of the plasma flow from the generator. Thus, the momentum lost by the input plasma is imparted to the electri-

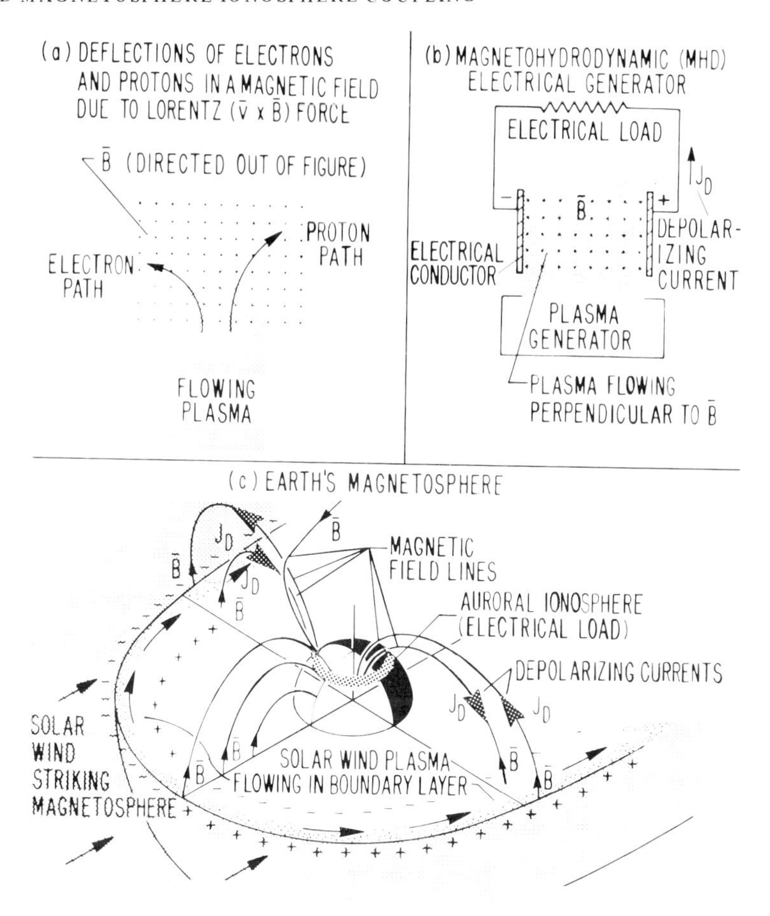

Figure 1. Upper left: Curved paths of electrons and protons in a magnetic field. Upper right: Electrostatic polarization of, and depolarizing current from, plasma forced to flow across a magnetic field. Bottom: Polarization of plasma in magnetospheric boundary layer and resulting depolarizing currents to the Earth's ionosphere.

cal load. Laboratory experiments demonstrating these momentum transfer processes in magnetized plasma have been performed by Baker and Hammel[1].

The lower part of Figure 1 illustrates how the above principles apply to the solar wind's loss of momentum to the equatorial sector of the magnetosphere. (The same principles apply elsewhere over the magnetosphere's surface and under the action of other processes to be mentioned below.) It has been observed with instrumented satellites (Hones et al. [2]; Rosenbauer et al.[3]; Eastman et al.[4]) that solar wind plasma penetrates through the magnetosphere's outer boundary (the magnetopause) and populates a region several hundred to several thousand kilometers thick with solar wind-like plasma. This region is called the boundary layer and is identified in Figure 1 by the decreasing density of dots inside the "magnetopause" (the solid curved line) with increasing penetration of the magnetosphere. As indicated by arrows, the plasma in the boundary layer flows inside and approximately tangentially to the magnetopause. The decreasing density of dots implies that the density of this solar wind plasma decreases with increasing penetration. It is also observed that the flow velocity of the plasma decreases with increasing penetration. Thus, with increasing penetration there is a decreasing momentum density of this penetrating solar wind plasma. Furthermore Eastman et al. [4] and Palmer and Hones [5] found that the boundary layer field lines at low latitudes are closed (i.e., both ends rooted at the Earth). Thus there must be a polarization electric field in the boundary layer to allow the flow of plasma across the magnetic field; the signs of the charge layers that produce the electric field are indicated by (+) and (-) signs in Figure 1 and are seen to be oppositely oriented with respect to the magnetopause on the morning (left) and evening (right) sides of the magnetosphere.

Besides the decreasing momentum density of the boundary layer plasma with increasing penetration into the magnetosphere (mentioned above) a further sign that this plasma is imparting its momentum to the magnetosphere is that the boundary layer field lines are observed to be strongly draped back toward the nightside showing that the magnetic field is acting against the plasma flow, i.e., that there is a polarizing current flowing across the boundary layer, inward on the dawn side and outward on the dusk side, making $\bar{J}_p \times \bar{B}$ sunward on both sides. These currents must be closed by depolarizing currents, J_D, flowing to the auroral ionosphere along the magnetic field lines in the directions indicated in Figure 1 by arrowheads drawn on sample field lines. These depolarizing currents, when they reach the conducting ionosphere, flow perpendicular to the field lines there. By their $\bar{J}_p \times \bar{B}$ force they transfer some of the momentum of the boundary layer plasma to the ionospheric plasma. Some of the momentum is also absorbed by the boundary layer magnetic field lines directly as the $\bar{J}_p \times \bar{B}$ force caused by the polarizing current stretches these closed flux tubes back along the flanks of the magnetosphere, adding to the magnetic flux in the magnetotail.

The addition of these flux tubes to the tail constitutes a storage of energy there. We shall see that the magnetic connection process that will be discussed next adds flux to the tail in a somewhat different manner. Release of the stored magnetic energy from the tail is an important consequence of these storage processes and will be discussed in a later section.

2.2 Transfer by connection of solar wind and magnetospheric magnetic field lines

The solar wind is threaded with magnetic field lines that it pulls out from the Sun. This interplanetary magnetic field (IMF), although its energy density is only about 1% of the solar wind's total energy density, strongly influences the coupling of solar wind plasma, momentum, and energy into the magnetosphere. It is thought to do this by becoming connected to the Earth's magnetic field. It thus provides paths for solar wind plasma to penetrate directly into some regions of the magnetosphere. Equally important, this connection of the IMF with the Earth's field provides a means whereby electric fields and magnetic stresses in the passing solar wind can be transmitted directly into the magnetosphere; momentum of the passing solar wind plasma can thus be imparted to the ionosphere in the same manner as was described above in our discussion of boundary layer processes.

The above discussions have illustrated that the basic solar wind-magnetosphere interaction is transfer of momentum from the solar wind to the magnetosphere. It is important to note that the transfer of momentum can be visualized as proceeding concurrently in two ways: (a) the force due to the depolarizing current flowing from the interaction region transfers momentum to the ionosphere, and (b) the polarizing current flowing within the interaction region causes a resistance to the plasma flow so that the magnetic field lines are stretched into the tail where they constitute stored energy. If the current flow were to stop, i.e., if the ionosphere suddenly became nonconducting, the interaction region would very quickly become fully polarized, acquiring an electrostatic field, $\bar{E} = -\bar{V} \times \bar{B}$ that would exactly compensate the Lorentz force on the plasma particles. The plasma would then flow unimpeded through the field and there would be no momentum transfer.

2.3 Influence of solar wind properties on interaction processes

Geomagnetic activity is the phenomenon most readily used as a measure of the solar wind-magnetosphere interaction. The activity is commonly characterized by an "index" measuring the degree of magnetic disturbance at a selected group of observatories. Since the basic interaction is transfer of momentum from the solar wind to the magnetosphere we expect that geomagnetic activity should increase with increasing momentum flux of the solar wind. If there are other solar wind parameters that enhance the coupling to the magnetosphere of the solar wind momentum flux they would be expected to increase the magnetic activity corresponding to a given momentum flux. We shall now mention briefly the results of one of many studies that have related geomagnetic activity to various solar wind parameters.

Svalgaard [6] related the am-index, an approximate index of global activity, to several solar wind parameters derived from measurements taken over nearly a complete solar cycle. Parameters treated separately in that study were: the solar wind momentum flux ($\propto nV^2$), where n is the number density of solar wind ions); BV, the influx of interplanetary magnetic field lines per unit length across the front of the magnetosphere; the angle, α , that the IMF makes with the Earth's field at the nose of the magnetosphere; the variability of the direction of the IMF; and the inclination of the Earth's dipole axis to the Earth-sun line. Svalgaard found that am varies linearly with BV and as $(nV^2)^{1/3}$. He also found a strong dependence on α and a weaker dependence on the variability of the IMF. Also there was a small but persistent tendency for am to be larger when the Earth's dipole was perpendicular to the solar wind flow direction.

The dependence of geomagnetic activity on α, with greater activity being associated with $\alpha > 90°$ (i.e., with southward IMF), that is illustrated in Svalgaard's study was first reported by Fairfield and Cahill [7] and has been substantiated in many other studies in the intervening decade. It is, of course, generally interpreted as strongly supporting the connection of the IMF to the Earth's field. In fact, this simple geometrical concept that connection of these fields occurs, and does so more efficiently with increasing southward relative orientation of the IMF, has had remarkable success in explaining a variety of observations. It is worthwhile to discuss some of these here.

The IMF usually displays two or four sectors corotating with the sun and having oppositely directed fields lying approximately in the ecliptic plane (Wilcox and Ness[8]). Away (also called positive or (+)) sectors contain field pointing away from the Sun along the "garden-hose" spiral at an angle that averages 135° to the east of the Sun. Toward (also called negative or (-)) sectors contain field pointing along the garden hose spiral at an angle that averages 45° to the west of the Sun. The field in these sectors also has randomly varying north and south components but these average close to zero. The 23.5° tilt of the Earth's axis with respect to the ecliptic plane causes the field in toward sectors of the IMF to assume a negative (southward) component relative to the Earth's field near spring equinox and northward component near fall equinox. These polarities are reversed for away sectors in the same seasons. Near the solstices no relative north or south components are caused. Russell and McPherron[9] pointed out that the occurrence of the southward components at the equinoxes due to this effect could be the cause of the long-recognized equinoctial maxima of geomagnetic activity. Their view was subsequently reinforced by Paulikas and Blake [10] who observed that trapped energetic electron fluxes at geosynchronous orbit, an indicator of geomagnetic activity, were correlated with passages of sector boundaries of the IMF. The electron fluxes were highest in the fall for away sectors and highest in the spring for toward sectors. Further support came from Sheeley _et al._ [11] who correlated geomagnetic activity with solar wind speed. They found, not surprisingly, that activity increased with increasing wind speed. Also, at a given solar wind speed the activity was greatest when the Earth was in toward sectors in the spring and in away sectors in the fall. These results of Sheeley _et al._ [11] are shown in Figure 2.

The IMF does not have to be exactly antiparallel to the Earth's field for connection to occur. Theories have been developed, e.g., Mozer _et al._[12] that predict how effectively arbitrarily oriented IMF will reconnect with the Earth's field. Several observations have suggested that the y-component of the IMF (i.e., the component perpendicular to

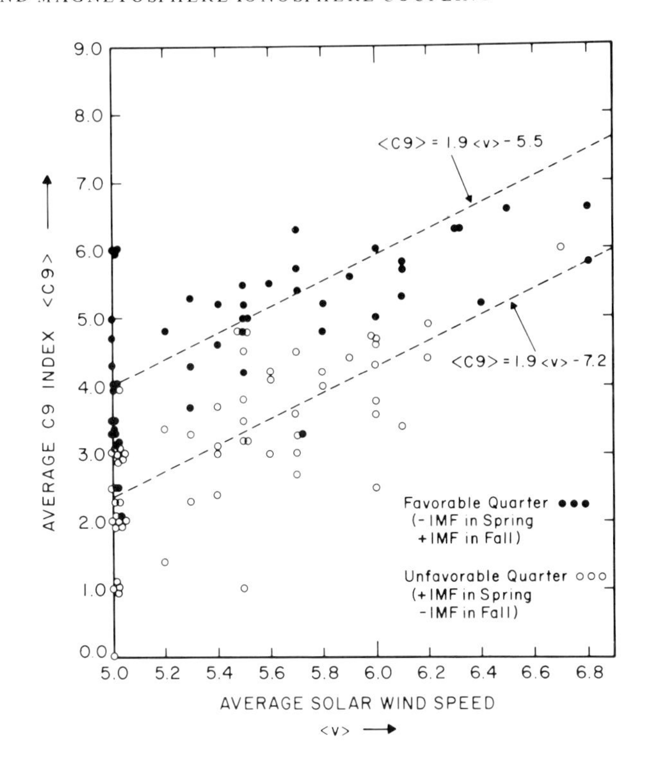

Figure 2. Average C9 index during passage of a given solar wind stream versus the average speed of that stream in units of 100 km s^{-1}. Closed circles are data taken in IMF toward-sectors in Spring and in away-sectors in Fall, conditions which are expected to provide a net southward component of the IMF. Open circles are data taken in IMF away-sectors in Spring and in toward-sectors in Fall, conditions which should provide a net northward IMF component (Sheeley et al.[11]).

the Earth-Sun axis) strongly influences the character, if not the magnitude, of the solar wind-magnetosphere interaction. As an example of this, Svalgaard [13, 14] deduced from analysis of ground magnetic records that an ionospheric current circles each magnetic pole at invariant latitude $\sim 10°$ away from the pole. The current direction as seen from above the poles is counterclockwise during away-sectors of the IMF and clockwise during toward-sectors. This phenomenon occurs so reliably that Svalgaard [15] used it to study the long-term behavior of the sector structure over four solar cycles. Friis-Christensen et al. [16], also studying ground magnetic records, were able to show that it is the y-component and not the x-component of the IMF that is associated with this current system (i.e., in away-sectors the y-component points 90° east of the Sun and in toward-sectors it points 90° west of the Sun). Independently Heppner [17], with polar satellite measurements, found that the dawn-to-dusk electric field over the polar caps is strongest on the dawn side of the northern polar cap and on the dusk side of the southern polar cap when the Earth is in an away-sector and reversed from this when Earth is in a toward-sector. Mozer _et al._ [12] found this same relationship in polar cap electric fields measured with balloons. They concluded that the magnetic effects studied by Svalgaard [13, 14] and by Friis-Christensen _et al._ [16] could be ascribed to the Hall currents flowing in response to the asymmetric polar cap electric fields and that all of these observations could be understood in terms of the magnetic connection model illustrated in Figure 3. The line of connection of the interplanetary and terrestrial fields is inclined at an angle that depends on the direction of the interplanetary field. The reconnection line coincides with the ecliptic plane when the IMF is purely southward but is tilted as indicated in the figure for pure positive or negative y-directed IMF. The white arrows indicate the resulting tendency for polar cap flow to be strongest (corresponding to the strongest electric field) over the dawn side of the northern polar cap for $B_y > 0$ and over the dusk side for $B_y < 0$.

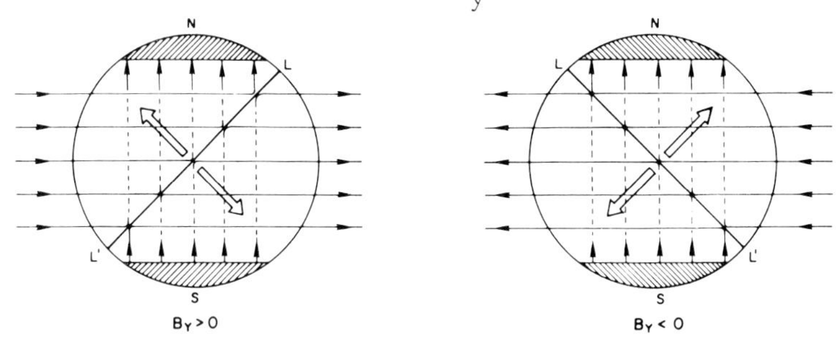

Figure 3. View from the Sun of the magnetopause convection due to the assumed reconnection between an azimuthal interplanetary magnetic field and the geomagnetic field for the cases $B_y > 0$ and $B_y < 0$. The solid lines are interplanetary magnetic field lines. The dashed lines are terrestrial magnetic field lines on the magnetopause. The hatched areas are the polar caps (Mozer _et al._ [12].

The final matter we shall discuss in this section is the dipolar component of the Sun's magnetic field, its observed 22 yr cycle of reversal (i.e., the solar-magnetic cycle) and the influence of this on geomagnetic activity. This observed 22 yr cycle in geomagnetic activity can be explained as another consequence of connection of the IMF to the Earth's field, even more subtle in its origin than the one just discussed. In a study of satellite measurements of the IMF over the period 1962 - 1968 Rosenberg and Coleman [18] found evidence for a distinct dominant polarity of the magnetic field. The dominant polarity of the field was inward (toward the Sun) at heliographic latitudes above the solar equatorial plane and outward at latitudes below the plane. Their analysis is illustrated in Figure 4. The data points show the number of days of negative polarity for each solar rotation. These numbers are greatest around September of each year when the Earth is at highest northern heliographic latitude (7.3°). These directions were in the same sense as the dipolar component of the Sun's field deduced from magnetographs of the polar regions of the Sun taken in the

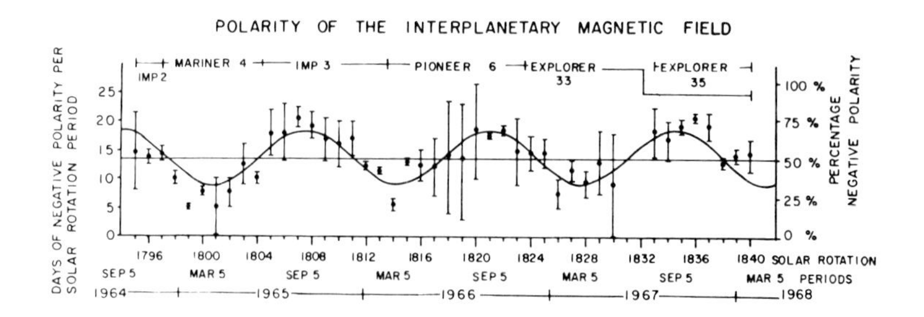

Figure 4. Polarity (toward or away) of the IMF. For each solar rotation the lower bar is the actual number of days of toward polarity. The upper bar is 27 minus the number of days of away polarity. The distance between the bars is the number of days of missing data. The dot is centered between the bars (Rosenberg and Coleman [18]).

same period. The latter showed that the dipolar component of the Sun's field had been inward over the northern hemisphere and outward over the southern hemisphere since the maximum of solar activity of sunspot cycle 19 in 1958.

Wilcox and Scherrer [19] extended this analysis over four solar cycles (1926 - 1971) by using the polarity of the IMF inferred by Svalgaard [15]. Their results generally agreed with those of Rosenberg and Coleman [18]. Furthermore, because of the much greater length of their analysis period they were able to show that the dominant polarity of the IMF changes each $\sim$10 yr along with the polarity changes at the Sun. The polarity changes within $\sim$2 yr after sunspot maximum. Thus the field was inward over the northern hemisphere from $\sim$1958 to $\sim$1971 (i.e., the last half of sunspot cycle 19 and the first half of cycle 20). It has been outward over the northern hemisphere for the last half of cycle 20 and into the beginning of the present cycle 21. The existence of the dominant outward polarity near the end of cycle 20 and the beginning of cycle 21 was beautifully demonstrated by Pioneer 11 measurements at high northern heliographic latitudes reported by Smith et al. [20]. At northern latitudes $\sim 16°$ the field was almost always outward and little evidence of sector structures was seen. Those observations suggest that a wavy azimuthal current sheet lies approximately in the Sun's magnetic equatorial plane and separates field lines from the two hemispheres

In a study of geomagnetic activity covering 80 yr from 1884 to 1964, Chernosky [21] found a 22 yr cycle having a peak of activity during the last half of the even-numbered sunspot cycles and during the first half of odd-numbered cycles. Svalgaard [6] recently reported another such study, covering the years 1868 - 1976, that yielded essentially identical results; these are depicted in Figure 5. An analysis by Svalgaard illustrates that the 22 yr cycle of geomagnetic activity can be understood as follows. During the last half of even cycles and the first half of odd cycles the outward polarity of the Sun's field in the northern hemisphere and its inward polarity in the southern hemisphere give rise

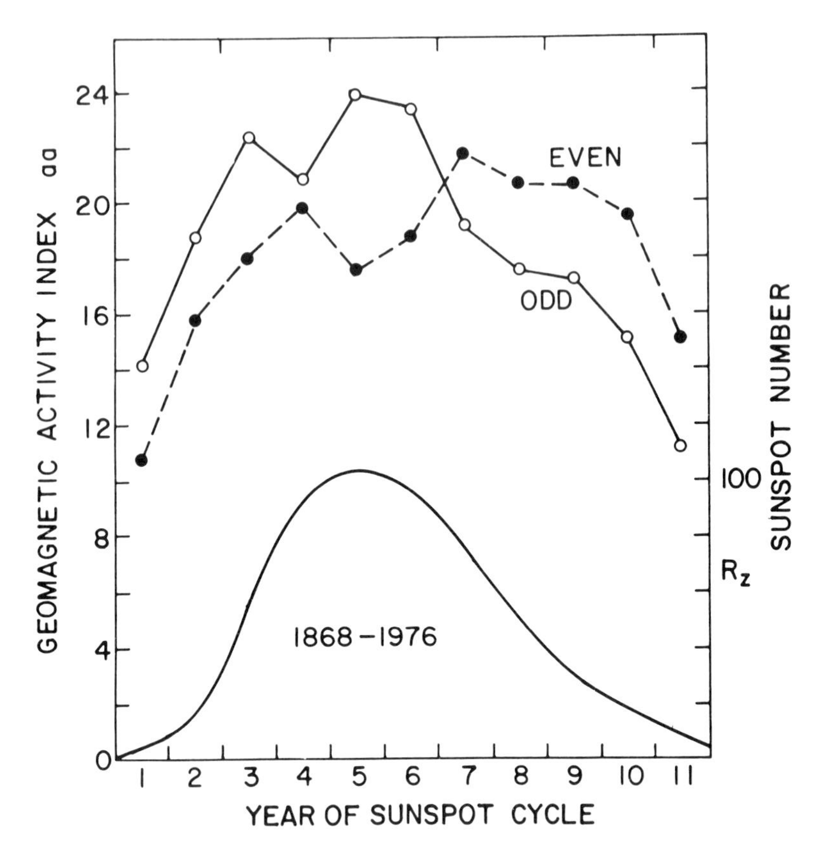

Figure 5. Variation of yearly means of geomagnetic activity as a function of phase within the sunspot cycle. Even-numbered (filled circles) and odd-numbered (open circles) cycles are considered separately. At the bottom is shown the average sunspot numbers for the solar cycles from 1868 to 1976, the interval covered by the other curves (Svalgaard [6].

to a net southward component (with respect to the Earth's field at the front of the magnetosphere) both around Spring equinox, when the Earth is at peak southerly heliocentric latitudes and around Fall equinox, when the Earth is at peak northerly latitudes. During the alternate 11 yr intervals a net northward component is presented at the Earth at the equinoxes.

We conclude this section with a caution to the reader. Despite the astounding ability of the magnetic connection concept to provide what appear to be logical interpretations of even the most subtle variations in geomagnetic activity, there remains a disturbing lack of observations of the direct consequences of connection (e.g., accelerated plasma, or energetic particles, or significant normal components of magnetic field) at the nose of the magnetosphere where the connection is supposed actually to take place.

3. THE MAGNETOTAIL: STORAGE AND RELEASE OF ENERGY

We noted in the last section that the transfer of solar wind momentum to the magnetosphere may be viewed as proceeding along two avenues concurrently, i.e., transfer to the ionosphere via the depolarizing currents flowing out of the interaction regions and transfer (and storage as magnetic energy) to the tail by the action of the polarizing currents within the interaction region. In this section we shall consider the storage and release of the energy in the magnetotail. We shall see that an appreciable fraction of the energy stored in the tail ultimately is transferred to the ionosphere. In the next section the transfer to the ionosphere from the tail and from the interaction regions will be discussed.

3.1 Energy storage in the magnetotail

The magnetotail (Ness [22]) is about 40 R_E (1 R_E = 6370 km) in diameter and is thought to extend 500 R_E or so downstream in the solar wind. The Earth's magnetic field, essentially dipolar out to geocentric distances of $\sim$5 R_E, assumes a tail-like stretched configuration beginning somewhere between $\sim$6 R_E and 10 R_E. Along the flanks of the magnetosphere, also, the field is strongly draped back, approaching a tail-like configuration. A plasma sheet lies in the midplane of the tail beyond $\sim$10 R_E and carries the electric current that divides the tail into a north lobe containing earthward-pointing magnetic field and a south lobe containing anti-earthward field (Figure 6).

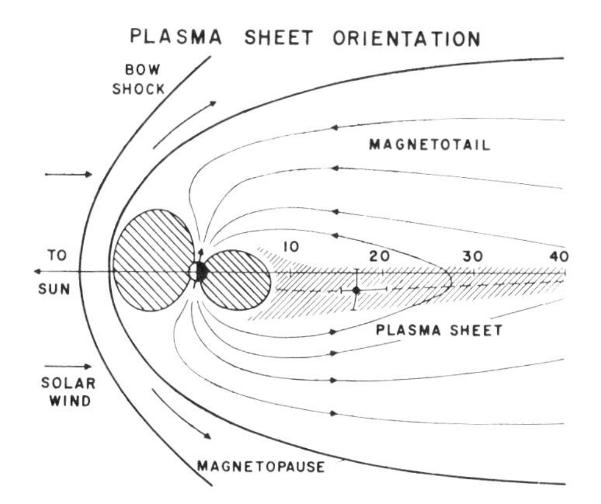

Figure 6. Sketched configuration of the magnetosphere in the solar magnetospheric noon-midnight meridional plane, showing bow shock, magnetopause, trapped radiation belts and plasma sheet (Bame et al. [34]).

Before continuing this discussion of the magnetotail it is enlightening to consider the magnitudes of some other magnetotail and magnetospheric parameters. These are listed in Table I and have been taken from several sources. The sources are indicated so that readers who are interested

Table I
Approximate magnitudes of some magnetotail and magnetospheric parameters

Quantity	Value	Source
Length	500 R_E	Svalgaard [31]
Diameter	40 R_E	
Average field strength	10 gammas (10^{-4}G)	
Energy content (magnetic and plasma)	7 x 10^{15} J	
Total cross-tail current	5 x 10^{7} A	Svalgaard [31]
Power applied continuously by solar wind to maintain the tail	3 x 10^{12}W	Svalgaard [31]
Solar wind energy flux onto the magnetosphere	1.6 x 10^{13} W	Svalgaard [31]
Total energy of Earth's field outside the solid Earth	10^{18} J	Willis [32]
Power dissipation to ionosphere and inner magnetosphere during a substorm	3 x 10^{11} W	Akasofu [33]

may look up the details of the calculations. Only a trivial fraction (0.007) of the energy of the Earth's magnetic field lies in the magnetotail. The power to maintain the tail is about 20% of the incident solar wind power. The power dissipation to the Earth and inner magnetosphere during a substorm is only about 10% of that required to maintain the tail. The last figure suggests that direct measurement of energy storage in the tail might be quite difficult to accomplish since such storage probably involves only small variations of the tail's total energy content at any instant; furthermore, both the tail field strength and the tail size may change to accommodate the changed energy content and it would seem almost a hopeless task, with one (or even a few) satellites, to determine how the tail's energy content is changing.

Caan et al.[23] used measurements of solar wind dynamic pressure and magnetic field together with simultaneous measurements of the tail lobe magnetic field to look for changes in tail energy density associated with the solar wind parameters. They found that enhancements of lobe energy density followed both southward turnings of the IMF and solar wind dynamic pressure increases. They also found that the lobe energy density decreased during substorm expansive phases. In a further attack on the tail energy storage (and release)problem Caan et al. [24] used a superposed epoch analysis technique to investigate the average relationship of the tail lobe magnetic energy density to the north-south component of the IMF and to substorm expansive phase. They found an enhanced lobe energy density to be associated with the expansive phase onset. Coincident with the expansive phase onset the tail lobe energy density starts a rapid decline.

Maezawa [25] observed associations of satellite crossings of the magnetotail boundary at distances of $\sim 30\ R_E$ to $70\ R_E$ behind the Earth with the occurrence of substorms. He found that the magnetotail radius increases during an interval of 1 to 2 h before the expansive phase onset. Shortly after the onset time the tail radius begins to decrease rapidly. He found that the average increment, ΔR of the tail radius is given by $0.5\ R_E < \Delta R < 4\ R_E$, the probable value being 1 to 2 R_E.

These in situ observations of tail changes support the concept of storage and release of magnetic energy there. But another avenue of investigation which is at least equally enlightening is the study of the variations of the diameter of the polar cap associated with substorms. The auroras (and the associated energy dissipation by precipitating particles and ionospheric currents) occur primarily in bands surrounding the north and south magnetic poles but displaced from them by $\sim 15°$ to $30°$ in magnetic latitude. These bands are called the auroral ovals. It is believed that the nightside of the ovals is magnetically connected to the tail plasma sheet and the dayside is connected to the boundary layer which, as we noted earlier, covers the sunward surface of the magnetosphere. The region of relative inactivity that comprises the central portion of the oval is connected to the tail lobes. When magnetic field is added to the tail lobes the diameter of the polar caps increases; variations of this diameter, therefore, are a measure of addition and loss of tail magnetic energy. Akasofu [26] has attempted to quantify this relationship by examining the diameter of the oval in auroral photographs taken by the U. S. Air Force DMSP satellites. He has found that periods of southward IMF do, in fact, lead to increased oval diameter, while periods of northward IMF lead to its decrease. When the oval is large, large areas of auroral activity (i.e., substorms) occur on the nightside, indicating that when there is much stored energy in the tail it is quite susceptible to being dissipated in auroral processes. When there is little stored energy there is little dissipation. Yet, even when auroral activity reaches very low values the area of the oval retains a finite size ($\sim 5 \times 10^6\ km^2$). This suggests that the magnetotail energy reservoir has a "ground level" below which it never falls. It may well be that this minimum level of stored energy is that maintained by the solar wind-boundary layer interaction when coupling by magnetic connection is at a minimum, and when the dynamic pressure of the solar wind is low.

3.2 Energy release from the magnetotail

Table I illustrates that dissipation of energy by the magnetotail must occur continuously, i.e., at a rate $\sim 3 \times 10^{12}$ W. This is the power that is used just to maintain the tail, analogous to the heat dissipated in the windings of a magnet by the current required to maintain the magnetic field. It is not known in any detail where or how the major part of this energy is dissipated. However, some of it ($\sim 10\%$ by Table I) is lost to the Earth and inner magnetosphere during substorms.

A magnetospheric substorm is characterized at Earth by a sudden brightening of auroras in the midnight sector of the auroral oval and subsequent development of a "bulge" of auroral brightness that spreads poleward into the polar cap and to the east and west along the oval. Intense westward currents (the westward electrojet) flow in the ionosphere near the bright auroras. These cause perturbations of the magnetic field at the ground underneath that are characterized, particularly, by decreases of the horizontal component of the main field called "negative bays." One theory holds that a substorm occurs when a part of the (westward flowing) cross-tail current that maintains the tail is somehow disrupted and becomes diverted down magnetic field lines and along a westward path through the ionosphere. This concept is illustrated in Figure 7.

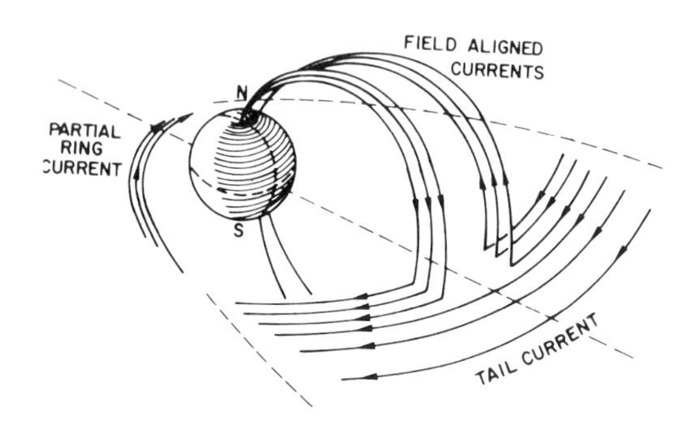

Figure 7. Currents within the magnetosphere during a magnetospheric substorm. The magnetotail current is disrupted and the magnetospheric currents establish a new circuit down the field lines to the ionosphere and back again to the tail. The intensity of the ring current becomes enhanced (Svalgaard [29]).

The disruption of the cross-tail current results in formation of a magnetic neutral line or x-line. The connection of magnetic field lines at the x-line allows the earthward portions of the field lines to contract rapidly Earthward, resulting in a convection of erstwhile plasma sheet plasma into the inner magnetosphere. The plasma is heated to energies from a few keV to a few tens of keV by magnetic compression as it is pushed inward to geocentric distances $\sim$5 to 6 R_E where it then drifts around the Earth to form a ring current. Also, some of the plasma is projected down field lines and precipitates into the ionosphere where it generates auroral light. Protons and electrons with energies of several hundred keV, sometimes as high as 1 MeV, are observed in the plasma sheet during some substorms. These are thought to be generated by strong electric fields induced by the magnetic variations near the x-line as the cross-tail current is disrupted.

4. INTERACTION OF THE OUTER MAGNETOSPHERE WITH THE IONOSPHERE

We have seen that in its interaction with the magnetosphere the solar wind is slowed by generating polarizing currents whose closing depolarizing currents flow along magnetic field lines and transfer the lost momentum to the ionosphere. These currents from the solar wind's initial interaction with the magnetosphere flow into the dayside portion of the auroral oval at the edge of the polar cap, since that is where field lines from the dayside surface of the magnetosphere reach the Earth. In like manner, stresses from the magnetotail, related to the energy changes therein (e.g., substorms), are transmitted by field aligned currents to the nightside of the auroral oval. Thus, if a good understanding of the ionosphere-outer magnetosphere interaction is to be achieved, it will be done in large part by developing a thorough understanding of the field aligned current generation, distribution, variations, and interactions with ionospheric plasmas.

Field aligned currents can be measured only at high altitudes (in and above the ionosphere) by rockets or satellites since their magnetic signatures at ground level are often indistinct and not reliably identified. The satellite and rocket measurements of the currents rely on the magnetic perturbations they cause transverse to the main field. The currents have been found to be distributed in sheets that trace, approximately, the curve of the auroral oval. Their distribution is illustrated for quiet and for more disturbed periods in Figure 8. The poleward-most ring has become known as the system 1 currents and the equatorward ring is known as the system 2 currents. Note that the system 1 currents flow into the ionosphere between 00 UT and 12 UT and out of the ionosphere from 12 UT to 24 UT. The system 2 currents have directions reversed from these. The two current systems are the same over the southern hemisphere with the same current directions relative to the ionosphere. The current densities in the region of measurement, a few hundred kilometers above the Earth, range from $\sim$0.3 to $\sim$2.5 mA m^{-2} (Iijima and Potemra [27]). Figure 8 illustrates another essential feature of the field aligned currents; that is, the underlying pattern does not change appreciably with increasing geomagnetic activity, it simply shifts to somewhat lower latitude (reflecting the increase in diameter of the auroral oval and thus in stored magnetotail energy as we discussed earlier). The total current flowing into the ionosphere is equal to that flowing out within experimental error. The current is $\sim 2.5 \times 10^6$ A during quiet times and $\sim 5 \times 10^6$ A for average disturbed times.

The system 1 currents border the polar cap; thus, those on the dayside are thought to connect to the sunward outermost regions of the magnetosphere. It is reasonable to suspect that these currents are the depolarizing currents from the magnetospheric boundary layer, especially so since it is found that boundary layer-like (i.e., low energy) plasma precipitates into the atmosphere along the system 1 current oval (McDiarmid et al. [28,29]). Eastman et al. [4] have shown that the direction and other features of the currents are consistent with this interpretation. The system 1 currents on the nightside of the polar cap are thought to flow on field lines that lie in the top and bottom surfaces of the magnetotail plasma sheet; the driving mechanism for these nightside system 1 currents is not known but no doubt it involves forces acting between the plasma and magnetic field in the plasma sheet.

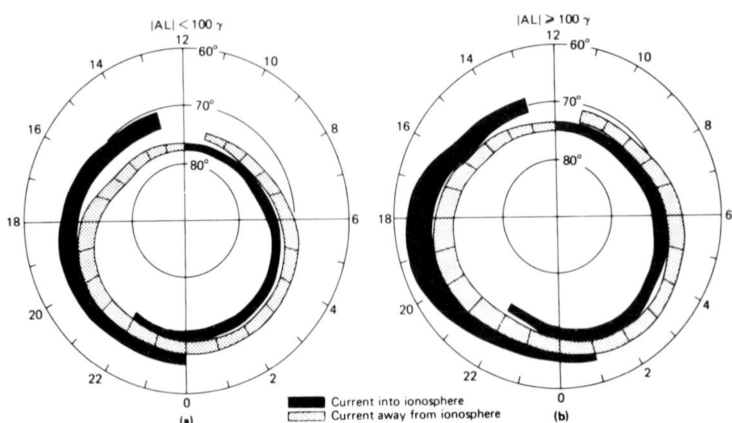

Figure 8. A summary of the distribution and flow directions of large-scale field-aligned currents determined from (a) data from 439 orbits of the Triad satellite during weakly disturbed conditions ($|AL| < 100\ \gamma$) and (b) data obtained from 366 Triad orbits during active periods ($|AL| \geq 100\ \gamma$) (Iijima and Potemra [27]).

The system 1 currents may also be the source of charge sheets that are the origin of the dawn-to-dusk electric field over the polar caps (mentioned earlier) that drives the tailward convection (i.e., $\bar{E} \times \bar{B}$ drift) of plasma there.

Above ∿150 km altitude currents perpendicular to $\bar{B}$ are prevented by lack of ion or electron collisions with neutral particles; here application of an electric field perpendicular to $\bar{B}$ simply causes an $(\bar{E} \times \bar{B})/B^2$ drift of ions and electrons that produces no net current. But below 150 km the ion-neutral collision frequency starts to become appreciable and the field aligned currents start to become diverted so as to pass horizontally through the ionosphere. In this region an electric field perpendicular to $\bar{B}$ will cause ion drift motion in two directions — in the $\bar{E}$ and $\bar{B}$ direction (Hall drift) and, as a consequence of the ion-neutral collisions, in the direction of $\bar{E}$ (Pedersen drift). The electron-neutral collision rate, however, remains inconsequential until altitudes below ∿80 km are reached. Thus perpendicular electric fields cause only a Hall drift of electrons in the region 80 to 150 km where ions have a substantial Pedersen drift, resulting in an altitude-dependent mixture of Hall and Pedersen currents in this region. Since the Hall currents are perpendicular to $\bar{E}$, no energy dissipation or voltage drop is associated with them. However, voltage drops and energy dissipation do result from the Pedersen currents. So the currents from the outer magnetosphere, diverted to flow horizontally through the auroral ionosphere as Pedersen currents, establish an electric field pattern there and, by means of their $\bar{J} \times \bar{B}$ forces, transfer the momentum of outer magnetospheric plasma to the ionospheric plasma and so to the neutral atmosphere at ionospheric heights. A detailed description of the ionospheric electric fields and current flow, and of their many consequences, is beyond the scope of this review but may be found, for example, in Banks [30].

5. SUMMARY

Momentum of the solar wind is transferred to the outer regions of the magnetosphere by direct entry of solar wind plasma into closed flux tubes at the magnetosphere's surface and by connection of the IMF to the terrestrial magnetic field. In both mechanisms electrical polarization of the solar wind plasma as it thus moves through an "interaction region" in the outer magnetosphere constitutes a source of electrical currents that cause a transfer of some of the lost momentum to the ionosphere and some to the magnetotail. The solar wind-magnetosphere interaction is, in several respects, very similar to the functioning of an MHD electrical generator. Various features of the solar wind influence the efficiency of the momentum coupling with the magnetosphere but the most important in this respect seems to be the direction (and to a lesser extent, the magnitude) of the IMF. This has been demonstrated in many convincing ways and seems to confirm that connection of the IMF to the terrestrial field occurs. But direct evidence of connection has yet to be found at the magnetosphere's Sunward surface where the process presumably occurs.

The currents from the interaction region flow along magnetic field lines to the polar ionosphere. These "field-aligned currents" cause field aligned electric fields in the few-thousand-kilometer altitude region that accelerate ionospheric particles to few-keV energies. Dispersed, and flowing horizontally in the lower ionosphere, the currents provide the $\bar{J} \times \bar{B}$ force that transfers solar wind momentum to ionospheric plasma and thence to the neutral gas.

Storage of solar wind energy in the magnetotail and its release during substorms can be sensed both by direct observations of the tail environment and by observations of changes of the auroral oval. Of the energy lost from the tail during a substorm, some is given to the Earth and inner magnetosphere and some is ejected downstream from the tail in a magnetic bubble or "plasmoid."

ACKNOWLEDGMENT

This work was done under the auspices of the U. S. Department of Energy.

REFERENCES

1. D. A. Baker and J. E. Hammel, Phys. of Fluids 8, 713, 1965.
2. E. W. Hones, Jr., J. R. Asbridge, S. J. Bame, M. D. Montgomery, S. Singer, and S. I. Akasofu, J. Geophys. Res. 77, 5503, 1972.
3. H. Rosenbauer, H. Grunwaldt, M. D. Montgomery, G. Paschmann, and N. Sckopke, J. Geophys. Res. 80, 2723, 1975.
4. T. E. Eastman, E. W. Hones, Jr., S. J. Bame, and J. R. Asbridge, Geophys. Res. Lett. 3, 685, 1976.
5. I. D. Palmer and E. W. Hones, Jr., J. Geophys. Res. 83, 2584, 1978.

6. L. Svalgaard, in: Coronal Holes and High Speed Wind Streams, ed. by J. B. Zirker, p. 371, Colorado Associated University Press, Boulder, 1977.
7. D. H. Fairfield and L. J. Cahill, Jr., J. Geophys. Res. 71, 155, 1966.
8. J. M. Wilcox and N. F. Ness, J. Geophys. Res. 70, 5793, 1965.
9. C. T. Russell and R. L. McPherron, J. Geophys. Res. 78, 92, 1973.
10. G. A. Paulikas and J. B. Blake, Geophys. Res. Lett. 3, 277, 1976.
11. N. R. Sheeley, Jr., J. R. Asbridge, S. J. Bame, and J. W. Harvey, Solar Physics 52, 485, 1977.
12. F. S. Mozer, W. D. Gonzalez, F. Bogott, M. C. Kelley, and S. Schutz, J. Geophys. Res. 79, 56, 1974.
13. L. Svalgaard, Geophys. Pap. R-6, 11 pp., Dan. Meterol. Inst., Copenhagen, August, 1968.
14. L. Svalgaard, J. Geophys. Res. 78, 2064, 1973.
15. L. Svalgaard, J. Geophys. Res. 77, 4027, 1972.
16. E. Friis-Christensen, K. Lassen, J. Wilhjelm, J. M. Wilcox, W. Gonzalez, and D. S. Colburn, J. Geophys. Res. 77, 3371, 1972.
17. J. P. Heppner, J. Geophys. Res. 77, 4877, 1972.
18. R. L. Rosenberg and P. J. Coleman, Jr., J. Geophys. Res. 74, 5611, 1969.
19. J. M. Wilcox and P. H. Scherrer, J. Geophys. Res. 77, 5385, 1972.
20. E. J. Smith, B. T. Tsurutani, and R. L. Rosenberg, J. Geophys. Res. 83, 717, 1978.
21. E. J. Chernosky, J. Geophys. Res. 71, 965, 1966.
22. N. F. Ness, J. Geophys. Res. 70, 2989, 1965.
23. M. N. Caan, R. L. McPherron, and C. T. Russell, J. Geophys. Res. 78, 8087, 1973.
24. M. N. Caan, R. L. McPherron, and C. T. Russell, J. Geophys. Res. 80, 191, 1975.
25. K. Maezawa, J. Geophys. Res. 80, 3543, 1975.
26. S. I. Akasofu, Planet. Space Sci. 23, 1349, 1975.
27. T. Iijima & T. A. Potemra, J. Geophys. Res. 83, 599, 1978
28. I. B. McDiamid, J. R. Burrows, and E. E. Budzinski, J. Geophys. Res. 80, 73, 1975.
29. I. B. McDiamid, J. R. Burrows, and E. E. Budzinski, J. Geophys. Res. 81, 221, 1976.
30. P. M. Banks, in Solar System Plasma Physics: A Twentieth Anniversary Overview, ed. by C. F. Kennel, L. J. Lanzerotti, and E. N. Parker, North Holland Publ. Co., 1977.
31. L. Svalgaard, in: Possible Relationships Between Solar Activity and Meteorological Phenomena, NASA SP-366, ed. by W. R. Bandeen and S. P. Maran, NASA Scientific and Technical Information Office, Washington, D.C., 1975.
32. D. M. Willis, J. Atmos. Terr. Phys. 38, 685, 1976.
33. S. I. Akasofu, Polar and Magnetospheric Substorms, p. 223, Springer-Verlag, New York, 1968.
34. S. J. Bame, J. R. Asbridge, H. E. Felthauser, E. W. Hones, and I. B. Strong, J. Geophys. Res. 72, 113, 1967.

A GEOMAGNETIC ACTIVITY RECURRENCE INDEX

H. H. Sargent III
Space Environment Services Center
325 South Broadway
Boulder, CO 80303 USA

ABSTRACT: A simple recurrence index for geomagnetic activity is constructed and index values are derived for a 110 yr period using Mayaud's aa-index. The results show a strong response every 11 yr, just before the minimum epoch of the 11 yr sunspot cycle. However, the strongest response (in terms of the duration, not amplitude) is shown to occur during alternate sunspot cycles each 22 yr, and in step with the American high plains droughts and other climate extremes.

1. RATIONALE

The Space Environment Services Center issues daily three-day forecasts of the 24-h A-index of geomagnetic activity as observed at Fredericksburg, Virginia. This mid-latitude index provides a good representation of the level of geomagnetic disturbance affecting the contiguous 48 states of the United States.

The most frequently used forecasting tool is the 27-day rotation chart, a version of the charting scheme developed by Bartels [1] to demonstrate recurrence or the repeatability of patterns in magnetic activity on a 27-day basis, consistent with solar rotation. Bartels charts used in geomagnetic forecasting are shown in Figure 1. These are simply stacked plots on which each new day's 24-h value is plotted as it is received. A new line is begun each 27 days, although each line is extended 8 days to emphasize continuity in the data. The forecaster looks at the last four or five rotations in order to determine whether recurrent patterns exist. Drifts in time (longer or shorter recurrence rates), as well as trends in the amplitudes of disturbances, can be established with these charts. This information, combined with knowledge of seasonal effects, is used to predict both the amplitude and timing of geomagnetic disturbances. Other data; i.e., solar flare data, coronal hole maps, interplanetary scintillation data, and measurements from satellites in Sun orbit are used, when available, to make additional modifications to the forecast — but each forecast evolves from the Bartels charts.

B.M. McCormac and T.A. Seliga (Eds.): Solar-Terrestrial Influences on Weather and Climate. 101-104.

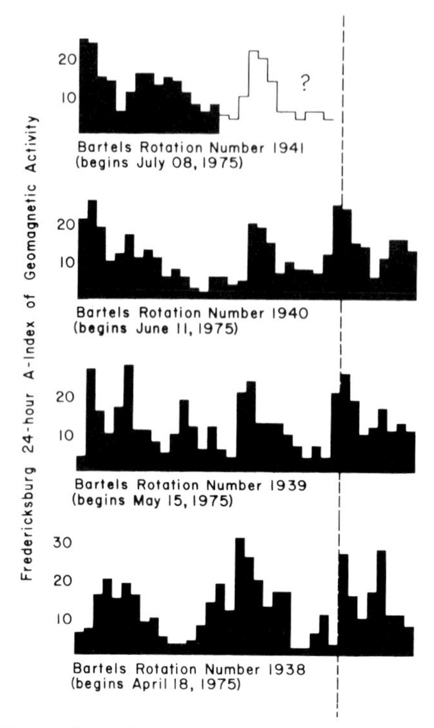

Figure 1. Bartels charts

Unfortunately, there are extensive periods during each sunspot cycle when recurrence simply does not work! Strong recurrent patterns in the Bartels charts are evidence of reasonably well-ordered and stable patterns in the interplanetary magnetic field structure. However, when major solar flares occur, shockwaves are formed which move outward through the interplanetary medium, destroying this order. When whole series of flares occur, the situation in the interplanetary medium can only be described as chaotic, and during these periods the Bartels charts are virtually useless. In order to identify those periods in the past when recurrence would have been a strong forecasting tool, a simple recurrence index was constructed using a published geomagnetic data set.

2. THE 27-DAY RECURRENCE INDEX

The data used for this study were the aa-index values of Mayaud [2], which are available from 1868 to the present. This index is recommended as a high quality consistent record of global geomagnetic activity. It is published in two 12-hourly average values per day for the entire period — thus, there are 54 values of the aa-index for each 27-day Bartels rotation.

The 27-day Recurrence Index was constructed in the following fashion:

1) The 54 values of Bartels Rotation 487 (beginning on January 12, 1968) were correlated with the 54 values of Bartels Rotation 488 (beginning on February 8, 1868). The correlation coefficient thus derived (determined to be 0.289) was assigned to the February 8th date (the first day of the second rotation — or, essentially, the midpoint of the 54-day period being examined). Next, Bartels Rotation 488 was correlated with Bartels Rotation 489 (beginning on March 6, 1868) and the correlation coefficient (0.384) was assigned to the March 6th date — and so on. Thus, each rotation was correlated with its immediate neighbor and a correlation coefficient was determined every 27 days.

2) The coefficients were all scaled by a factor of 100 (to make whole numbers of the set).

3) The entire set was smoothed using a simple 15-Bartels rotation running average. This smoothing function is nearly equivalent in length to the modified 13-mo running average traditionally used to smooth the monthly mean sunspot numbers. The resulting 27-day Recurrence Index is shown in Figure 2, along with the smoothed monthly mean sunspot numbers.

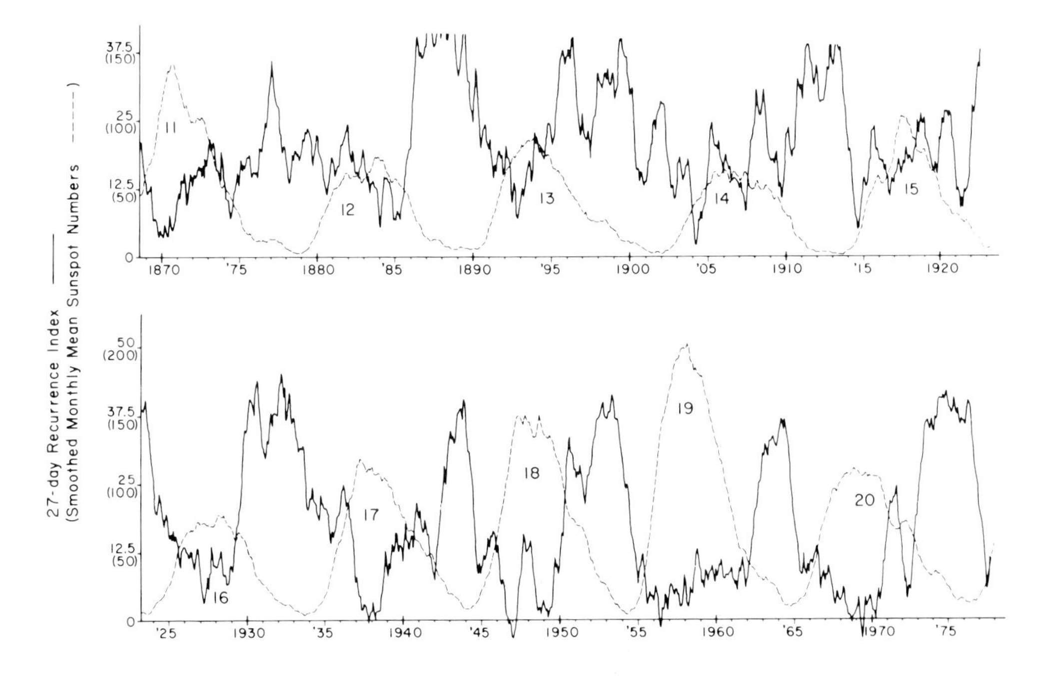

Figure 2. 27-day Recurrence Index and smoothed monthly mean sunspot numbers from 1868 to the present.

3. EVIDENT RELATION TO CLIMATIC PATTERNS

The signal-to-noise ratio of the 27-day Recurrence Index is surprisingly good, especially during the past 55 yr. Indeed, during that period, the response can almost be termed "binary" (in the sense that the signal is either "ON" or "OFF"). It should be noted that the index responds every 11 yr, prior to sunspot minimum; however, the duration of the response happens to be doubled each 22 yr. Interestingly, the double duration response of the Recurrence Index occurs at just those periods that Roberts [3] identified with the high plains droughts in the United States. There is even a hint of still more extensive response each 44 yr, which matches well-known North American climate extreme periods around the years 1888, 1932, and 1976.

Those periods of extended duration in the Recurrence Index are believed to be periods when the interplanetary structure was relatively well-ordered, stable and undisturbed for several years at a time. The high plains droughts are evidence of periods when the atmospheric circulation patterns over North America are *also* reasonably well-ordered, stable, and undisturbed for a similar interval. This does not appear to be mere coincidence, and the observation begs for a physical explanation.

REFERENCES

1. J. Bartels, *Terrestrial Magnetism and Atmospheric Electricity*, **37**, 1, 1932.
2. P. N. Mayaud, A Hundred Year Series of Geomagnetic Data 1868 – 1967, *IAGA Bulletin No. 33*, Meppel, Holland, 1973.
3. W. O. Roberts, *Possible Relationships Between Solar Activity and Meteorological Phenomena*, NASA SP-366, NASA, Washington, D.C., 1975.

EFFECTS OF SOLAR FLARES ON THE ATMOSPHERIC CIRCULATION

Cornelius J. E. Schuurmans
Royal Netherlands Meteorological Institute
Wilhelminalaan 10, De Bilt, The Netherlands

ABSTRACT. Strong solar flares cause atmospheric circulation changes (or alternatively changes in the mass and temperature distribution) at middle and high latitudes, starting less than 12 h after the solar eruption. This early effect lasts approximately one day, is strongest in winter and most pronounced at certain geographical locations. Delayed effects are reported having a maximum 2 to 4 days after a flare. In general, the connection between these effects is insufficiently known. This also applies to the possible role of solar flares in solar-climatic relationships. The observed early effect in the troposphere is plausibly explained by strong convergence of air over large areas in the lower stratosphere. However, a mechanism causing this convergence is still unknown. The early effect itself by changing the baroclinic stability in the mid-latitude troposphere could lead to some of the observed delayed effects.

1. INTRODUCTION

The question of the influence of solar activity on the circulation of the lower atmosphere is still a matter of controversy. In itself such a statement is not very fruitful or productive. Progress in this field is severely hampered by the lack of a quantitative theory, but our poor knowledge of the characteristics of the effect(s) is an equally serious drawback. In view of this it was felt appropriate to concentrate on one of the proposed effects and further analyze its nature. The subject is the effect of solar flares on the atmospheric circulation. The information in this paper is mainly based on a comparison of results described in a monograph by the present author in 1969 [1], with the results of subsequent publications on the subject. The effect is introduced in Section 2, while in Sections 3 and 4 the confirmation or nonconfirmation of it is discussed. In Sections 5 and 6 the aftereffects are treated. Finally in Section 7 some theoretical ideas concerning the cause and development of the flare influence are advanced.

B.M. McCormac and T.A. Seliga (Eds.): Solar-Terrestrial Influences on Weather and Climate. 105-118.

2. SOLAR FLARE EFFECT

In 1969 the results of an extensive study, relating solar flare outbursts to changes in the Earth's lower atmosphere, were published [1,2]. The study was based on a sample of 81 strong solar flares, which occurred during the period July 1957 - December 1959. The atmospheric parameter most extensively analyzed was the height of the 500 mb level. The statistical part of the study concentrated on the 24 h height change of the 500 mb level for various time lags with respect to the moment of a flare outburst. The investigation was global in coverage, except for the equatorial belt. It turned out that the largest mean height changes of the 500 mb level occurred very soon after the flare outburst, on average within 6 h, while the duration of the influence appeared to be restricted to the first 24 h.

A map of the maximum effect is reproduced here as Figure 1. It shows the average height change of the 500 mb level between the first radiosonde observation after a flare and the observation 24 h earlier, divided by ($\sigma N^{-\frac{1}{2}}$), N being equal to 81 and σ the standard deviation of the 24 h height changes. The first radiosonde observation after a flare varied in time from less than 1 to nearly 12 h. On the average, the map refers to a time of approximately 6 h after a flare.

Figure 1. Mean 500 mb height difference between the first aerological observation after a flare and the observation 24 h earlier, divided by the standard error of the mean.

The map shows a rather chaotic pattern consisting of cells of mean height rise, alternated by cells of mean height fall. This pattern is dominant at mid-latitudes, while at the polar regions negative changes prevail. At low latitudes, mean height changes are very small. They nowhere reach a value of two times the standard deviation ($\sigma N^{-\frac{1}{2}}$), althought this parameter itself is very small in the low latitude belt. At middle and high latitudes the number of gridpoints with a t-value ($t = \overline{\Delta H}/\sigma N^{-\frac{1}{2}}$) exceeding certain limits is larger than may be expected by chance. This has been verified by comparison with random samples (R) and by successive widening of the network in order to eliminate spatial correlations between gridpoint values. Pertinent figures on these tests are given in Tables I and II.

After these tests it was concluded that the pattern shown in Figure 1 constitutes a real effect, well above the synoptic noise at the level concerned.

Table I. TESTS

sample	epoch	number of t's where $\|t\|>1.96$ (5% level) number	%	number of t's where $\|t\|>2.58$ (1% level) number	%
R I		45	4.4	7	0.7
R II		71	7.0	10	1.0
R III		38	3.7	4	0.4
R IV		60	5.9	10	1.0
Flares	n = - 1	26	2.6	1	0.1
Flares	n = + 1	84	8.2	23	2.3
Flares	n = + 2	82	8.0	18	1.8
Expectation		51	5.0	10	1.0

Table II. Percentage of grid points where $|t|>1.96$ (5% level) for various network densities

N_t = number of grid points	R I	R II	R III	R I + R II + R III	solar flares
1020	4.4	7.0	3.7	5.0	8.2
521	4.6	7.5	4.0	5.4	8.4
262	3.4	7.2	3.8	4.8	8.0
128	5.6	5.2	3.9	4.9	7.3
66	6.1	5.0	5.6	5.6	8.6

Data for the Southern Hemisphere were analyzed in the same way, the only difference being the number of flare key days which was only 56 in this case. The resulting pattern looked very similar to the one shown in Figure 1: an alternating pattern at mid-latitudes dominated somewhat by the cells of height rise with negative height changes at the polar regions.

3. DIRECT CONFIRMATION

3.1 General Aspects

Similar and directly comparable studies as the one described above were until recently limited to the European region. For this area two analyses on independent flare samples were already published, one by Duell and Duell in 1948 [3] and one by Valnicek in 1953 [4]. In both studies mean height rises were shown to occur, as an early effect after solar flares fully comparable with the cell of height rise over Europe in our study. However, no statistical significance tests were applied in these early studies. Recently, however, Zerefos [5] performed an independent statistical check of the influence of solar flares on the heights of pressure levels in the atmosphere. His analysis was based on a sample of 33 flares from the period 1956 to 1969 and covered the whole of North America and parts of the surroundings.

Zerefos fully confirmed the alternating pattern of mean height rises and height decreases of tropospheric constant pressure levels. He also found that at 7 to 8% of the stations used in his study the t-value of 2 is exceeded. The overall agreement between Zerefos' results and those of our own analysis is striking and in my view convincingly enough to regard the effects as a real phenomenon related to the occurrence of solar outbursts.

3.2 Specific Characteristics

Apparently time is ripe to inquire further than the general aspects. Do comparable analyses agree on certain specific properties of the effect? Four specific points will be discussed here. First, the earliness and duration of the effect. The first aerological observation after a flare showing the largest effect (see Section 2) was confirmed by Zerefos' study; secondly, the distribution with height. In our study we found the largest mean height changes at the 300 mbar level, but in view of the fact that the standard deviation at this pressure level is also a maximum the significance of the effect does not differ very much from that at the 500 mb level.

Below 500 and above 300 mb the statistical significance drops very quickly as is also shown by Zerefos. Thirdly, the geographical distribution is a point of very great importance which, due to the lack of comparable hemispheric or global studies, cannot be completely covered. As far as the latitudinal distribution is concerned, we may compare Figure 2 and Figure 3, which do indicate the two things already mentioned: preponderance of the height rises at the (lower) mid-latitudes and height fall at subpolar to polar latitudes. A major difference between the two results appears to be the latitude where the zonal average changes from positive to negative: 65° N in our case and 52° N in Zerefos' case. However, we should bear in mind that the curves are not strictly comparable in view of the fact that Zerefos's study covered only the American sector of the Northern Hemisphere.

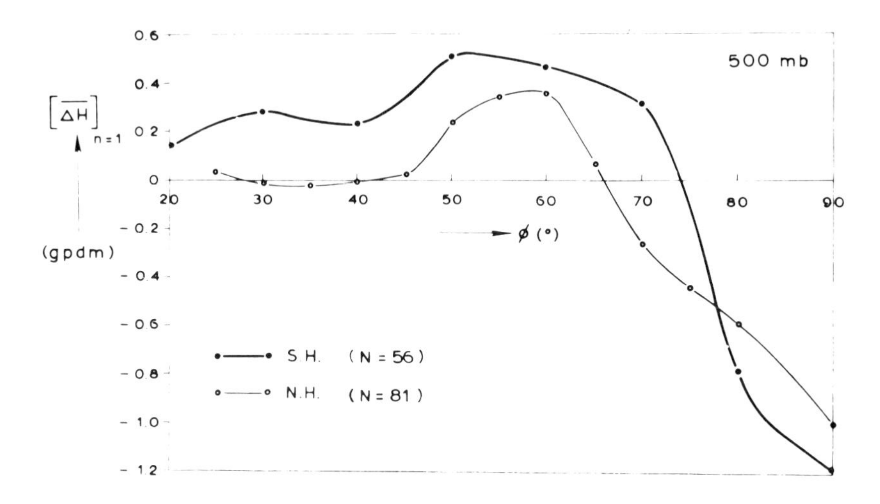

Figure 2. Zonal averages of the mean 500 mb height changes between the first aerological observation after a flare and the observation 24 h earlier (in geopotential decameters).

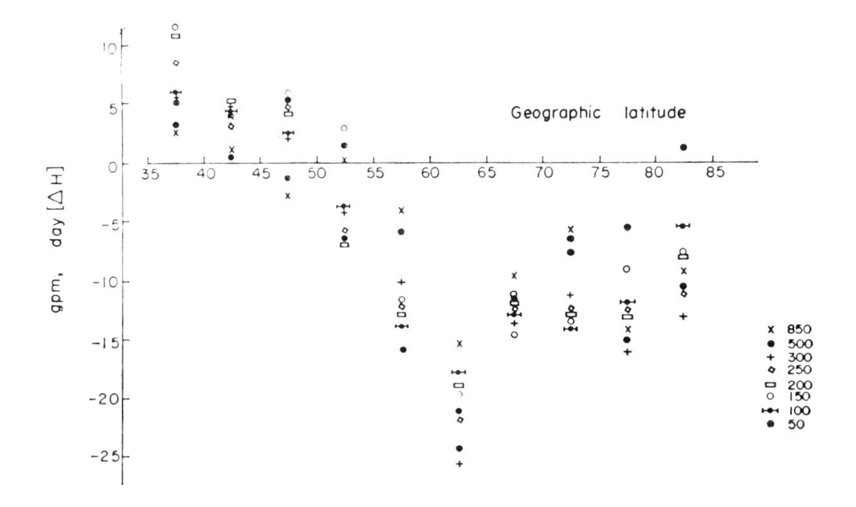

Figure 3. The same as Figure 2, but for an independent sample of 33 flares and for all indicated pressure levels (after Zerefos, 1975).

Focusing on the mid-latitudes we may inquire about the longitudinal positions of the cells of height rise and height fall. Is there a preference, are they possible fixed? A comparison of our results with the older ones of Duell and Duell [3] and those of Valnicek [4] would indicate that Western Europe is a preferred region for height rises. Although very little could be done in comparing other results, I have prepared Figure 4 showing the variation with longitude of the extremes in mean height change in the latitude belt 40 to 70° N. The comparison is based on independent flare samples for the IGY (July 1957 through December 1958), for 1959 and for the sample used by Zerefos [5]. Since Zerefos did not publish the values of mean height change itself, but only values divided by $\sigma N^{-\frac{1}{2}}$, extremes in his map could only be indicated by its longitude (arrows). From the curves and the arrows it is difficult to conclude whether or not the longitudinal variation of the solar flare effect is fixed. For Europe, the Atlantic and the Eastern part of North America, the data support each other. Clearly this subject has to be investigated further before definite conclusions can be drawn. Finally, the seasonal variation of the effect is considered. In our own study the four seasons were analyzed separately and significant effects were found in each of them. However, the winter season showed by far the strongest response. Zerefos' work seems to confirm this, although he only subdivided his sample into two subsamples: cold and hot season.

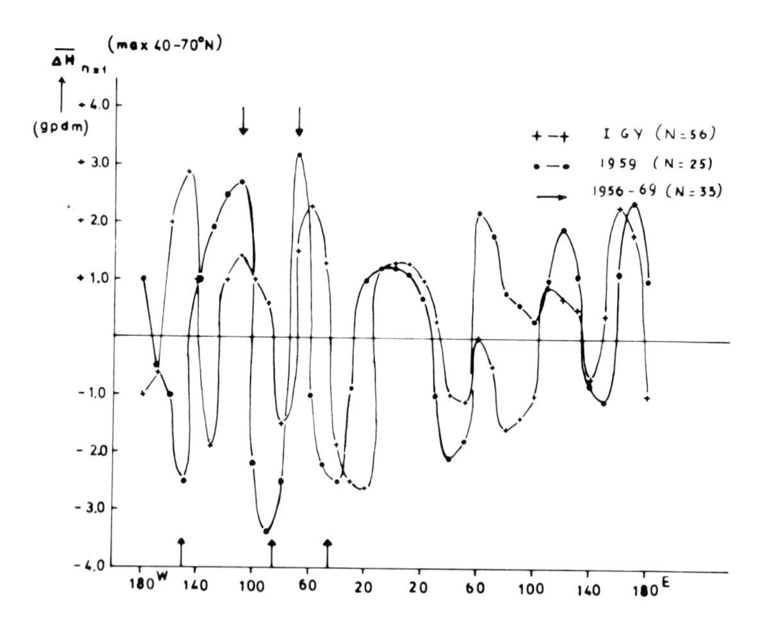

Figure 4. Extremes of mean 500 mb height change after solar flares in the latitude belt 40 to 70° N as a function of geographical longitude; IGY and 1959 flare samples analyzed by Schuurmans and 1956 to 1969 flare samples by Zerefos (1975).

If we do the same, the predominance of the cold season over the warm season is confirmed, but not as overwhelmingly as we expected. This conclusion may be verified by comparing Figures 5 and 6. Interestingly enough, comparison of these figures also adds to the earlier suggestion that the longitudinal variation of the effect is probably rather fixed, not only from year to year but also during the year.

Figure 5. Cells of mean 500 mb height rise after a solar flare greater than 1 gpdm, for the summer half-year.

4. INDIRECT CONFIRMATION

Olson et al. [6] investigated the correlation between the occurrence of solar flares and the so-called vorticity area index (VAI) at the 500 mb level over the Northern Hemisphere. This index is defined as the area over which the absolute vorticity exceeds a certain value. Actually 24 h changes of the VAI were studied. Their superposed epoch analysis was based on a sample of 94 flares, all occurring during the cold season over the period 1955 to 1969. From related studies based on geomagnetic key days the authors expected a large decrease of the VAI 3 to 4 days after a flare, but they were quite surprised to observe an additional peak of VAI-increase on the first and second day of a flare. It seems reasonable to assume that this peak of VAI-increase is somehow related to the early flare effect discussed in Sections 2 and 3. As a matter of fact the mean 500 mb height changes shown in Figure 1 do give rise to an increase of the area of strong vorticity over the hemisphere. Namely, at mid-latitudes positive and negative changes of relative vorticity occur, their effect on VAI probably cancelling out, but in the polar cap area relative vorticity is strongly increased (and necessarily

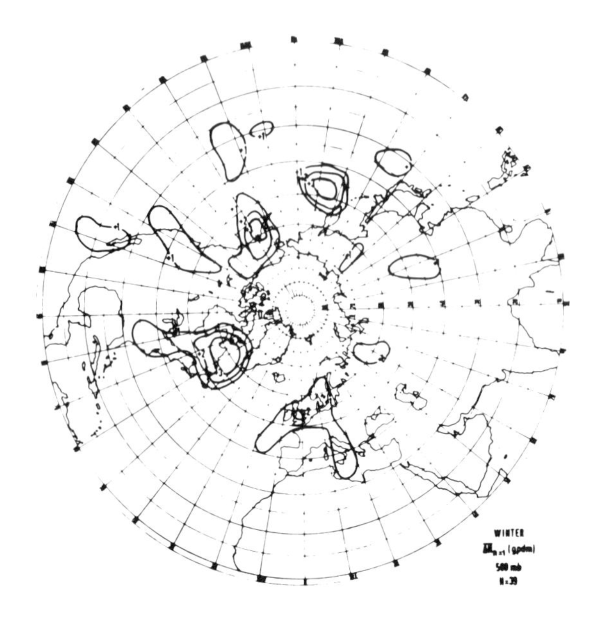

Figure 6. Cells of mean 500 mb height rise after a solar flare greater than 1 gpdm, for the winter half-year.

also absolute vorticity). According to the authors [6] the effects on VAI are also found for the warm season sample of flares, although weaker. This would agree with the fourth specific property of the early flare effect discussed in Section 3.

5. LATE EFFECT

The analysis of Olson et al. clearly suggests that the influence of solar flares on the atmospheric circulation consists of a sequence of events. To a first approximation we may distinguish between an early and a delayed or late effect. It is of interest to see if there are other reports of late effects of solar flares to be found in the literature. Our search for such claims has resulted in the following list (Table III). Although the effects described are very dissimilar, there exists some agreement as to the time lag of the maximum which on the average seems to be 2 to 4 days after the flare. It is not clear what other conclusions can be drawn from such a list. Certainly there are important interrelations between the effects mentioned but these are insufficiently known to draw up an integrated picture. Also the connection between the early and the late effects is very poorly known and described.

Nevertheless from the indications available some evidence emerges that the flare effect starts in a number of hours after the flare, maintains a certain type of influence for a day or so and changes the atmosphere

Table III. Reported late effects of solar flares

Author	No. of flare days	Period	Location	Effect	Time lag (days after flare)
Duell and Duell [3]	51	1936-41	Central Europe (3 stations)	Maximum of interdiurnal pressure change	2 - 4
				Maximum of pressure	4 - 6
Valnicek [7]	69	1936 (33) 1947 (36)	Eastern Atlantic & Europe	SE displacement of polar front; change from zonal to meridional circulation	2 - 5
Kubishkin [8]	41	1956-61	Northern Hemisphere (43 stations) Africa (9 stations)	Extreme of surface pressure*	3 - 4
Schuurmans [1]	81	1957-59	Eastern Atlantic & Europe	Maximum of meridional and blocking type circulations	2 - 3
Olson, Roberts, and Zerefos [6]	94	1955-69	Northern Hemisphere ($\phi > 20°$ N)	Significant decrease of vorticity area index	3 - 4
Reiter [9]	103	1967-69	Central Europe	Maximum of spherics	2
Reiter [9]	20-30	1967-69	Zugspitze and Wank peaks	Maximum of potential gradient and air-earth current	1 - 3
Reiter [9]	15-25	1957-59	Zugspitze	Maximum of potential gradient and air-earth current	1 - 2
Reiter [10]	181	1967-75	Zugspitze	Maximum of intrusions of stratospheric air into the troposphere	2

*Results not confirmed by Stolov and Spar [11] and Stolov and Shapiro [12] on North American and Arctic and Antarctic stations, respectively.

in such a way through this influence that it is forced to develop in a certain specific direction. In doing so, after 2 to 4 days a circulation type is produced which gives rise to a large number of interrelated atmospheric phenomena.

6. SOLAR-CLIMATIC RELATIONS

If the flare effect is to be considered a real phenomenon, one may inquire into its role in establishing solar-climatic relationships. However, this problem can be explored only if the flare effect itself and, more importantly its delayed effects, are better investigated. Somehow one has the feeling that flares occur too rarely to be completely responsible for such phenomena as the 22 yr cycle in the occurrence of droughts [13]. One aspect, however, may be of importance here; namely, the possibility that the flare effect initiates and favors the occurrence of blocking circulations. In view of the strong persistence of blocking systems, one flare event may exert its influence on the atmosphere for several weeks, although the actual intervention lasted only a number of hours.

Clearly we still know too little about the exact nature of the flare effect to say something beyond speculation. One may, however, conclude that the relation solar flares-atmospheric blocking deserves special attention in view of the fact that over the last century solar activity and the occurrence of blocking in the Atlantic-European area shows a parallel upward trend [14]. As shown in Figure 7, the average yearly number of meridional and/or blocking type circulations is 161 in the period 1928 to 1975 (last 47 yr) and 154 in the period 1881 to 1927 (first 47 yr). The difference of 7 is statistically significant.

7. THEORETICAL CONSIDERATIONS

7.1 The Early Effect

Physical mechanisms to explain the observed solar flare effects still have to be discovered. Nearly 10 yr ago this author [1] proposed a chain of events starting with solar flare protons entering the Earth's atmosphere to lower stratospheric depths and creating *in situ* a heat sink at the tropopause level. Neither the physical processes involved, nor the heat sink itself have been confirmed as yet. In order to explain the observed early flare effect in the troposphere it was supposed that the lower stratospheric heat sink would give rise to isallobaric convergence of air leading to a circulation system as sketched in Figure 8. From a meteorological point of view such a scheme is acceptable as an explanation of the strong height rises of tropospheric pressure levels. The reasoning is as follows. Convergence in the stratosphere causes compression and anticyclogenesis in the troposphere. The vertical motion profile will have a maximum at the mid-troposphere, near 500 mb. This downward motion causes an adiabatic temperature rise, which also will be largest near 500 mb. Maximum temperature changes at 500 mb give rise to

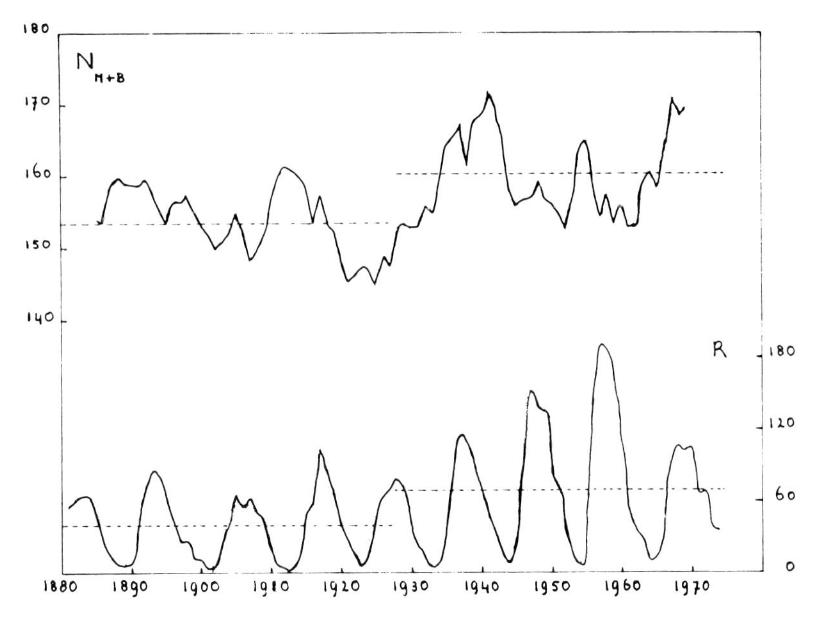

Figure 7. Yearly mean sunspot number (lower curve) and the yearly number of meridional and blocking type circulations in the Atlantic-European area for the period 1881 to 1975. Horizontal lines indicate the averages for the first and second halves of the period.

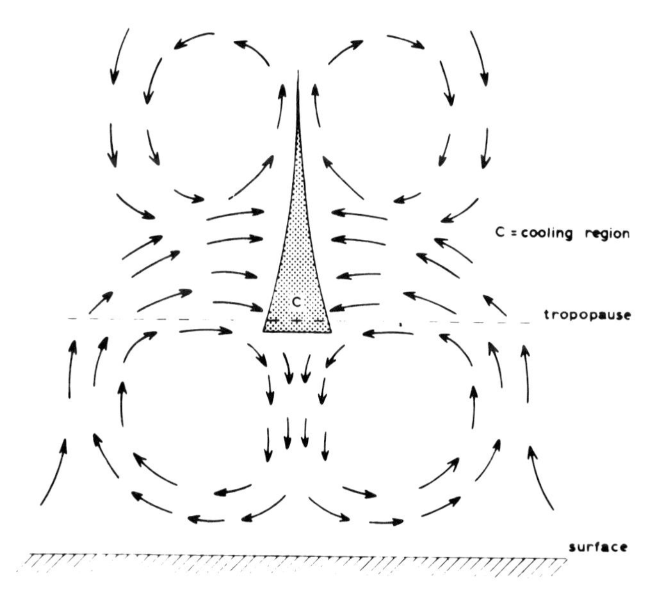

Figure 8. Possible circulation system initiated and maintained by a heat sink or other source of convergence at the tropopause and lower stratospheric levels.

maximum height changes at 300 mb and this is exactly what has been observed.

In order to give an estimate of the necessary convergence in the lower stratosphere we use the observed temperature change at 500 mb after a flare. From the observed height changes reported before we easily compute that on the average this temperature change is of the order of 1°C. Hartman [15] observed a 24 h temperature rise (area-mean of the Northern Hemisphere) of the 500 mb level after the solar flare of June 1, 1960 amounting to 4°C. We therefore may safely assume that in an individual flare case the temperature rise at the 500 mb level is of the order of 1 to 5°C. If we furthermore assume that the effect is established in a time interval of 6 h, which is on the average the case, the downward motion we need is 0.5 to 2.5 cm s^{-1}. With zero vertical motion in the lower stratosphere the computed convergence is of the order of 10^{-6} to 5 x 10^{-6} s^{-1}. This is a rather strong convergence but not unrealistically strong. Of course, by this convergence process we have only explained the height rises after solar flares, and not the large areas of height drop. The latter, however, might very well be the necessary counterparts of the development, demanded by continuity. A strong argument for giving priority to the height increases and not to the height decreases is the complete lack of vertical variation of the latter. It is not clear at all why such a mechanism, if it really exists, should work out stronger in winter than in summer. One possible reason could perhaps be found in the difference in the stratospheric circulation regime between winter and summer.

7.2 The Late Effect

As indicated earlier, late effects of solar flares may be an atmospheric consequence of the early flare effect. At this stage one can only speculate about the atmospheric processes involved. From Table III one may infer that the long waves in the westerlies somehow are changed, either by wave number, amplitude or phase. Such changes might be a consequence of changes in static stability of the atmospheric layers. As shown by Figure 9 the critical wavelength (L) for baroclinic development is strongly dependent on static stability (σ). The curves shown are computed for the two level model from linear theory as discussed by Holton [16]. Since L roughly increases with the square root of σ a fourfold change in static stability can double the wavelength of the most unstable wave. In other words, an increasing static stability stabilizes the shorter waves, prohibiting for instance the development of the short wave extratropical cyclones. The decrease of the area with strong cyclonic vorticity as observed by Olson _et al._ [6] would fit into this picture. Remaining questions, however, are does the early flare effect constitute an increase of static stability? And if so: is this increase large enough to explain a significant change in L_c? The answer to the first question is: yes, in the areas of height rise the static stability is increased while in the areas of height drop no important change in static stability occurs, at least on the average. The answer to the quantitative question is much more difficult. It strongly depends on the initial situation whether or not a certain change in the vertical temperature profile changes the static stability significantly or not. In general the changes will be more significant when saturated adiabatic conditions at relatively low surface temperatures prevail, i.e., in winter time, over the oceans.

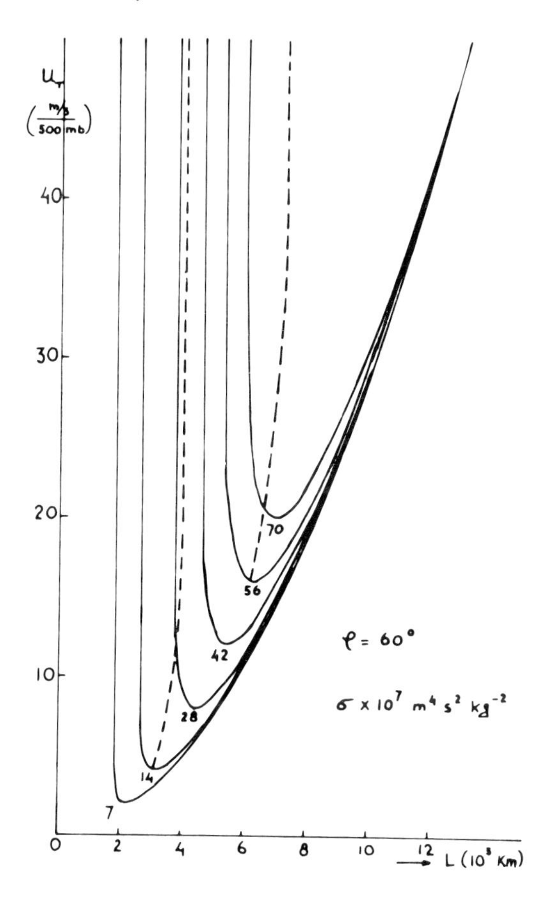

Figure 9. Stability curves for a two level baroclinic model atmosphere for different values of the static stability σ; L is the critical wavelength and U_T the vertical windshear in the layer between 750 and 250 mb.

REFERENCES

1. C. J. E. Schuurmans, The influence of solar flares on the tropospheric circulation, Royal Netherlands Meteorological Institute, Mededelingen en Verhandelingen, No. 92, 1969.

2. C.J. E. Schuurmans and A. H. Oort, Pure and Applied Geophysics, 73, 30, 1969.

3. B. Duell and G. Duell, Smithsonian Miscellaneous Collections, 110, no. 8, 1948.
4. Valnicek, B., Bulletin Astronomical Institute of Czechoslovakia, 4, 179, 1953.
5. C. S. Zerefos, Planet. Space Sci., 23, 1035, 1975.
6. R. H. Olson, W. O. Roberts, and C. S. Zerefos, Nature, 257, 113, 1975.
7. B. Valnicek, Bulletin Astonomical Institute of Czechoslovakia, 3, 71, 1952.
8. V. V. Kubishkin, Astronomical Journal Academy of Sciences USSR, 43, 374, 1966.
9. R. Reiter, Pure and Applied Geophys., 86, 142, 1971.
10. R. Reiter and M. Littfass, Archiv für Meteorologie, Geophysik und Bioklimatologie, Series A, 26, 127, 1977.
11. H. L. Stolov and J. Spar, J. Atmos. Sci., 25, 126, 1968.
12. H. L. Stolov and R. Shapiro, J. Atmos. Sci., 26, 1355, 1969.
13. J. M. Mitchell and C. W. Stockton, 1979, this volume.
14. C. J. E. Schuurmans, Proceedings of the WMO/IAMAP Symposium on Long-term Climatic Fluctuations, Norwich, August 1975.
15. W. Hartmann, Meteorologische Rundschau, 16, 150, 1963.
16. J. R. Holton, An introduction to dynamic meteorology, Ch. 10, Academic Press, New York, 1972.

ENERGY DEPOSITION IN THE EARTH'S ATMOSPHERE DUE TO IMPACT OF SOLAR ACTIVITY-GENERATED DISTURBANCES

S. T. Wu and L. C. Kan
The University of Alabama in Huntsville
Huntsville, AL 35807 USA

E. Tandberg-Hanssen
Space Sciences Laboratory/NASA
Marshall Space Flight Center, AL 35812 USA

M. Dryer
Space Environment Laboratory/ERL
National Oceanic and Atmospheric Administration
Boulder, CO 80302 USA

ABSTRACT. Energy deposition in and dynamic responses of the terrestrial atmosphere to solar flare-generated shocks and other physical processes such as particle precipitation and local heating are investigated self-consistently in the context of hydrodynamics. The problem is treated as an initial boundary-value problem. The undisturbed terrestrial atmosphere is the steady-state solution of hydrodynamic equations with appropriate boundary conditions corresponding to a realistic atmospheric model from 50,000 ($\sim$10 R_E) to 120 km. A solar event (such as the impact of a flare-generated shock wave on the magnetosphere or other physical processes) is considered to initiate an impulsive increase of density, temperature and/or velocity of long duration at 50,000 km. The model follows this perturbation down to the lower boundary (i.e., 120 km) as time progresses. Numerical results are presented for energy flux and propagation trajectories of the disturbances for various initial pulses in relation to different physical processes. We find that the amount of energy flux and the time required to reach to 120 km depend on the strength of the initial disturbances as well as on the physical origin of the disturbance. The amount of energy flux is of the order of 10^{-1} to few erg $cm^{-2}s^{-1}$, which is comparable to the day-side cusp auroral emission. The results further show that gravity waves are excited by a density pulse.

1. INTRODUCTION

Recently, the idea that variable solar activity influences the weather and climate on the Earth has attracted much attention in the scientific community. The new evidence has been reviewed by Roberts and Olson [1], King [2], and Wilcox [3]. It has been suggested that some changes in

B.M. McCormac and T.A. Seliga (Eds.): Solar-Terrestrial Influences on Weather and Climate. 119-124.

the Earth's weather pattern may be linked to certain features of flow of solar-wind plasma past the Earth environment (Cole [4] and Willis [5]). The latter authors investigated the possible mechanisms in which the magnetosphere serves as an active or passive medium between the solar activity and meteorological phenomenon. It is extremely difficult to construct a general model for the link: solar activity-magnetosphere-atmosphere; however, a limited model for this link is possible. In the present study, we describe such a model, and present some results on energy deposition into the Earth's atmosphere due to solar activity-generated disturbances.

2. MODEL

In order to avoid unnecessary complexity while still retaining the basic physical process of the problem, we consider that the solar-generated disturbance is guided downward along the magnetospheric cleft. The possibility of such a confined propagation of the disturbance is due to the fact that the magnetic pressure is much larger than the plasma pressure, i.e., the present model is limited to the case of a low plasma-β flow; β being the ratio of the plasma pressure to magnetic pressure

$$\text{(i.e., } p/\frac{B^2}{8\Pi}\text{)}.$$

The appropriate equations of this problem can be written as follows:

continuity:

$$\frac{\partial \rho}{\partial t} + \frac{\partial}{\partial s}(\rho u) + \frac{\rho u}{A}\frac{dA}{ds} = 0, \tag{1}$$

momentum:

$$\frac{\partial}{\partial t}(\rho u) + \frac{\partial}{\partial s}(\rho u^2 + p) + \rho u^2 \frac{1}{A}\frac{dA}{ds} + \rho g(s) = 0 \tag{2}$$

energy:

$$\frac{\partial}{\partial t}\left[\frac{p}{\gamma-1} + \frac{\rho u^2}{2}\right] + \frac{\partial}{\partial s}\left(\frac{\gamma}{\gamma-1}\, pu + \frac{1}{2}\rho u^3\right) + \rho u g(s) + \left(\frac{\gamma}{\gamma-1}\, p + \frac{1}{2}\rho u^2\right)\frac{u}{A}\frac{dA}{ds} = 0, \tag{3}$$

where ρ, u, A, p, and γ are the density, velocity, cross section area, pressure and ratio of specific heats, respectively, and all quantities are functions of space and time. The independent variables are *s* and *t*, where *s* represents the distance measured from the Earth's surface and *t* is time. Finally, gravity is given by

$$g(s) = G\left(\frac{R_E}{s}\right)^2 \quad . \; . \; . \tag{4}$$

where G is the Earth's gravitational constant and R_E the Earth's radius. An ideal gas law is used, and the cross-section area is $A \sim s^2$.

In order to seek time-dependent numerical solutions for the above set of equations (Equations (1) - (3)) for the disturbed atmosphere, we first determine the steady state solution. This is done by considering an initially quiet atmosphere (u = 0 at t = 0 everywhere), and by setting

$$\frac{\partial}{\partial t} = 0$$

in Equations (1) - (3) and obtaining a separate integration of the steady-state equations, specifying ρ, T, and u at the outer and inner boundaries: $\rho = 4.34 \times 10^{-23}$ gm cm^{-3}, T = 420°K at 50,000 km and $\rho = 2.34 \times 10^{-11}$ gm cm^{-3}, T = 334°K at 120 km. These conditions were selected according to the observational data with thermopause temperature 1000°K given by Banks and Kockarts [6]. After we have obtained the steady-state conditions, we seek the time-dependent solution by using the modified Lax-Wendroff method and the detailed computation procedures can be found in reference [7].

3. RESULTS

We specify the solar activity-generated disturbances at the outer boundary (50,000 km $\approx$ 10 R_E above the Earth) and trace the disturbances downward to the lower atmosphere. The numerical results of the computations for the propagation of the disturbances are summarized for three specific physical causes; i.e., (i) a density pulse; (ii) a temperature pulse; and (iii) shock waves. From an observational point of view, the density pulse may be identified with particle precipitation; the temperature pulse corresponds to local heating and the shock waves are solar flare-generated shocks.

Figure 1 shows the calculated trajectories resulting from density, temperature and shock wave pulses of long duration; i.e., the pulse has been left on during the entire calculation. It is clearly indicated that the shock wave results in the fastest propagation, followed by the temperature pulse; a density pulse leads to the slowest propagation. Also, we notice that the stronger the pulse, the faster is the propagation. Figure 2 shows the time-sequence for the energy fluxes at 300 km, for four different pulses, the long tails are due to the infinitely long duration of the pulses. Figure 3 gives a snapshot at t = 1500 s of the spatial structures of the energy fluxes resulting from these four different pulses. We observe that the disturbance generated by a shock-wave pulse of Mach number 20 has already reached lower atmosphere, and the energy flux deposited beyond 1000 km is due to the reflected wave. This situation is shown in Figure 4, where we have plotted the distribution of plasma flow velocities at different time steps. Notice that at t = 1440 s, the velocity is negative, i.e., the direction of velocity is upward, indicating the reflected wave. In the meantime, the downward propagation of the disturbance has reached to a height of a few hundred km, which cannot be discerned in the figure.

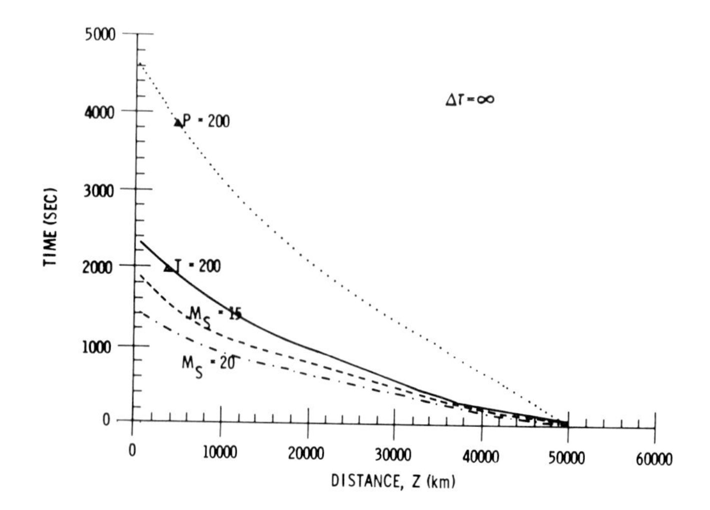

Figure 1. Calculated trajectories of propagation of disturbances initiated at 50,000 km ($\sim$10 R_E) resulting from density temperature and shock wave pulses.

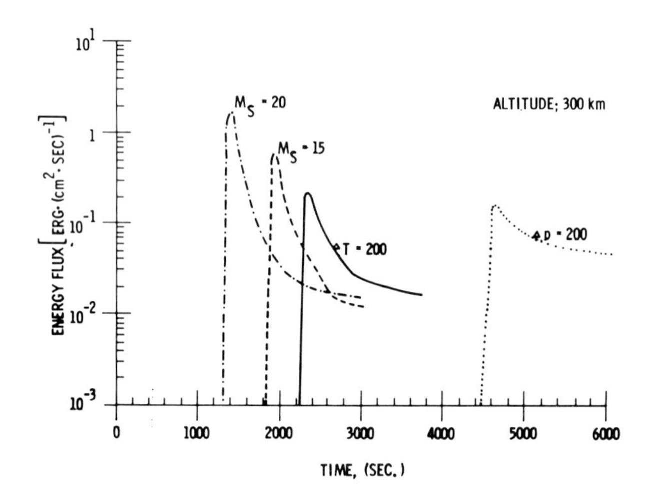

Figure 2. Energy flux vs time at 300 km resulting from density, temperature and shock wave pulses.

4. DISCUSSION

From the present calculations, several results emerge: (i) The amount of energy flux resulting from these pulses is of the same order of magnitude as the observed day-side cusp auroral emissions (i.e., a few erg $cm^{-2}s^{-1}$ [8]). (ii) The disturbance resulting from a density pulse shows a second hump, which we take as an indication of the

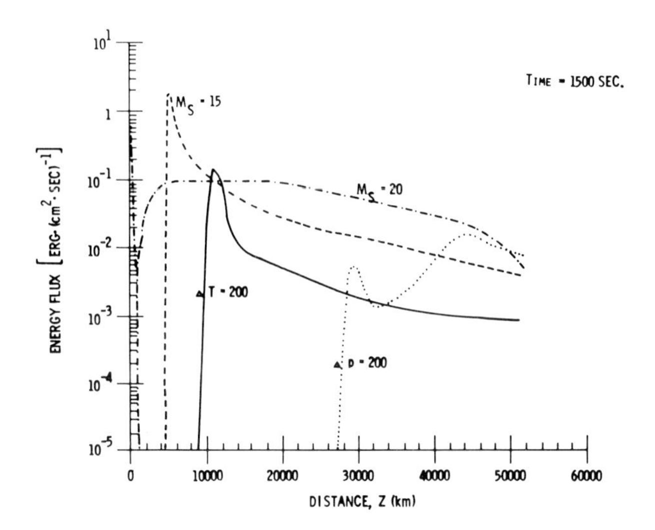

Figure 3. Energy flux distribution at T = 1500 s resulting from density, temperature and shock wave pulses.

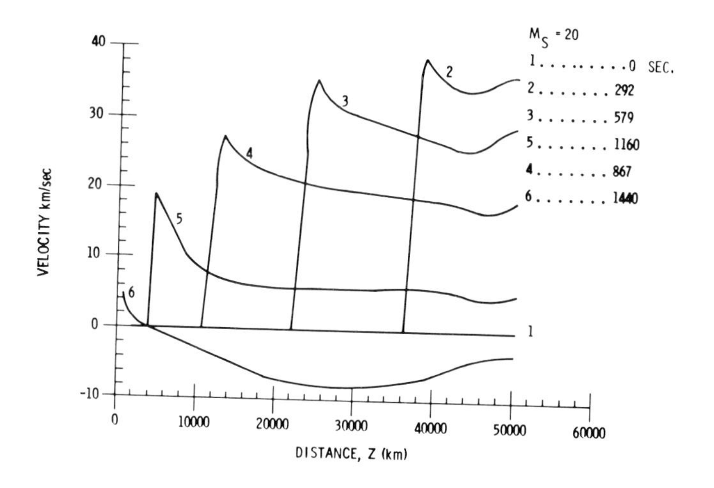

Figure 4. Distribution of plasma-flow velocities at various times resulting from a shock wave pulse (M_S = 20).

excitation of a gravity wave. If this is correct, it provides the physical mechanism to modify the dynamics of the lower atmosphere, and may ultimately lead to an understanding of solar-terrestrial influences on weather and climate.

We realize that the present model is still in its infancy; significant physical aspects have been neglected. But, we believe it has the merit of giving a systematic approach to the problem from a dynamical point of view.

ACKNOWLEDGMENT

The work done by STW and LCK was supported by NASA/Marshall Space Flight Center under Contract NAS8-28097.

REFERENCES

1. W. O. Roberts and R. H. Olson, Rev. Geophys. Space Phys., 11, 731, 1973.
2. J. W. King, Astron. Aeron., 13, 10, 1975
3. J. M. Wilcox, J. Atmos. Terr. Phys., 37, 237, 1975.
4. K. D. Cole, J. Atmos. Terr. Phys., 37, 939, 1975.
5. D. M. Willis, J. Atmos. Terr. Phys., 38, 685, 1976.
6. P. M. Banks and G. Kockarts, Aeronomy Part B, Academic Press, New York and London, 1973.
7. S. T. Wu, Murray Dryer, and S. M. Han, Solar Phys., 49, 187, 1976.
8. D. S. Evans, private communication, 1978.

EVIDENCE OF A 22-YEAR RHYTHM OF DROUGHT IN THE WESTERN UNITED STATES RELATED TO THE HALE SOLAR CYCLE SINCE THE 17TH CENTURY

J. Murray Mitchell, Jr.,

NOAA Environmental Data and Information Service,
Silver Spring, MD 20910 U.S.A.

Charles W. Stockton and David M. Meko

Laboratory of Tree-Ring Research, University of Arizona
Tucson, AZ 85721 U.S.A.

ABSTRACT. Families of Drought Area Indices (DAI) have been derived from tree-ring data for the entire U.S. west of the Mississippi River, for each year back to either 1700 or 1600 A.D., depending on the data base used. Each DAI is expressed in terms of the relative area in which the Palmer Drought Severity Index (PDSI) lies below a specified threshold value between -1 (mild drought) and -4 (extreme drought). Three families of DAI are considered in the analysis reported in this paper. Each DAI family is based on reconstructions from a selection of between 40 and 65 tree-ring sites ranging from Canada to Mexico and from the west coast to the Plains states. Variance spectrum analysis of the DAI series shows a concentration of variance at periods near 22 yr, at significance levels ranging from 5 to 0.1% (relative to a pink noise continuum). Band-pass filters tuned to periods near 22 yr are used in a form of harmonic dial analysis proposed by G. W. Brier to verify the extent of phase locking between the drought area variations and the Hale sunspot cycle since 1700 A.D. Phase locking is confirmed at significance levels of order 1 to 0.1% for all DAI families and all drought severity limits except extreme drought (PDSI < -4). A tendency is found for the amplitude of the 22 yr drought rhythm to vary systematically in parallel with the amplitude (envelope) of the Hale sunspot cycle, on the Gleissberg time scale of about 90 yr. This relationship is statistically significant between the 5 and 1% levels, and is independent of the phase locking found within the Hale cycle. The DAI series that extend back to 1600 A.D. reflect a well-defined 22 yr drought rhythm during the early stages of the Maunder Minimum of solar activity, but a very weak rhythm near the end of the Maunder Minimum. The average areal extent of drought was relatively low, and remained low for a prolonged period, during the Maunder Minimum. This analysis strongly supports earlier evidence of a 22 yr drought rhythm, or "cycle," in the U.S. and suggests that the drought rhythm is in some manner controlled by long-term solar variability directly or indirectly

B.M. McCormac and T.A. Seliga (Eds.): Solar-Terrestrial Influences on Weather and Climate. 125-143.

related to solar magnetic effects. The solar control is best described as a modulation of terrestrial drought-inducing mechanisms, such that it alternately encourages and discourages the development of major continental droughts which are set up by evolutionary climatic processes unrelated to solar activity.

1. INTRODUCTION

This paper is in the nature of a progress report on a joint research effort begun in 1976 to reconstruct the history of drought in the western two-thirds of the U. S. by means of "proxy" tree-ring data, and to seek to verify early indications of a *bona fide* drought "cycle" and assertions of its connection to solar cycles. Inasmuch as the research results to date have not yet been submitted for formal publication elsewhere, we briefly describe the nature of the data involved before proceeding to a discussion of our analysis of the data. In this paper we confine ourselves to some of our principal results only, and leave consideration of a number of corroborative studies and further results to other papers now in preparation.

2. THE DROUGHT AREA INDICES (DAI)

Some years ago, two of us (C. S. and D. M.), following the technique outlined by Fritts *et al.* [1], adopted a multivariate approach to reconstructing drought patterns in the western U. S. (west of the Mississippi River) from the extensive collection of tree-ring indices at the University of Arizona Laboratory of Tree Ring Research. On the drought side, values of the Palmer Drought Severity Index (PDSI) [2] were obtained from the NOAA National Climatic Center for a total of 208 state climatological divisions between 1931 and 1970, and these were averaged into 40 regions identified in Figure 1A. On the tree-ring index side, indices from 40 tree sites known to be sensitive to local climatic variations of one kind or another (depending on tree specie and exposure [3]) were chosen on the basis of sufficient tree age and a reasonably dispersed geographical distribution (Figure 1B). Eigenvector analysis was then performed on the field of PDSI drought data between 1931 and 1970, and independently on the field of tree-ring index data between 1700 and 1962 (the last year of tree core measurements available). In each of the two fields a few of the eigenvectors were chosen that accounted for a reasonably high fraction of the total variance of that field. The eigenvector coefficients for the period of overlap between the two fields (1931 to 1962) were then related to each other by means of canonical analysis [4,5]. This canonical "bridge" between the PDSI drought patterns and the tree-ring index patterns was then used to reconstruct the PDSI value for each of the 40 climatic regions in Figure 1A, for each year from 1700 to 1962. Then four DAI were evaluated for each year in that period. The first DAI is simply the count for a given year of the number of climatic regions (out of 40) in which the reconstructed PDSI was less than -4 (the threshold of "extreme" drought by the Palmer definition). The other DAI are the corresponding counts of climatic regions in which the PDSI was less than -3 ("severe" drought or worse", less than -2 ("moderate" drought or worse), and less than -1 ("mild" drought or worse). A general description of the procedures used, and a plot of the series of the DAI values (PDSI < -1) along with comparable values for the wet side of the Palmer index (PDSI > +1), has been published by Stockton and Meko [6].

Subsequent to the first drought reconstruction indicated above, two other reconstructions were undertaken, using the same data field on the drought side but different choices of tree-ring sites for which tree-ring indices are available back to the year 1600 A.D. These additional reconstructions, identified in this paper as S-50 and F-65 [7] (Figures 1C and 1D), were made essentially for two reasons. First, special interest has developed in the drought situation during the 17th century (see later). Second, verification of the robustness of drought reconstructions from a single choice of tree-ring indices was considered highly desirable to ensure that the information transfer from tree rings to drought statistics is relatively insensitive to the choice of tree sites. (A gratifyingly high degree of robustness has been confirmed in this manner, at least with respect to DAI variations on time scales appreciably longer than 1 or 2 yr which is our main concern here.)

In all, we have three "families" of DAI's, with four members of each family, one for each of the four PDSI limits previously noted. Further information on these DAI reconstructions is summarized in Table I. Here, as in later sections of this paper, the DAI families are distinguished by a letter-number combination in which the number identifies the number of tree-ring sites used in that reconstruction. The original DAI family that is described in [6] is identified here as family S-40. A plot of three members of this DAI family is shown by the vertical bars in Figure 2 (for PDSI < -1, < -2, and < -3).

3. VARIANCE SPECTRA OF THE DAI

Variance spectrum analysis has been performed on most of the 12 DAI series identified above. In each case, a more or less prominent concentration of variance is found around a period of 22 yr. This is illustrated by four spectra shown in Figure 3. Three of these spectra are of the three series plotted in Figure 2. The fourth spectrum is from another DAI family, F-65 (PDSI < -1), to compare with family S-40 (PDSI < -1). In testing the statistical significance of the spectral peaks near 22 yr, it is necessary to specify a realistic null-hypothesis continuum to compare with each spectrum. The low-lag autocovariance structure of the DAI series makes it abundantly clear that the appropriate null continuum is neither white noise nor red noise [8] but rather a mixture of both which (following the terminology of Tukey and others) is best described as "pink noise." This is as one might

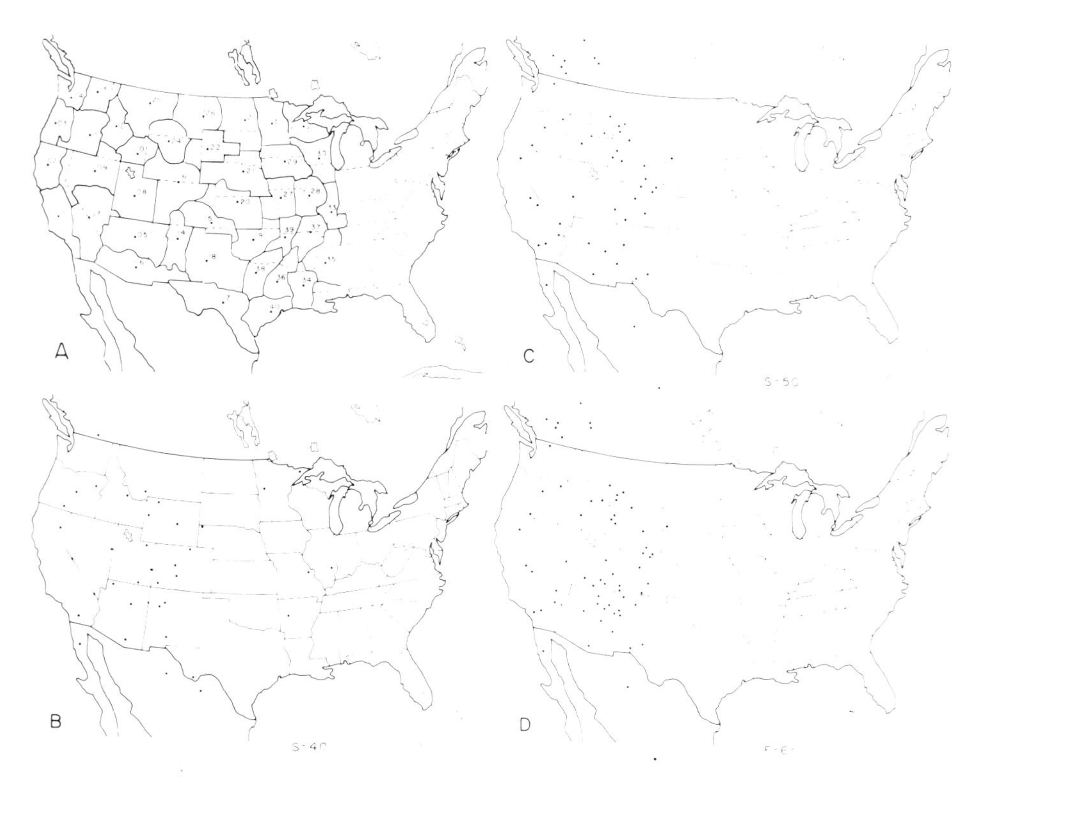

Figure 1. Domain of drought analysis on which this study is based. 1(A): Areas with solid boundaries and interior numbers identify the 40 climatic regions in which Palmer Drought Severity Index was observed or reconstructed. 1(B): Locations of 40 tree-ring data sites used to reconstruct drought data in DAI Family S-40 beginning 1700 A.D. 1(C): Locations of 50 tree-ring data sites used to reconstruct drought data in DAI Family S-50 beginning 1601 A.D. 1(D): Locations of 65 tree-ring data sites used to reconstruct drought data in DAI Family F-65 beginning 1600 A.D.

Table I

Drought Area Reconstructions

DAI FAMILY	TREE RING INDICES Period of Record	Data Points	EVs# Retained	Variance Explained	DROUGHT INDICES (PDSI) Period of Record	Data Points	EVs# Retained	Variance Explained
S-40	1700-1962	40	6	45%	1931-1970	40	5	75%
S-50	1601-1962	50	2	31%	1931-1970	40	7	83%
F-65	1600-1962	65	3	41%	1931-1970	40	5	75%

DAI FAMILY	CANONICAL "BRIDGE" Terms Retained[@]	Variance in Common*	DAI VARIANCE EXPLAINED* PDSI LIMIT < -1	< -2	< -3	< -4
S-40	5	64%	72%	72%	74%	78%
S-50	6	56%	67%	76%	79%	75%
F-65	5	65%	69%	70%	79%	74%

EVs = Eigenvectors (principal components) PDSI = Palmer Drought Severity Index

* In overlap period 1931-1962 (32 yr) DAI = Drought Area Index

[@] The maximum number of terms retained in the canonical correlation "bridge" cannot exceed the least number of EVs retained in either set of indices [5]. Appearances to the contrary in this Table are explained as follows. In all DAI families the drought EV coefficients (amplitudes) for year t are related to the tree-ring EV coefficients for both year t and year t+1. In the case of Family S-50, the drought coefficients for year t are additionally related to the tree-ring coefficients for year t-1. For justufication of this lagging procedure see [10].

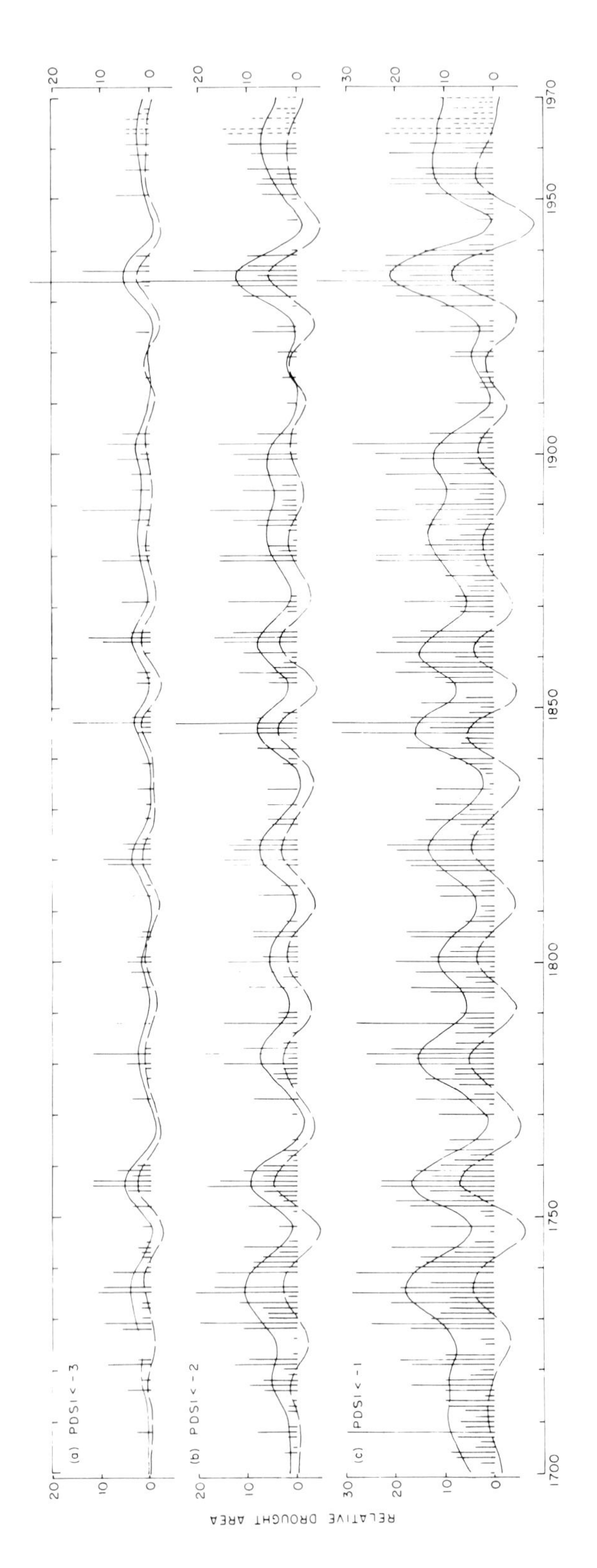

Figure 2. Chronology of drought areas (number of regions out of 40) for three Palmer Drought Severity Index limits, based on DAI Family S-40 reconstruction, 1700 to 1962 A.D., and on observed values after 1962. Vertical bars denote original reconstructions. Lower wavy line in each panel denotes series after filtering with band-pass Filter 1 (see text). Upper wavy line in each panel denotes series after filtering with low-pass filter.

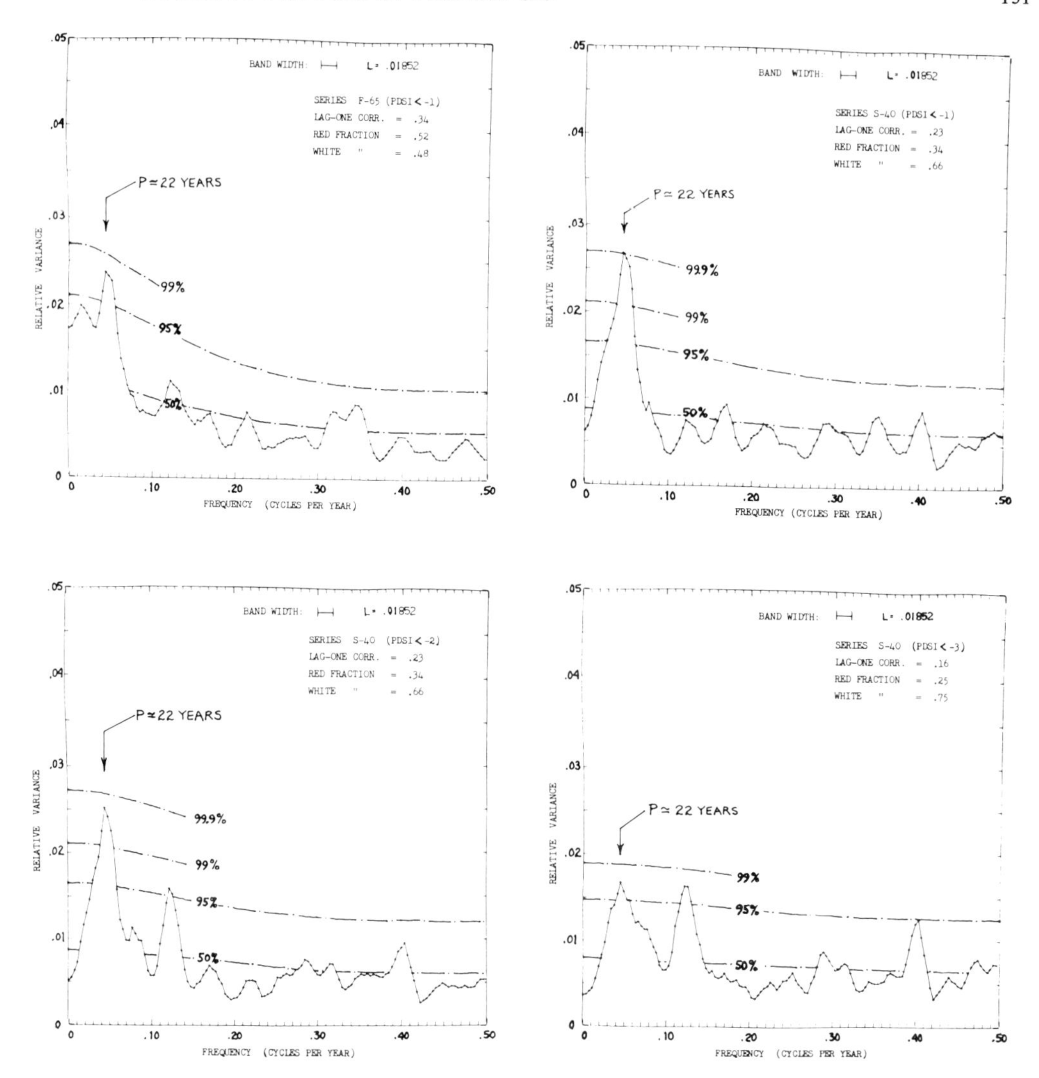

Figure 3. Variance spectra of four selected Drought Area Index series. Upper left, Series F-65 (PDSI < -1); upper right, Series S-40 (PDSI < -1); lower left, Series S-40 (PDSI < -2); lower right, Series S-40 (PDSI < -3). Background curves indicate fitted pink noise continua and associated confidence limits.

expect, inasmuch as the PDSI reflects carryover storage of subsurface moisture that makes it behave as a first-order Markov process in time, and inasmuch as estimates of PDSI from tree-ring data introduce errors of estimate of PDSI that are likely to be random from year to year (e.g., white noise).

On the basis of the low-lag autocovariances of each DAI series, pink noise spectra were fitted to the spectra in Figure 4 and in all cases these were found to give very satisfactory comparisons with the observed spectra at periods well removed from 22 yr. It is then a simple matter to test the statistical significance of the spectral peaks near 22 yr by comparing their amplitudes with certain multiples of the fitted pink noise continua corresponding to preselected significance points in the Chi-square/degrees-of-freedom distribution with degrees of freedom in this case lying between 9 and 10.

In the case of all DAI series thus far examined, the concentration of variance near 22 yr has been found by means of the above procedure to exceed the 95% confidence limit ($P = 0.05$). In one case, that for DAI series S-40 (PDSI < -1), the confidence limit reached is as high as 99.9% ($P = 0.001$). These results imply that spectral peaks as large as those shown in Figure 3 near 22 yr would be highly unlikely to arise by chance in data having the same statistical properties as the DAI but no rhythmic component. We conclude from the variance spectrum analysis that there is strong statistical evidence of a pulse in the extent of drought in the western U. S. that has averaged close to 22 yr in period since 1700 A.D.

4. PHASE LOCKING WITH THE HALE SUNSPOT CYCLE

When considering possible explanations for a 22 yr rhythm in drought it is no surprise that we would have thought to look at the familiar Hale sunspot cycle, and try to devise a suitable test of the consistency with which the drought rhythm has marched in step with the Hale solar cycle since 1700 A.D. The test we have employed for this purpose is one originally conceived by Brier [9] which is ideally suited to a situation in which one may be dealing with an imprecise rhythm rather than with a strict periodicity, and in which the independent variable (sunspot cycles) is likewise an imprecise rhythm. The starting point for the analysis is to filter the DAI series with a band-pass filter designed to pass with little loss of fidelity any variation in the series not only at a selected period of variation but also at any other periods between about half that selected period and about twice that selected period. In this way, even a very irregular quasi-periodicity or rhythm can be followed as a function of time, to reveal any changes of phase, period, and/or amplitude when and as they occur.

Following the methods of Brier [9], two band-pass filters were developed for this analysis which have somewhat different response profiles as shown in Figure 4. The result of applying one of these filters to the DAI series (that identified as Filter 1) is shown in Figure 2 by the lower member of the pair of wavy curves superposed on each DAI series. This gives some indication of the flexibility of the filter in reflecting changes of both phase (or wavelength) and amplitude of the drought rhythm in the series.

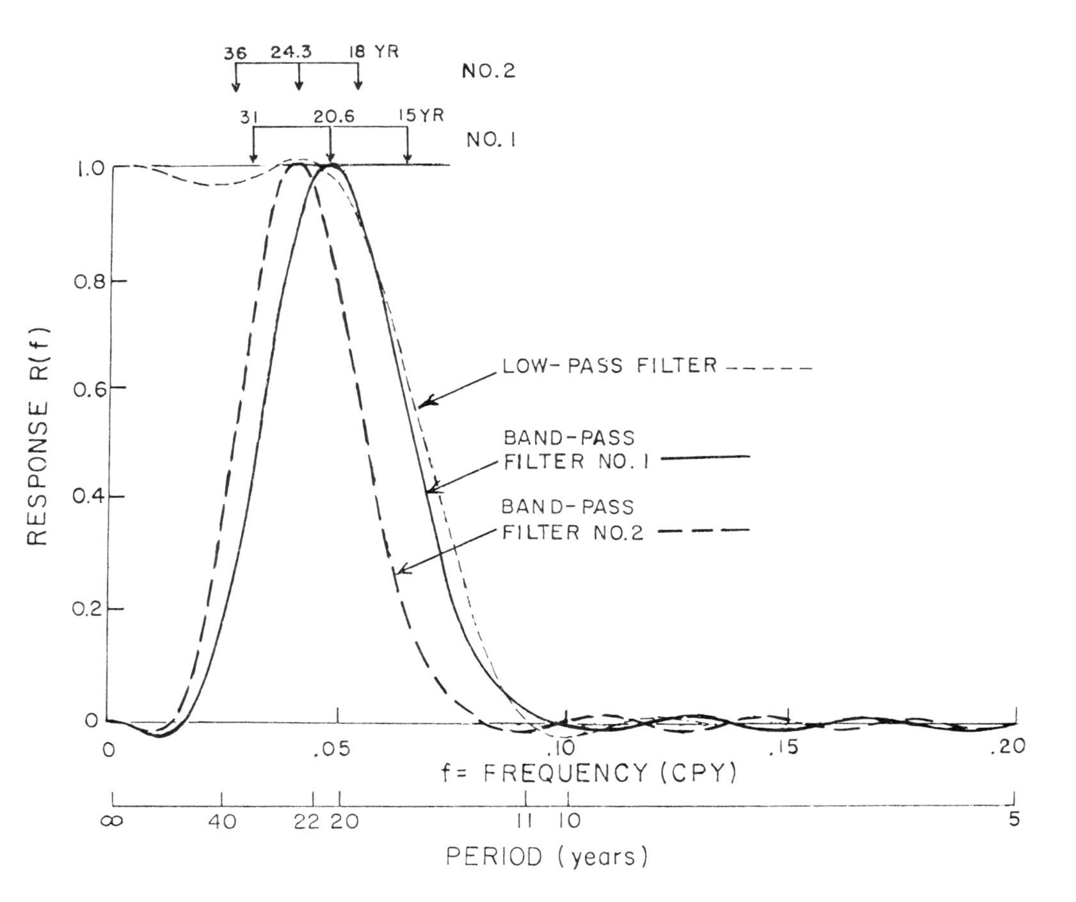

Figure 4. Response functions of band-pass filter 1 (heavy solid curve), band-pass filter 2 (heavy dashed curve), and low-pass filter applied in Fig. 2 (light dashed curve). Peak response and half-response points are indicated for both band-pass filters at top (period, in years).

The next step in the analysis is to pick out the year in which each maximum of drought area occurs, as defined by the band-pass filtered version of the DAI series, and also the local amplitude of the drought maximum as defined by the filter. The year of each maximum is related to the dates of the two successive Hale sunspot minima that bracket it, and the timing of the drought maximum is converted to an angular measure (in degrees azimuth) that describes its phase relationship to the Hale sunspot cycle. This process is repeated for each drought maximum between 1700 and 1962.

The third step in the analysis is to set up a harmonic dial in which the zero azimuth is the time at which each Hale sunspot cycle begins, and in which time progresses around the dial at such a rate that the next passage through the zero azimuth coincides with the beginning of the next Hale sunspot cycle. (The rate of time progression may vary from one Hale cycle to the next if the length of the cycle changes.) In this framework, each drought maximum is plotted as a point on the dial, at an azimuth appropriate to its time of occurrence relative to the beginning and ending dates of the Hale cycle involved, and at a distance from the origin of the dial that is proportional to its amplitude. The result of this procedure resembles a cluster of bullet holes in a rifle range target (see Figure 5).

The final step in the analysis is to calculate the position of the centroid of the drought maxima as positioned in the harmonic dial, and to test the statistical significance of the distance of the centroid from the origin of the dial. If there were no systematic relationship between drought events and the phase of the Hale sunspot cycle, one would expect the drought maxima to distribute themselves more or less uniformly around the origin. In fact, however, there is a consistent tendency for the drought maxima to cluster toward the first quadrant of the dial, or in other words to be favored to occur in the first few years following Hale sunspot minima. The significance of this clustering tendency can be assessed by means of a straightforward analysis of variance involving the position of the centroid. An F-test is used. Because the length of the band-pass filter used here is longer than the mean interval between successive drought maxima, however, a downward correction to the degrees of freedom associated with the within-groups variance is necessary when applying the F-test. At worst this amounts to a reduction of about 45% for the filters used here, which is the correction we have conservatively adopted throughout.

It was noted earlier that two somewhat different band-pass filters were used with the DAI series. The harmonic dial analyses were repeated on each DAI series using both of these filters. One filter was designed to have a peak response near 20.6 yr, and the other, near 24.3 yr. If the filters themselves were responsible for the rhythmic appearance of the series to which they were applied, rather than an inherent rhythm in the series themselves which the filters would of necessity tend to follow, then the application of one filter in the

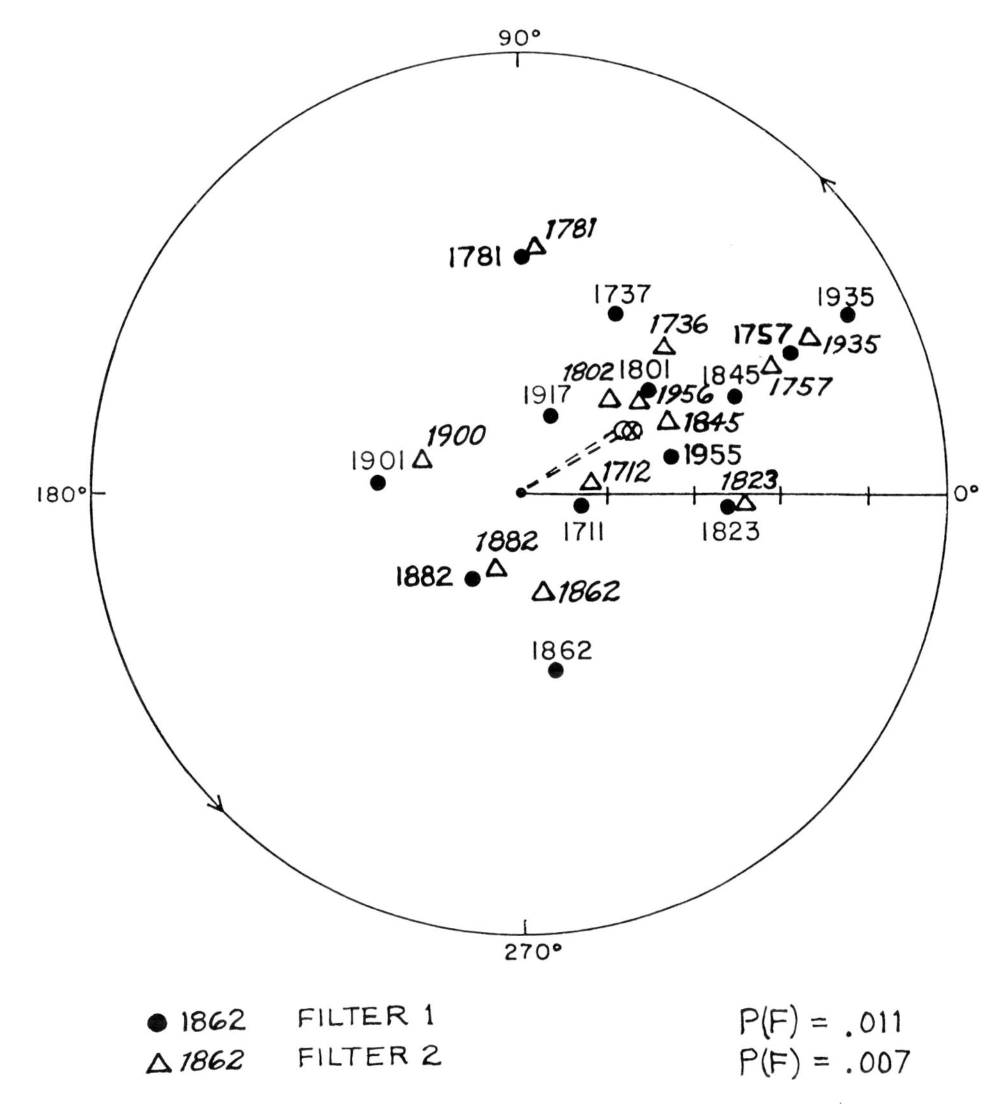

Figure 5. Harmonic dial analysis of DAI Series S-40 (PDSI < -1), referred to Hale sunspot cycle phase (see text). Results are shown for Filter 1 (solid circles) and for Filter 2 (open triangles). Dates of each DAI maximum are indicated. Centroids of data with Filter 1 (circle with cross) and Filter 2 (open circle) are indicated with dashed lines to origin. Beginning and ending of each Hale sunspot cycle occurs at azimuth marked 0°. Time advances counterclockwise.

harmonic dial analyses would result in a rather systematic clockwise drift of about 30° between successive drought maxima, whereas the application of the other filter would result in a similar counterclockwise drift. In fact, both filters result in very similar dial plots, with almost identical centroid positions. An example of this is shown in Figure 5. This is a clear indication that the results of the harmonic dial analyses are not very sensitive to the choice of filter used, and no artificial periodicity introduced by either filter is the explanation of the clustering found. The results of these analyses for all DAI series, which are detailed in Table II, indicate that Filter 2 (with peak response near 24.3 yr) is slightly more efficient than Filter 1 (with peak response near 20.6 yr) in capturing the phase locking that evidently exists in all 12 DAI series.

Table II

Harmonic dial analysis for individual DAI series

PDSI LIMIT	DAI FAMILY	FILTER	NUMBER MAXIMA	CENTROID DIR	(LAG*)	MAG	F-VALUE	SIGNIFICANCE P(F)#
	S-40	1	13	29°	(1.8)	2.87	6.48	0.011
	S-50	1	13	42°	(2.6)	3.02	8.92	0.004
-1	F-65	1	13	42°	(2.6)	3.81	12.42	0.001
	S-40	2	12	32°	(2.0)	2.80	7.70	0.007
	S-50	2	12	42°	(2.6)	2.95	10.74	0.002
	F-65	2	12	41°	(2.5)	3.93	22.39	0.0001
	S-40	1	13	26°	(1.6)	1.95	6.85	0.009
	S-50	1	14	31°	(1.9)	1.83	6.10	0.012
-2	F-65	1	14	34°	(2.1)	1.88	6.67	0.009
	S-40	2	12	27°	(1.7)	1.94	8.58	0.005
	S-50	2	12	29°	(1.8)	1.99	9.52	0.003
	F-65	2	13	32°	(2.0)	1.87	9.09	0.003
	S-40	1	13	24°	(1.5)	0.93	5.98	0.014
	S-50	1	14	21°	(1.3)	0.97	4.58	0.029
-3	F-65	1	14	29°	(1.8)	0.95	4.56	0.030
	S-40	2	12	22°	(1.4)	0.86	6.56	0.012
	S-50	2	12	23°	(1.4)	1.00	6.09	0.015
	F-65	2	12	27°	(1.7)	1.01	6.06	0.015
	S-40	1	12	358°	(-0.1)	0.39	3.44	0.066
	S-50	1	12	17°	(1.0)	0.43	2.99	0.088
-4	F-65	1	14	35°	(2.1)	0.18	0.53	0.600
	S-40	2	11	355°	(-0.3)	0.39	4.24	0.043
	S-50	2	12	17°	(1.0)	0.37	3.59	0.060
	F-65	2	13	19°	(1.2)	0.17	0.65	0.540

*Lag in nominal years after Hale sunspot minimum
#Corrected for filter overlap

We conclude from this analysis that there is strong evidence of a systematic phase locking between the 22 yr rhythm of large-scale drought in the western U. S. and the Hale sunspot cycle since 1700 A.D. The phase relationship involved is such that maxima of drought area are most favored to occur within the first 2 or 3 yr following minima in the Hale sunspot cycle (i.e., following alternate 11 yr sunspot minima including the minimum reached in 1976). The statistical significance of this result is assessed as being better than 1% for PDSI < -1 and < -2, about 1 or 2% for PDSI < -3, and about 10% for PDSI < -4.

5. LONG-TERM MODULATION OF DROUGHT RHYTHM AMPLITUDE

In examining a number of the DAI series after band-pass filtering had been applied for purposes of the harmonic dial analyses, discussed above, it appeared that the amplitudes of successive drought maxima were changing with time in a manner that bore some resemblance to the changes of amplitude of the sunpot cycle on the time scale of the Gleissberg cycle. To test whether this apparent relationship might be a real one we examined each DAI series in terms of the coefficient of correlation between the amplitude of each drought maximum (after filtering by both band-pass filters individually) and the envelope of the Hale sunspot cycle at the time of that drought maximum. For this purpose we omitted the drought maximum dating around 1955 because this came too close to the end of the DAI series (1962) to permit a satisfactory estimate of its amplitude by means of the filters used. The results of this correlation analysis are shown for all DAI series, for both filters, in Table III. Also shown in Table III are the correlations for averages of the DAI maxima for each PDSI limit, and those for averages involving more than one PDSI limit. These averages were derived first by rescaling all drought maxima since 1700 for a given DAI and a given filter so that the average drought maximum amplitude for that DAI and filter was equal to one, and then by pooling the results for more than one DAI. In the case of averages involving more than one PDSI limit, the drought amplitudes (after rescaling) were weighted in inverse proportion to the PDSI limit in each case. This means, for example, that in the grand average on the bottom line of Table III, the average was derived by weighting the PDSI < -1 average by 1, the PDSI < -2 average by 1/2, the PDSI < -3 average by 1/3, and the PDSI <-4 average by 1/4 before pooling. This was done because the higher (negative) PDSI limits appeared to be noisier than the lower. Table III shows that these "weighted means" have appreciably better correlations with the Hale sunspot envelope than all but a handful of individual DAI series. It is apparent that the means based on Filter 1 correlate better than those based on Filter 2 with the sunspot envelope, where the significance levels of the correlations reach between 1 and 2%, and about 5%, respectively.

The "weighted means" involving all PDSI limits, based on each of the two filters, are plotted as time series in Figure 6 along with the Hale sunspot cycle envelope since 1600 A.D. Some interesting conclusions are suggested by this figure. First, one can see a tendency for the epochs of large-amplitude drought cycles to lead those of large-amplitude sunspot cycles by 10 or 20 yr. A similar lag relationship may also exist between low-amplitude epochs in the two variables but this is less clear.

Table III

Correlation between drought cycle amplitude and envelope of Hale sunspot cycle* 1700 - 1962 A.D.

PDSI LIMIT	DAI FAMILY	FILTER 1 NUMBER MAXIMA	FILTER 1 CORRELATION COEFFICIENT	FILTER 1 SIG.# LEVEL	FILTER 2 NUMBER MAXIMA	FILTER 2 CORRELATION COEFFICIENT	FILTER 2 SIG.# LEVEL
-1	S-40	12	0.645	0.05	11	0.548	-
	S-50	12	0.683	0.02	11	0.685	0.02
	F-65	12	0.365	-	11	0.209	-
	All @	13	0.652	0.02	13	0.517	-
-2	S-40	12	0.555	-	11	0.557	-
	S-50	12	0.571	-	11	0.554	-
	F-65	12	0.550	-	12	0.386	-
	All @	12	0.583	0.05	13	0.518	-
-3	S-40	12	0.522	-	11	0.554	-
	S-50	12	0.544	-	11	0.634	0.05
	F-65	12	0.555	-	11	0.456	-
	All @	12	0.600	0.05	12	0.518	-
-4	S-40	11	0.511	-	10	0.506	-
	S-50	11	0.770	0.01	11	0.776	0.005
	F-65	12	0.338	-	11	0.371	-
	All @	12	0.639	0.05	12	0.589	0.05
			Weighted Means				
-1 to -2	All @	13	0.682	0.02	13	0.522	-
-1 to -3	All @	13	0.697	0.01	13	0.560	0.05
-1 to -4	All @	13	0.712	0.01	13	0.588	0.05

*Envelope of Hale sunspot numbers filtered by Filter 1
#Two-tailed test (significance 0.05 or higher)
@Average of all normalized DAI for designated PDSI limit(s)

Also of interest here is the indication that the amplitude of the 22 yr drought rhythm was quite large going into the Maunder Minimum but that it decreased to a very low value near the end of the Maunder Minimum (about 1700 A.D.). The drought rhythm amplitude was nearly as low again around 1900 A.D. The amplitude of the Dust Bowl drought of the 1930's stands out in this analysis as probably the most widespread drought event, as well as one of the most extreme events on the Palmer drought severity scale, to have occurred in the western U. S. since 1600 A.D. It may or may not be significant that the 1930's were followed within 20 yr by the highest sunspot numbers ever recorded.

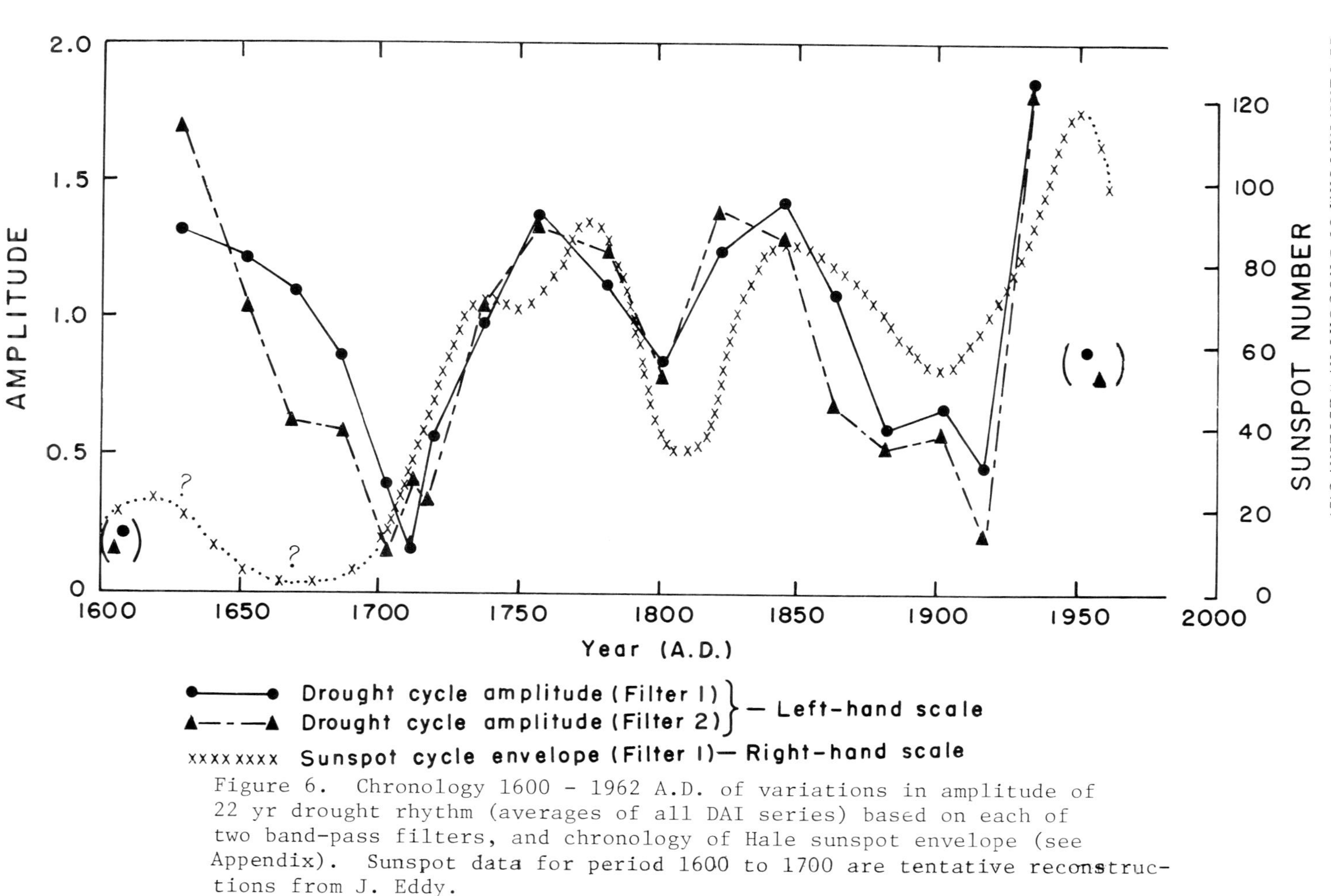

Figure 6. Chronology 1600 - 1962 A.D. of variations in amplitude of 22 yr drought rhythm (averages of all DAI series) based on each of two band-pass filters, and chronology of Hale sunspot envelope (see Appendix). Sunspot data for period 1600 to 1700 are tentative reconstructions from J. Eddy.

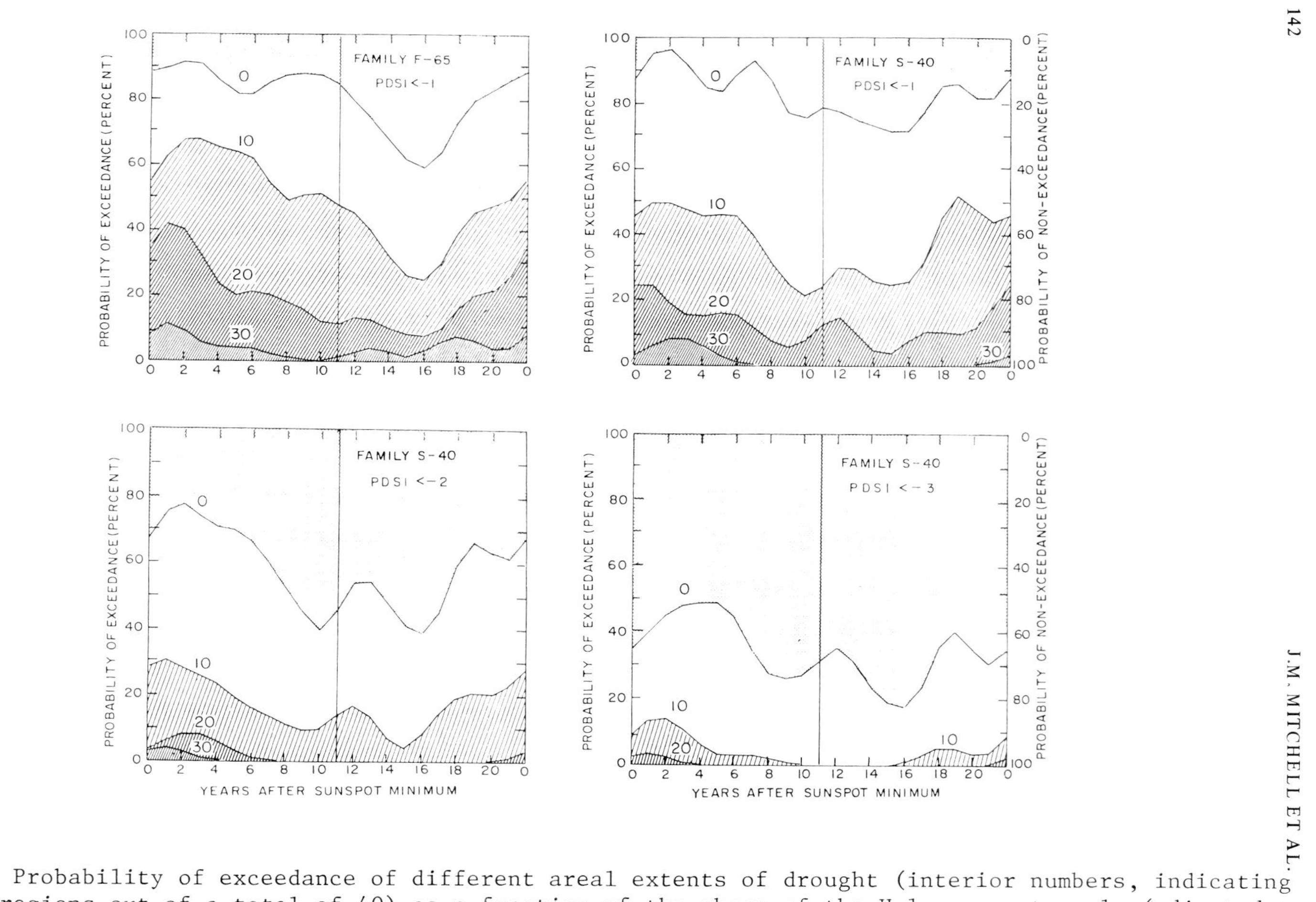

Figure 7. Probability of exceedance of different areal extents of drought (interior numbers, indicating number of regions out of a total of 40) as a function of the phase of the Hale sunspot cycle (adjusted to its average length of about 22 yr). Based on data for period 1700 - 1962 A.D., for each of four DAI series indicated. All curves have been smoothed as a function of time by a binomial moving average with weights 1-4-6-4-1 over 16.

REFERENCES

1. H. C. Fritts, T. J. Blasing, B. P. Hayden, and J. E. Kutzbach, J. Appl. Meteorol., 10, 845, 1971.
2. W. C. Palmer, Meteorological Drought, U. S. Weather Bureau Research Paper No. 45, Washington, D.C., 1965.
3. H. C. Fritts, Tree Rings and Climate, Academic Press, London-New York-San Francisco, 1976.
4. H. R. Glahn, J. Atmos. Sci., 25, 23, 1968.
5. T. J. Blasing, SIAM-SIMS Conference Series, No. 5, Society for Industrial and Applied Mathematics, Philadelphia, 211, 1978.
6. C. W. Stockton and D. M. Meko, Weatherwise, 28, 244, 1975.
7. H. C. Fritts and D. J. Shatz, Tree-Ring Bulletin, 35, 31, 1975.
8. D. L. Gilman, F. J. Fuglister, and J. M. Mitchell, Jr., J. Atmos. Sci. 20, 182, 1963.
9. G. W. Brier, Some Statistical Aspects of Long-Term Fluctuations in Solar and Atmospheric Phenomena, in: Solar Variations, Climatic Change, and Related Geophysical Problems, ed. by F. N. Furness, New York Academy of Sciences, New York, 173, 1961.
10. C. W. Stockton, Long Term Streamflow Records Reconstructed from Tree Rings, Pap. Lab. Tree-Ring Res. No. 5, Univ. Arizona Press, Tucson, 1975.

APPENDIX

A. DERIVATION OF HALE CYCLE MINIMUM DATES AND SUNSPOT ENVELOPE

For purposes of this paper, the Hale sunspot cycle is defined in terms of the Zurich Relative Sunspot Number, with alternate (even) 11 yr cycle spot numbers considered positive and intermediate (odd) 11 yr cycle spot numbers considered negative. The series of annual mean sunspot numbers defined in this manner was then subjected to analysis in two respects. Where the sign of the spots changed from positive to negative, the intercept with zero was identified as the end of one Hale cycle and the beginning of the next. This resulted in the following list of dates of Hale sunspot minima which were used in all harmonic dial analyses cited in Section 4:

(1694.0)	1798.4	1913.2
1711.5	1823.2	1933.4
1733.2	1843.4	1954.1
1755.5	1867.2	1976.8
1775.3	1889.4	

The Hale sunspot series was then filtered by means of Filter 1 described in the text. The maxima and minima (with sign changes to positive) of this filtered version of the series were used to define the envelope of the Hale cycle as a smooth function of time, shown in Figure 6. This led to a much less ambiguous measure of the long-term variations of sunspot cycle amplitude than use of absolute maxima of sunspot counts, or the like.

AVAILABLE POTENTIAL ENERGY IN THE MIDDLE ATMOSPHERE AS IT RELATES TO THE SUN-WEATHER EFFECT

M.F. Larsen
Atmospheric Sciences Department and School of Electrical Engineering
Cornell University
Ithaca NY 14853, U.S.A.

M.C. Kelley
School of Electrical Engineering
Cornell University
Ithaca NY 14853, U.S.A.

ABSTRACT

We calculate the potential energy available in the middle atmosphere and compare it to the energy needed to account for the Sun-weather effect.

1. INTRODUCTION

Historically, the field of Sun-weather physics developed as a series of correlation studies which indicated that a tropospheric effect could be seen in response to a solar perturbation. The reluctance of the general scientific community in accepting these results has stemmed from the nature of the energy considerations and the lack of clear physical mechanisms to test against theory. How can a solar signal affect the tropospheric circulation when the energy density of the latter is orders of magnitude higher than that of the signal? The mechanisms most recently proposed rely on various triggering effects in which the solar signal would tap a source of energy already present in the atmosphere. These triggered energy sources however, must also be evaluated to determine if they are capable of supplying the necessary energy to cause a Sun-weather effect of the magnitude seen in the statistical studies and as a means of determining which of the proposed mechanisms seem most promising.

There are at least three major mechanisms proposed to date which are amenable to such testing:

1) The possible effects of the solar input on planetary wave propagation

B.M. McCormac and T.A. Seliga (Eds.): Solar-Terrestrial Influences on Weather and Climate. 145-147.

either in the form of tides or the reflection of Rossby waves in the middle atmosphere;
2) The effect of the modulation of the atmospheric electric field by variations in solar activity on the condensation of water vapor in the troposphere; and
3) The possible effects of heating due to the release of stored chemical energy onto variations in ozone in the stratosphere as linked to variable solar activity.

We propose here a fourth mechanism to be added to the above list and then carry out the required energy calculation.

2. DESCRIPTION AND TEST OF THE MECHANISM

The mechanism involves the generation of neutral atmospheric waves in the upper or middle atmosphere with a downward component of group velocity. Although the amplitude of neutral atmospheric waves varies with altitude as $\exp(Z/2H)$, where H is the scale height ($\approx$8 km) of the atmosphere and Z the altitude, this exponential decrease of the wave amplitude with decreasing altitude could be partially offset if the waves were capable of tapping the free energy inherently present in the middle atmosphere. The source for the waves could, for example, be ion drag and Joule and particle heating in the auroral zone or ozone variations in the middle atmosphere. The most important question to be answered then is whether or not enough free energy exists in the middle atmosphere to account for the changes in the tropospheric circulation seen in the statistical studies.

Free energy exists in the atmosphere whenever constant density and constant pressure surfaces are misaligned. Since density is not measured on a routine basis, this "misalignment" is detected as a horizontal temperature gradient. This is created on the largest scale in the system by differential heating between the poles and the equator. In fact, this source of free energy is the driving force for our tropospheric weather which can be viewed as an instability in the zonal flow. This same source of energy can also be related to energy available in a sheared medium since for a flow in approximate geostrophic balance, a horizontal temperature gradient necessarily implies the existence of a vertical shear in the wind (Holton, 1972). The question then becomes, exactly what is the magnitude of the available energy which can be derived from the middle atmosphere.

To answer this question, we can use the result derived by [1] in which he found that the available potential energy (APE) in the atmosphere is related to the variance of temperature on a constant pressure surface. Specifically, the APE is given by the integral of the expression

$$1/2\ \overline{T}(\Gamma_d-\overline{\Gamma})^{-1}\overline{(T'/\overline{T})^2}\ dP \qquad (1)$$

through a vertical column. Here $\overline{T}$ is the mean temperature in the layer,

P is the pressure, T' is the local deviation in temperature, Γ_d is the dry adiabatic lapse rate, and Γ is the mean lapse rate in the layer. To evaluate the integral, we use the values given in CIRA (1972) for the latitude band 60° to 70° N in January. Integrating between 10 and 90 km altitude, we find that the energy per unit area in the column is $2.8 \times 10^5 Jm^{-2}$. If we assume that the wave which is generated has a latitudinal and meridional wavelength of approximately 2×10^3 km, then the energy available to the wave as it propagates through the middle atmosphere is 11.2×10^{17}J. For various reasons which are not well understood, usually only 10% of the APE in the troposphere appears as the short wave structures responsible for most low and high pressure features at mid-latitudes [2]. Using this same percentage then, the accessible energy is the order of 1.12×10^{17}J. This can be compared with the value of 10^{17}J calculated by Willis [3] as necessary to account for the Sun-weather effect.

3. THEORETICAL AND EXPERIMENTAL TESTS FOR THE MECHANISM

In view of the energy arguments which seem very favorable, it is worthwhile to pursue the Sun-weather problem from this point of view. The problem of understanding how waves redistributed the energy in a sheared flow is interesting not only in terms of Sun-weather physics, but also in terms of understanding the dynamics of the middle and upper atmosphere. We have been working on these questions from the viewpoint of instability theory, starting with the theory of gravity wave propagation in a medium with vertical and horizontal shear. The work looks very promising but to date is not complete.

ACKNOWLEDGMENTS

This research was sponsored by the Atmospheric Sciences Division of the National Science Foundation under Grant ATM-77-04518.

REFERENCES

1. E. N. Lorenz, Energy and Numerical Weather Predictions, Tellus, 12, 364, 1960.
2. J. R. Holton, An Introduction to Dynamic Meteorology, Academic Press, New York, 1972.
3. D. M. Willis, The Energetics of Sun-Weather Relationships - Magnetospheric Processes, J. Atmos. Ter.. Phys., 38, 685, 1976.

INFLUENCE OF THE SOLAR MAGNETIC FIELD ON TROPOSPHERIC CIRCULATION

John M. Wilcox

Institute for Plasma Research
Via Crespi, Stanford University
Stanford, CA 94305, U.S.A.

ABSTRACT. The solar and interplanetary magnetic sector structure has its source in a large-scale photospheric magnetic field structure that is carried away from the Sun by the solar wind to form a warped equatorial current sheet in the heliosphere. A complete rotation of this current sheet is carried past the Earth by the solar wind every 27 days, and several terrestrial effects result. The area of wintertime troughs in the northern hemisphere as measured using the vorticity area index (VAI) reaches a minimum about one day after the current sheet passes the Earth (a sector boundary). The percentage depth of this minimum is larger for the areas with the most intense tropospheric circulation. Certain passages of the current sheet that are observed in the interplanetary medium to be particularly active have an influence on the trough area that is twice as large as the average effect. A quantitative analysis indicates that forecasting using the Limited Area Fine Mesh grid is significantly less accurate during the two days following the passage of the current sheet. Individual troughs crossing 180° longitude when the interplanetary magnetic field is directed away from the Sun (i.e., at the present time the current sheet is south of the Earth) are significantly larger than troughs crossing 180° longitude when the field is toward the Sun. Possible physical mechanisms include a solar phasing influence on planetary waves leading to constructive or destructive interference, an observed influence of the heliospheric current sheet on the vertical electric field in the troposphere, and sector-related changes in solar EUV intensity which may influence O_3 concentrations and the reaction rates of atmospheric chemical processes.

The large-scale interplanetary magnetic field in the heliosphere can be described in terms of an electric current sheet that is warped northward and southward of the solar equator as shown in Figure 1[1]. This current sheet has its source in an observed structure in the photospheric magnetic field [2], and therefore makes one complete rotation past the Earth in the solar rotation period of 27 days. The structure is usually rather stable from one rotation to the next, but there is enought evolution so that a power spectrum of the interplanetary magnetic fields shows

B.M. McCormac and T.A. Seliga (Eds.): Solar-Terrestrial Influences on Weather and Climate. 149-159.

Figure 1. A painting of the warped heliospheric current sheet related to the solar source of the meteorological effects described in this review.

a rather broad peak centered at 27 days. The lack of a sharp periodicity in the interplanetary magnetic field helps to lessen the possibility of a chance coincidence between a 27 day solar structure and a 27 day meteorological periodicity.

At the present time the region of the heliosphere northward of the current sheet has interplanetary magnetic field directed away from the Sun, while the region southward of the current sheet has interplanetary field directed toward the Sun. The polarities change every 11 yr, about 2 yr after the maximum of the 11 yr sunspot cycle [3].

In the Sun-weather analyses to be described, the time at which the warped current sheet is observed to be carried past the Earth by the solar wind gives the phase information, or the zero day in Figure 2. It is unlikely that the current sheet as such has significant terrestrial effects but it provides a convenient, precise time marker. Presumably the physical causes are to be associated with other quantities that are organized in terms of the sector structure (the location of the current sheet)[4].

We may note several advantages in using the heliospheric warped current sheet in Sun-weather investigations:

1) It is unquestionably of solar origin, yet its time of arrival on Earth is precisely observed.

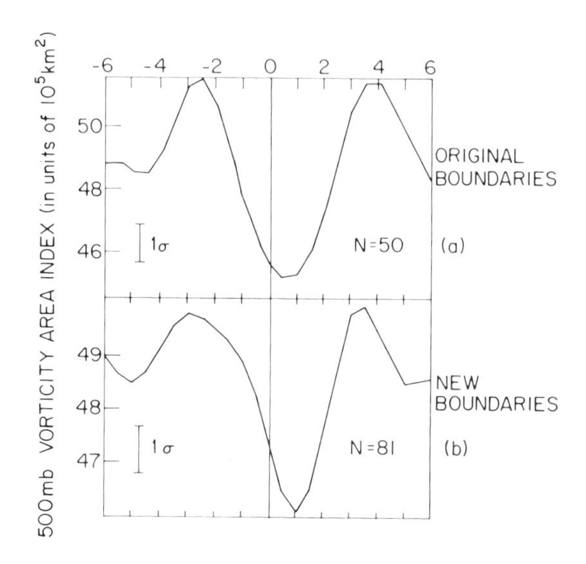

Figure 2. The change in the VAI computed for the northern hemisphere as the heliospheric current sheet shown in Figure 1 is carried past the Earth by the solar wind. The ordinate is an approximation to the area of a dozen or so troughs usually present in the northern hemisphere. The zero day is when the current sheet is carried past the Earth. Figure A shows the 50 cases studied first, and Figure B shows 81 new cases [7].

2) It is a well-defined physical structure related to observed magnetic fields on the Sun. It is a much more precise description of the changing Sun than is the sunspot number.

3) Since several research groups have already published Sun-weather investigations using the heliospheric warped current sheet (i.e., the time of sector boundary passages), a new investigation can readily be compared with and fitted into the fabric of knowledge established by previous workers. This is a much better situation than with most of the Sun-weather literature, which tends to consist of isolated fragmentary claims that are difficult to evaluate and to intercompare. An Atlas of the observed times of sector boundary passages is available [5].

4) A subset of boundary passages observed to be particularly active (see below) has yielded a larger influence on the area of low-pressure troughs, and might similarly be expected to give a larger amplitude of other Sun-weather effects.

The suggestion that the solar and interplanetary magnetic structure be used in Sun-weather research was first made by Markson [6]. A few years later Wilcox et al. [7] obtained results typified by Figure 2, which shows that the area of wintertime low-pressure troughs in the northern hemisphere reached a minimum about one day after the warped current sheet was carried past the Earth by the solar wind. The area of the troughs was represented by a vorticity area index (VAI) defined as the area in km^2 in which the absolute vorticity (circulation per unit area) exceeded a value of $20 \times 10^{-5}s^{-1}$, corresponding to a large well-formed trough [8]. In order to increase the influence of large circulations, the area in which the vorticity exceeded $24 \times 10^{-5}s^{-1}$ was added in computing the VAI. The VAI was computed from the National Meteorological Center grid in the geostrophic approximation using the four points adjacent to each grid point. This is an objective calculation; no interpretations or personal judgments are required.

The initial published results shown in Figure 2A for 50 sector boundary passages were reproduced in an analysis of 81 new additional boundaries as shown in Figure 2B. The most important investigation of the significance of the result shown in Figure 2 was by Hines and Halevy [9], who investigated many ways in which the effect could have resulted from various aspects of the data such as long term trends and regularities. They concluded that they were obliged to accept the validity of the claim shown in Figure 2 and to seek a physical explanation.

We will next show that the percentage depth of the minimum shown in Figure 2 is larger for areas in the troposphere having the most intense circulation. Figure 3 shows the results of an exercise in which computations similar to those of Figure 2 were performed except that the value of vorticity used in computing the VAI was a parameter, and also the northern hemisphere was divided into 8 zones of latitude such that each zone had approximately the same number of grid points in the National Meteorological Center octagonal grid. If we focus our attention on the latitude band 43 to 51° near which most of the troughs circulate we see a monotonic increase in fractional depth of the minimum VAI with increasing vorticity. The same trend is found in the other latitude bands.

Svestka et al. [10] have pointed out that some sector boundary passages are observed to be followed by a stream of protons with energy a few MeV that lasts for a few days. Figure 4 shows an analysis similar to that of Figure 2 of the average northern hemispheric VAI as a function of time near a sector boundary passage. The solid line represents 18 boundaries that were followed by streams of MeV protons as tabulated in [10], while the dashed line represents the 60 other boundary passages observed during the same winters (1963-69). We note that the minimum VAI associated with the MeV proton boundaries is almost twice as deep as the minimum associated with the other boundaries.

It is by no means clear that it is the presence of the MeV protons that causes the enhanced meteorological response shown in Figure 4. The sectors containing the MeV proton streams are found to be more active in almost every way. For example, the solid line in Figure 5 shows that the MeV proton boundaries were followed by a large increase in the mag-

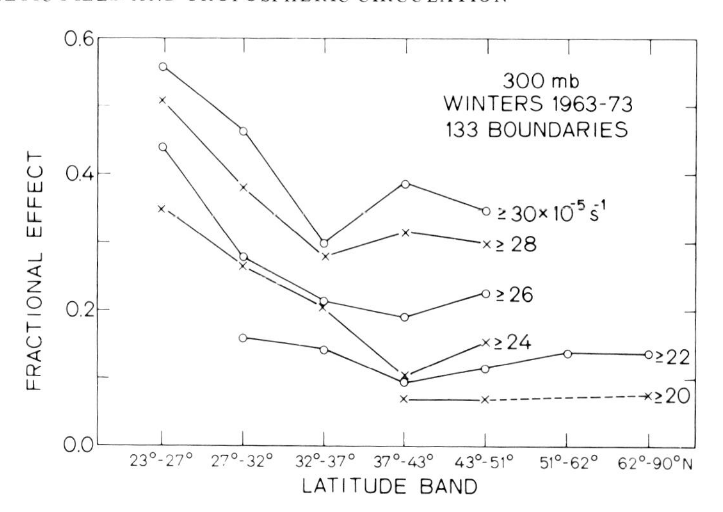

Figure 3. The fractional depth of the sector-related minimum in VAI as a function of the vorticity value used in computing the index.

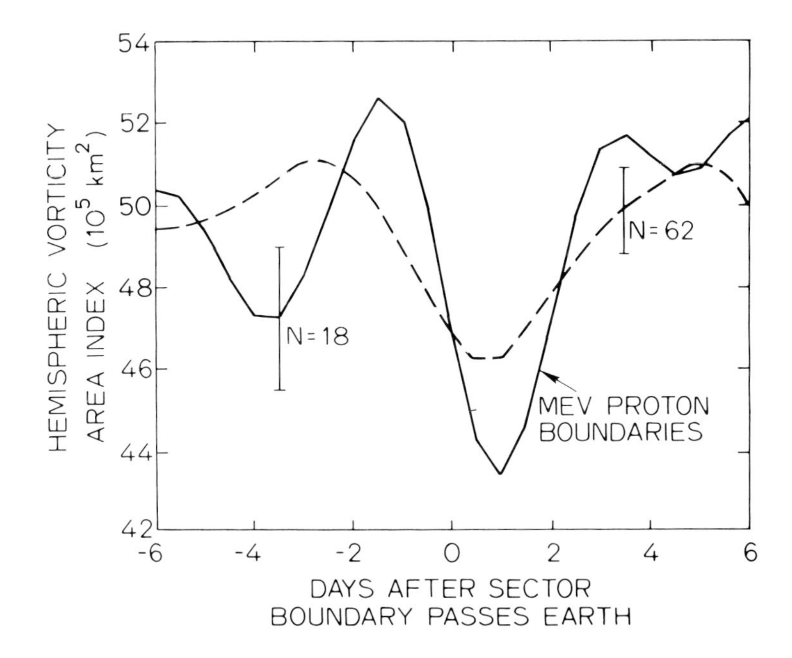

Figure 4. Similar to Figure 2, but computed for sector boundary passages accompanied by MeV proton streams (solid line) and for all other boundary passages in the same winters (dashed line).

nitude of the interplanetary magnetic field, while the other boundaries in the same winters showed no change in the field magnitude. A similar result is found in the response of geomagnetic activity to the two

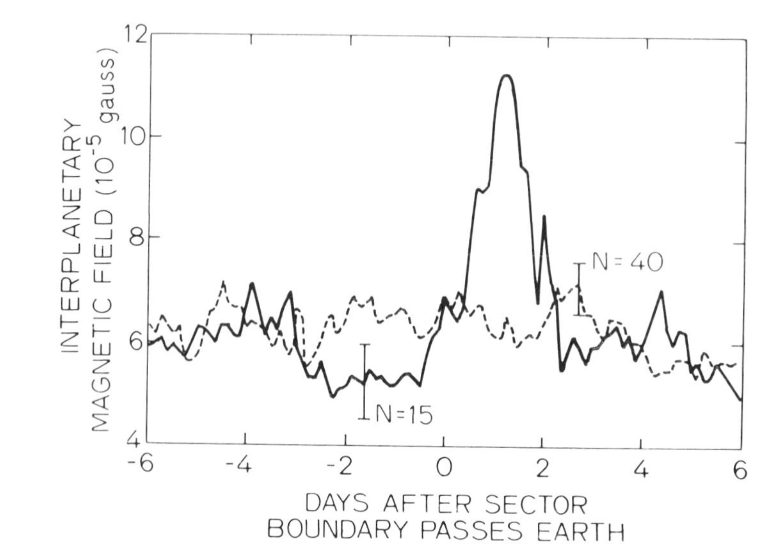

Figure 5. The variation of the magnitude of the interplanetary magnetic field after boundary passages with MeV proton streams (solid line) as compared with other boundary passages in the same winters (dashed line).

classes of sector boundary passages. These new results may prove helpful in defining the elusive physical mechanisms involved in Sun-weather effects. The enhanced meteorological response associated with the MeV proton boundary passages might be looked for in the other meteorological responses to the sector structure that have and will be investigated.

An analysis by Larsen and Kelley [11] used the recently available Limited Area Fine Mesh Grid over the continental Units States. They reproduced the minimum in VAI shown in Figure 2 in new data. Their most interesting result was a quantitative analysis of the accuracy of forecasting in a time frame related to the solar and interplanetary sector structure. In Figure 6 the abscissa is keyed to the times at which the warped current sheet was carried past the Earth by the solar wind, and the ordinate is a measure of the accuracy of forecasts made at particular times (for details of this computation see [11]). We see in Figure 6 that the forecasting accuracy is significantly decreased during the two days following the passage of the sector boundary. The appreciable decrease in forecasting accuracy near the time of sector boundary passage suggests that Sun-weather influences may have a significant and important impact on forecasting in spite of the elusive nature of the Sun-weather effects at the present time.

It is appropriate to note that the work of Larsen and Kelley [11] was only possible because they had access to daily data that had been archived by the Atmospheric Sciences Department at Pennsylvania State University. When attempting to perform similar investigations one often finds that the requisite daily data have not been archived — only monthly

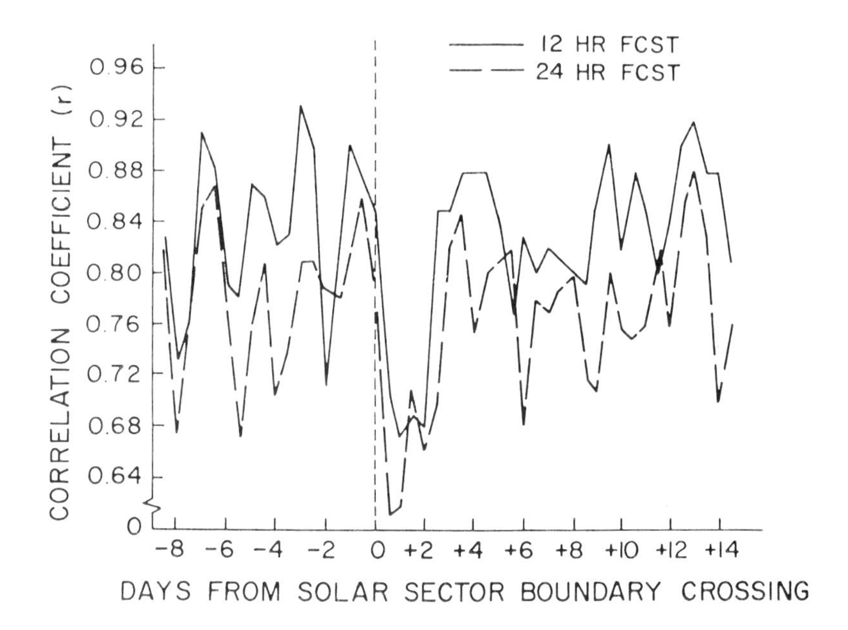

Figure 6. An analysis of the accuracy of forecasts of tropospheric vorticity in a time frame related to the passage of the warped heliospheric current sheet. The ordinate is a cross correlation between the forecast and observed vorticity indices. During the first two days after the passage of the current sheet the correlation coefficient appears to be systematically lower by about 0.15 units [11].

means are preserved. In the absence of daily data extending over several years it is impossible to look for Sun-weather relationships. In planning for future spacecraft and ground-based observations it is important that daily values of the observations be archived so as not to exclude before the fact the possibility of establishing and understanding Sun-weather relationships.

The vertical electric field in the troposphere is a measure of the electric potential between the ionosphere and the Earth's surface. Figure 7 shows a convincing variation of the vertical field at the top of Zugspitze as measured by Reiter [12]. Figure 7 shows that during a complete sunspot cycle the vertical electric field increases after the time of boundary passage. The increase is larger at toward-away boundaries and in years of maximum sunspot activity. The air-Earth current density decreases before the time of the boundary. We note that the influence of the sector structure on the VAI also appears to be larger during years of maximum sunspot activity.

Park [13] found that the vertical electric field observed near the southern geomagnetic pole at the Soviet station Vostok showed a minimum a few days after the time of boundary passage. Local winter showed a

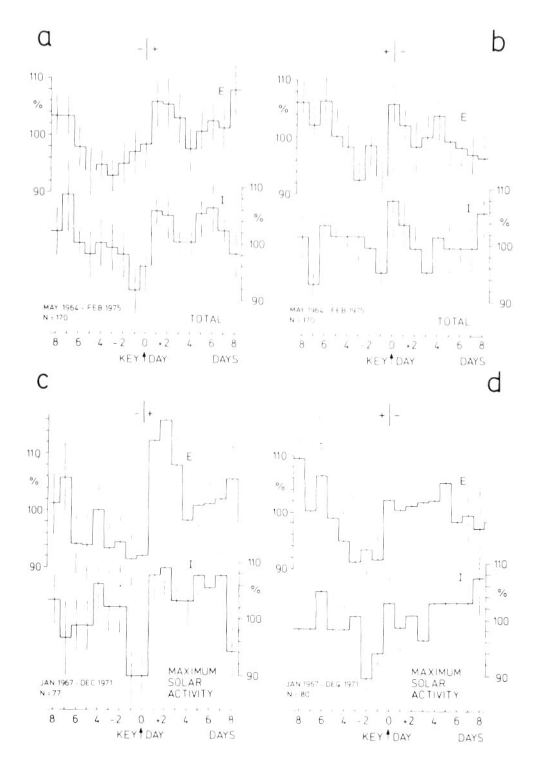

Figure 7. Superposed epoch analysis of the atmospheric electric field E and the air-Earth current density I recorded at Zugspitze Peak Observatory near the time of sector boundary passage [12].

larger amplitude than local summer. It is not clear why a minimum was found at Vostok and a maximum at Zugspitze. Further observations at various points around the world are in progress to try to understand this. In any case, it is plausible that changes in the atmospheric electric field may have important meteorological consequences such as influencing the processes by which clouds are formed [6]. Deformation of a bank of cirrus clouds can have an important influence on the atmospheric heat budget as discussed by Roberts and Olson [8].

An important clue to the changes in atmospheric circulation associated with the sector structure is found in the results of Reiter and Littfass [14] shown in Figure 8. This figure shows the concentration of Be7 and O_3 observed at Zugspitze with relation to the sector structure. Be7 is predominantly originating from the straosphere by reaction of cosmic rays with atmospheric N_2. An increased concentration of Be7 observed on a mountain peak at 3 km altitude indicates an influx of stratospheric air into the lower troposphere. Figure 8 shows that such influxes tend to occur a few days after the passage of a sector boundary. The effect is seen for both away-toward and toward-away boundaries, but with the largest amplitude in away-toward boundaries. Theoretical cal-

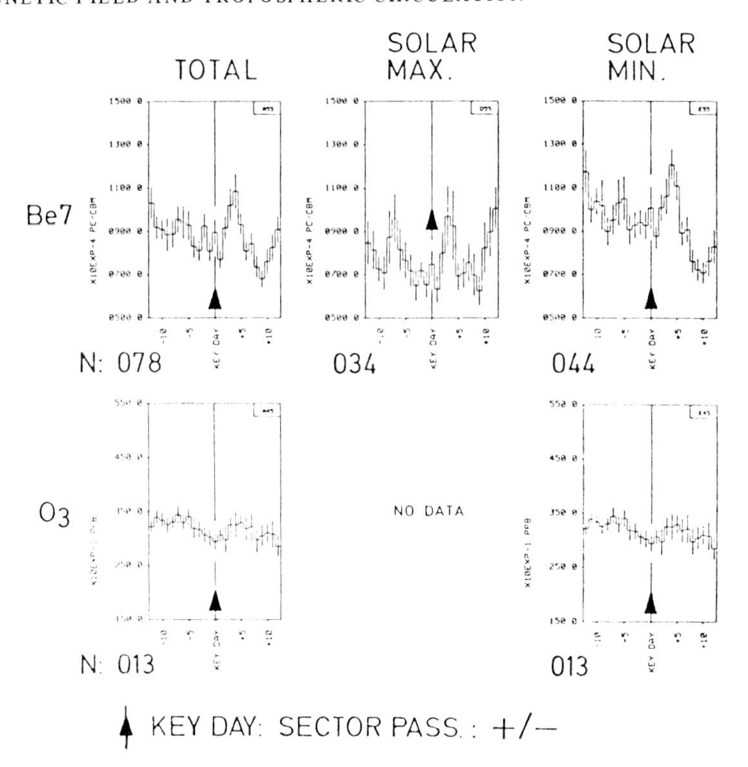

Figure 8. Superposed epoch analysis of Be7 and of O_3 concentrations observed at Zugspitze Peak Observatory near the times of sector boundary passage [14].

culations relating atmospheric chemistry to solar activity may need to take account of such circulations in order to arrive at a realistic description of the actual atmosphere.

It appears that in the approximate area of the Gulf of Alaska the direction of the interplanetary magnetic field toward or away from the Sun influences the area of troughs. On the day that wintertime troughs crossed 180°W longitude they were divided into two classes [15]. In Figure 9 the solid line is the average VAI of troughs for which the interplanetary magnetic field was directed away from the Sun when they were crossing 180°W, while the dashed line similarly represents troughs associated with interplanetary field directed toward the Sun. During the first five days of measurement the "away" troughs are significantly larger on the average than the "toward" troughs. A series of approximately 1,000 similar analyses using random assignments of troughs to the away or toward class indicates that the effect is significant at better than the 99% level. The effect presists through most of the winters 1951 - 1973, and can be found in each of the months October to March analyzed separately. The physical mechanism may relate to the topological condition that an interplanetary magnetic field line directed

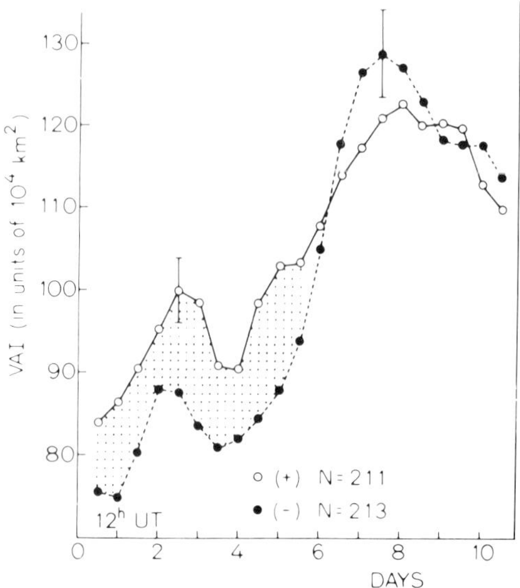

Figure 9. Low-pressure troughs near the Gulf of Alaska when the interplanetary magnetic field is directed away from the Sun have a significantly larger area than troughs associated with field directed toward the Sun. The zero day is the day on which the troughs crossed 180°W longitude [15].

away from the Sun may merge with the geomagnetic field line going into the northern polar region, as illustrated by the observation of Yeager and Frank [16] that the flux of solar electrons of a few hundred eV into the northern polar cap is an order of magnitude larger when the interplanetary field is directed away from the Sun as compared with toward the Sun.

ACKNOWLEDGMENTS

This work was supported in part by the Office of Naval Research under Contract N00014-76-C-0207, by the National Aeronautics and Space Administration under Grant NGR 05-020-559 and Contract NAS5-24420, by the Atmospheric Sciences Section of the National Science Foundation under Grant ATM77-20580 and by the Max C. Fleischmann Foundation.

REFERENCES

1. L. Svalgaard and J. M. Wilcox, Nature, 262, 766, 1976.
2. L. Svalgaard, J. M. Wilcox, P. H. Scherrer, and R. Howard, Solar Physics, 45, 83, 1975.
3. J. M. Wilcox and P. H. Scherrer, J. Geophys. Res., 77, 5385, 1972.
4. J. M. Wilcox, Space Sci. Rev., 8, 258, 1968.

5. L.Svalgaard, Stanford University Institute for Plasma Research Report No. 629, An Atlas of Interplanetary Sector Structure 1957-1974, 1975 and Report No. 648, Interplanetary Sector Structure 1947-1975, 1976.
6. R. Markson, Private communication, 1969; Pure and Applied Geophysics (PAGEOPH) 84, 161, 1971.
7. J. M. Wilcox, P. H. Scherrer, L. Svalgaard, W. O. Roberts, R. Olson, and R. L. Jenne, J. Atmos. Sci., 31, 581, 1974; J. M. Wilcox, L. Svalgaard, and P. H. Scherrer, J. Atmos. Sci., 33, 1113, 1976.
8. W. O. Roberts and R. Olson, Rev. Geophys. Space Phys., 11, 731, 1973.
9. C. O. Hines and I. Halevy, Nature, 258, 313, 1975; J. Atmos. Sci., 34, 382, 1977.
10. Z. Svestka, J. T. Fritzova-Svestkova, H. W. Dodson-Prince, and E. R. Hedeman, Solar Phys., 50, 491, 1976.
11. M. F. Larsen and M. C. Kelley, Geophys. Res. Lets., 4, 337, 1977.
12. R. Reiter, J. Atmos. Terr. Phys., 39, 95, 1977.
13. C. G. Park, Geophys. Res. Lets., 3, 475, 1976.
14. R. Reiter and M. Littfass, Arch. Met. Geophys. Biokl., Ser. A, 26, 12, 1977.
15. P. B. Duffy, J. M. Wilcox, W. O. Roberts, and R. Olson, E⊕S, 58, 1220, 1977.
16. D. M. Yeager and L. A. Frank, J. Geophys. Res., 81, 3966, 1976.

AN EXPERIMENTAL SEARCH FOR CAUSAL MECHANISMS IN SUN/WEATHER-CLIMATIC RELATIONSHIPS

Richard A. Goldberg
Laboratory for Atmospheric Sciences
NASA/Goddard Space Flight Center
Greenbelt, MD 20771, USA

ABSTRACT. The stratosphere and lower mesosphere are strong candidate regions for the occurrence of Sun/weather coupling processes. The approach presented here investigates the role of the middle atmosphere as a buffer region; i.e., where external energy inputs are modulated, filtered, amplified, and/or reflected. An ongoing experimental program is presented. It is designed to investigate atmospheric responses to the energetics of high latitude corpuscular radiations, which are closely related to solar activity. Many Sun/weather relationships show their best correlation at high latitudes during local winter, when direct solar insolation is least competitive, thereby qualifying the high latitude atmosphere as a primary region for causal-mechanism search. Satellite data (Nimbus IV BUV O_3) are currently under study to measure O_3 response to high latitude energetics. Furthermore, two complex and highly coordinated rocket-balloon programs (Aurorozone I and II) have been conducted at Poker Flat, Alaska to measure the direct penetration of energetic radiations to stratospheric depths, and their effect on the local electrical and neutral atmospheric properties. Preliminary findings exhibit an unexpected depletion in O_3 during two modest X-ray auroral events, and display ionic effects which could imply significant changes in the local aeronomy. Future plans for continued studies incorporating the Spacelab are also discussed.

1. INTRODUCTION

The concept that certain terrestrial weather patterns and climatological behavior are induced by solar phenomena is based principally on statistical evidence. These include, for example, variability with solar activity [1], solar magnetic boundary passage [2,3] and the single and double sunspot cycles [4,5]. To understand such statistical relationships we must determine the physical processes at work and trace the steps between cause and effect. A full comprehension of the physics and chemistry responsible for the observed correlations is essential in developing a base on which to build a capability for prediction of weather and especially climate.

B.M. McCormac and T.A. Seliga (Eds.): Solar-Terrestrial Influences on Weather and Climate. 161-173.

Consideration of the energetics involved [6,7] demonstrates that the energy (primarily in the form of corpuscular radiation) supplied to the Earth's atmosphere by isolated solar events, does not compete with the total solar irradiance and is inadequate to feed a typical tropospheric storm system. The disparity between the corpuscular radiation and solar insolation becomes less severe when we recognize that shielding by the geomagnetic field forces the major deposition of corpuscular radiation to focus at high latitudes where solar insolation effects become minimal, particularly under local wintertime conditions. It can be shown that locally at high latitudes, the ratio of corpuscular related radiation to solar insolation can significantly increase [8]. It may not be surprising then, that many Sun/weather correlations improve at higher latitudes, and are most pronounced during local winter conditions.

There are several categories of energetic radiation reaching various altitudes at high latitudes. Galactic cosmic rays, which are regulated by solar activity, normally dominate the electrical properties of the atmosphere between 30 and 5 km. Solar proton events (PCA's) are relatively infrequent, but can enhance ionization by orders of magnitude near 30 km. Also, relativistic electron precipitation events (REP's) are less energetic but more frequent, and may induce changes and effects in O_3 for example, of a more persistent and cumulative nature [9]. Finally, there are local electron precipitation events within the auroral zone of very frequent occurrence, and these often produce significant X-ray fluxes by the bremsstrahlung process. The conversion of electron energy to X-ray photons permits penetration of the energy to lower depths within the atmosphere, thereby causing effects well below the absorption height for the parent electrons.

Recent observations of stratospheric O_3 in connection with particle injection events have already suggested some possible candidate mechanisms. Weeks *et al.* [10] and, more recently, Heath *et al.* [11] have observed a depletion of O_3 below 60 km during solar proton events. On the other hand, evidence for an X-ray bremsstrahlung enhancement of O_3 near 50 km, following a diffuse auroral event, has been reported by Maeda and Heath [12]. Fluctuations in O_3 have been suggested to affect tropospheric weather through UV filtering and stratospheric heating modulations, which in turn induce circulation pattern changes [13].

This paper provides a description of and preliminary results from an ongoing program to study mesospheric and stratospheric responses to the above-mentioned corpuscular radiations induced by solar activity. Satellite data are used to map correlations between O_3 variability and high latitude energetic phenomena. Two rocket/balloon programs have already been conducted to measure local changes of O_3 and electrical parameters *in situ*, while simultaneously studying the parent electron distributions responsible for the incoming radiations. Plans for mapping the global input of electron precipitation events from the Spacelab are also discussed.

2. APPROACH AND PRELIMINARY RESULTS

2.1 Satellite-Nimbus IV Backscattered Ultraviolet (BUV) Experiment

The Nimbus IV BUV experiment is a scanning ultraviolet (UV) spectrophotometer designed to measure the atmospherically backscattered UV radiations from the Sun. Twelve distinct wavelengths are monitored; each radiance is subject to a unique absorption characteristic by atmospheric O_3, both as a function of atmospheric depth and O_3 concentration. It is therefore possible to map changes in total O_3 and to produce height profiles of O_3 concentration by inversion techniques combining the various wavelengths.

Heath et al. [11] have already demonstrated that the technique is sufficiently sensitive to detect changes in O_3 induced by a PCA event. Figure 1 shows their reported response of total atmospheric O_3 above 4 mb for three geographical zonal regions. These data represent daily zonal means. It is apparent from the Figure that during the great PCA events of August 1972, which began near day 217 (August 4), a significant depletion of O_3 ($\sim$20%) occurred at high latitudes and persisted well beyond the termination of the event. The effect is reduced in the 55 to 65°N zone, and is not apparent in the tropical zone. Heath (private communication) has also indicated that similar effects were seen, but with much larger fluctuations, in the southern (local winter) high latitude hemisphere. These results are even more striking when one realizes that geographical zonal means tend to smear effects induced by corpuscular radiations under geomagnetic control.

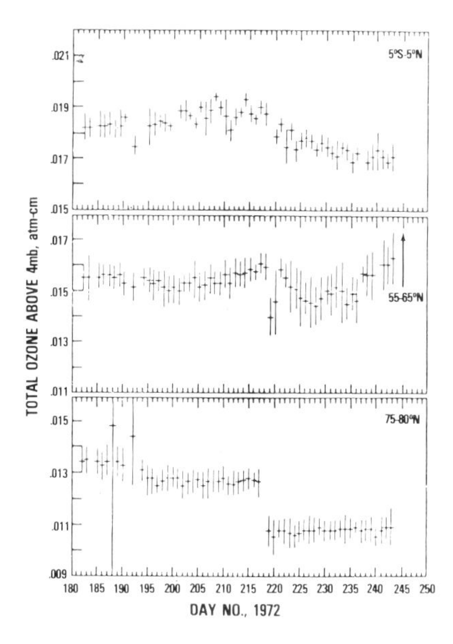

Figure 1. Daily, zonally averaged total O_3 above 4 mb pressure for tropical, mid and high latitude zones, during July - August 1972 (after Heath et al. [11]).

A 7 to 8 year data base for BUV O_3 now exists, so that high powered statistical techniques may permit retrieval of information concerning O_3 response to PCA's, REP's, and other high latitude phenomena. The successful use of the data base inherently depends on the quality of the data itself. An Ozone Processing Team (OPT) has been formed at the Goddard Space Flight Center under the direction of A. Fleig, specifically to optimize the quality of the Nimbus IV BUV data set. Part of this study has demonstrated that the occurrence of background effects, induced in part by energetic precipitating particles, can distort the irradiance values measured by the instrument. This is particularly important at high latitudes where background effects must be appropriately corrected for before the remaining fluctuations can be attributed to O_3.

Under the OPT program (and with E. G. Stassinopolous) we have developed a software package to analyze the dark current count rates for the BUV experiment aboard Nimbus IV. This package permits analysis of the BUV data on both a spatial and temporal scale from block sizes of 3° LAT X 5° LONG worldwide, and from 1 to N passes. Averaging, including the statistical significance of any chosen data set, can be established. In addition, spatial coordinates for formatting include geographic, magnetic, magnetic B-L and magnetic time. Solar and geophysical conditions are established through the use of worldwide indices including Ap, AE, Dst, solar sector boundary passage, and 10.7 cm flux. The instantaneous or time integrated spectral distribution of particles trapped in the Earth's radiation belts is available for correlation. Finally, geomagnetic shielding effects on high latitude orbit positions and corresponding solar flare proton intensities can be established. It is therefore possible to assemble and group the BUV data in any desired averaging or sequential mode, and to relate the data to concurrent solar and geophysical processes.

This package is currently in use to evaluate background current characteristics, and compare modulations in the dark current with high latitude phenomena and with the magnetospheric particle environment. We are now beginning an analysis of BUV O_3 data with the same software package. The software should also be applicable to projected data sets expected from future satellite BUV experiments aboard Nimbus G and Tiros N.

2.2 Rocket/Balloon - Operations Aurorozone I and II

The BUV satellite data are extremely useful for correlating atmospheric O_3 response with specific types of geomagnetic phenomena, on both a local and global scale. However, the BUV suffers from restrictions which limit the scope of the technique. It cannot provide O_3 response under nighttime conditions. Also, it cannot measure parameters associated with the characteristics of the geomagnetic event, the direct energy deposition at stratospheric altitudes, nor the atmospheric chemical and physical processes leading to the observed modulations in O_3.

These processes can only be studied in depth through highly coordinated and sequenced _in-situ_ soundings of the appropriate parameters, aboard rocket and balloon platforms.

For example, recent results have been obtained during Aurorozone I (September 1976) and II (March 1978). These rocket-balloon experiments were launched from Poker Flat Research Range, Alaska and each involved a large number of launches. Scientists from several universities and laboratories coordinated their efforts to make the programs possible. The institutions represented include NASA/Goddard Space Flight Center, Naval Research Laboratory, University of Denver, University of Texas at El Paso, the Pennsylvania State University, Utah State University, University of Washington, Atmospheric Sciences Laboratory/White Sands Missile Range, and SRI International.

Aurorozone I was the first concerted experimental effort to trace X-ray auroral bremsstrahlung radiation into the lower mesosphere and stratosphere to study its direct effect on O_3 electrical conductivity, ionization, ion mobility, and on the meteorological structure. The rocket flights included 2 Nike Tomahawks to measure auroral particle and X-ray energetic structure, 7 chemiluminescent and optical ozonesondes, 7 conductivity and Gerdien probes to measure stratospheric electrical parameters, and 17 datasondes to measure temperatures, densities and winds. These flights were coordinated with 5 large balloon flights to measure X-rays, conductivity and cosmic rays near 40 km. In addition, 4 balloon ozonesondes were employed to measure O_3 up to 30 km. The nearby Chatanika radar backscatter facility was also coordinated to predict, verify, and map electron particle energetics during the Nike Tomahawk flight periods.

Figure 2 depicts the chronological logistics for Aurorozone I. Two events of interest occurred on September 21 and 23. Each of these moderately intense X-ray auroral events was studied with a nighttime launch sequence. The results shown here demonstrate the characteristics of each event, and show the associated changes in the selected middle atmospheric parameters.

The electron density contour maps for each auroral event were obtained by the Chatanika radar facility (Vondrak and Wickwar, SRI International) during the Nike Tomahawk flights, and are depicted in Figure 3 along with the rocket trajectories. The intensity of the auroral events is measured by the magnitude of the electron density contours. From this, we observe that the September 21 event was reasonably intense during the early upleg portion of the flight (which is in the spatial vicinity above where stratospheric soundings were made), but gradually faded to non-auroral conditions down-range. The September 23 event was more intense, and extended down-range beyond the downleg portion of the rocket flight. This behavior was also observed with riometers and visually, with all sky cameras, at Poker Flat and Ft. Yukon ($\sim$200 km down-range).

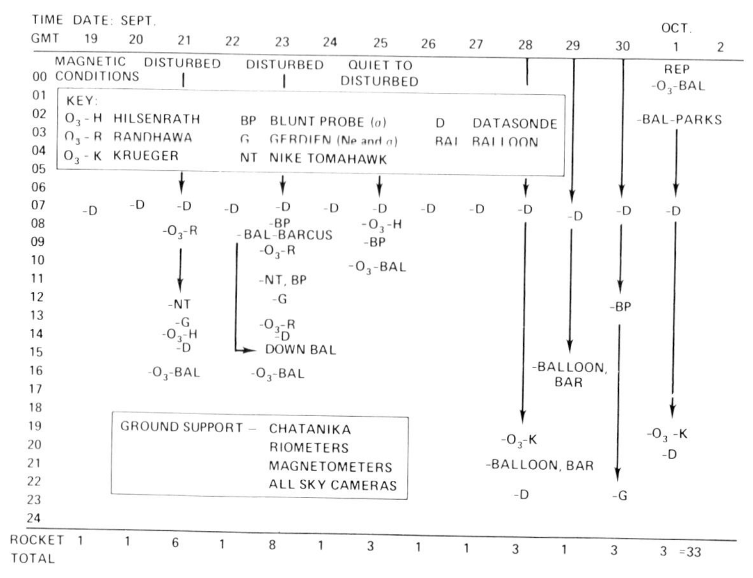

Figure 2. Aurorozone I - Chronological launch sequence.

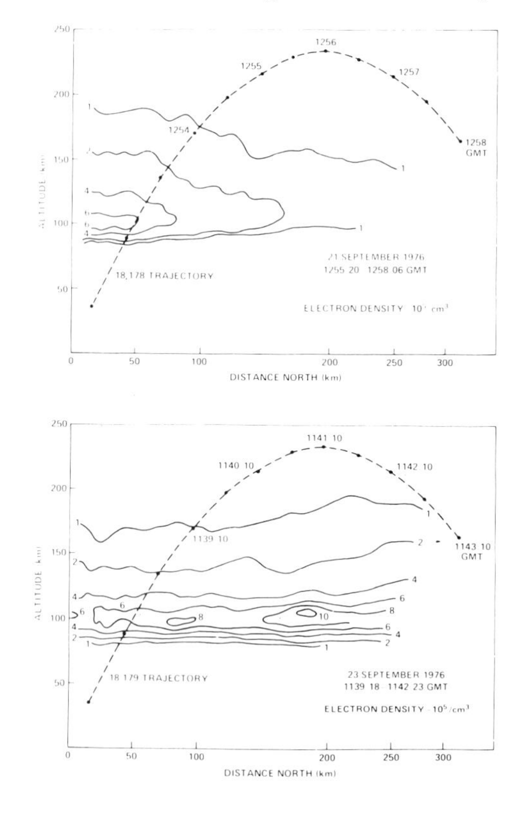

Figure 3. Electron density contour maps (Chatanika) and superimposed Nike Tomahawk rocket trajectories during Aurorozone I.

The integral particle fluxes for electrons > 15, 25, 40, and 100 keV were measured with Geiger Mueller detectors (Goldberg and Jones, GSFC) on each Nike Tomahawk. These in-situ measurements verified the Chatanika result, and also showed that the more energetic (> 40 keV) component of the electron flux sustained a greater increase than the low energy component during the second event relative to the first event. Further examination of the auroral ionosphere was obtained from pulsed plasma probes (Szuszczewicz, NRL) flown on each Nike Tomahawk.

Also aboard was an upward-looking X-ray detector (NaI scintillator) with energy discrimination in 4 channels above 5 keV. Figure 4 compares for the nighttime events the X-ray energy deposition in the atmosphere by 5 to 40 keV bremsstrahlung X-rays between 80 and 30 km (Barcus, University of Denver) with an extrapolation to higher altitudes. From the figure we see that the properties of the events were similar and provided X-ray radiation which peaked near 60 km. The ratio of the energy depositions as a function of altitude demonstrates that the second event possessed a strong high energy component in accord with the measurements of the parent electron distribution. Above 50 km, the ratio between the second and first event is about 2.

The O_3 results were obtained by both the chemiluminescent (night) and optical (day) sounding techniques. Figure 5 shows the GSFC results for the O_3 mixing ratio. The two nighttime profiles were obtained on September 21 following the first auroral event and on September 25, which was used as a control quiet night (Hilsenrath, GSFC). Also shown are profile points for two daytime soundings made one week later (Krueger, GSFC). The results indicate an early nighttime O_3 enhancement, which may be caused by O_3 production from atomic oxygen attachment when the Sun sets. They then show an O_3 depletion to near daytime values above 50 km following the auroral event. Further verification of this effect was found in the two chemiluminescent O_3 soundings (Randhawa, ASL) during the September 23 event. In both cases, a significant depletion (more than 25%) was seen above 50 km. The above results are unexpected, since energy arguments using neutral photochemical models cannot account for the observed depletion.

The effects of X-ray and particle ionizing radiations are illustrated in Figure 6 (Mitchell, University of Texas, El Paso), which displays a sequence of three conductivity soundings during the September 23 event. The profiles at 2200 AST (before) and 0137 AST (during) were made with blunt probes, and at 0220 AST with a Gerdien probe. This latter instrument measures conductivity and ionization simultaneously permitting an estimate of ion mobility and, hence, ion size.

Below 50 km, all three profiles are parallel and nearly equal, reflecting the dominance of cosmic ray ionization in this region. The increased conductivity above 50 km is due to ion production by the X-rays and energetic electrons. The lower altitude of the ion ledge found on the second sounding is expected from the X-ray measurement in Figure 4.

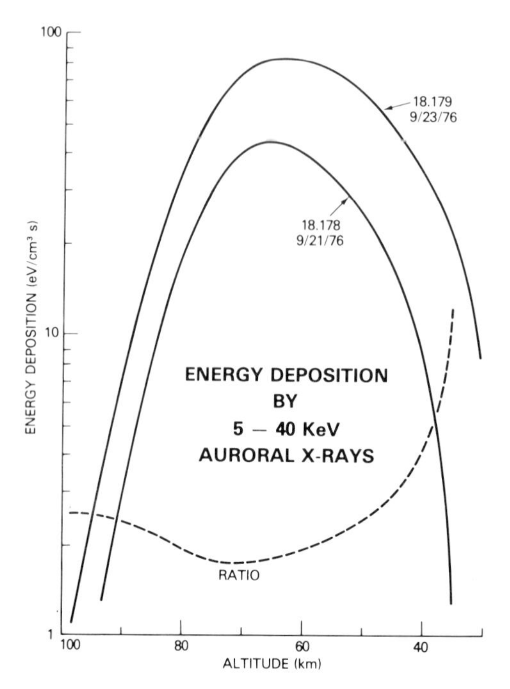

Figure 4. Atmospheric deposition of 5 to 40 keV bremsstrahlung X-rays during September 21 and 23, 1976 events for Aurorozone I.

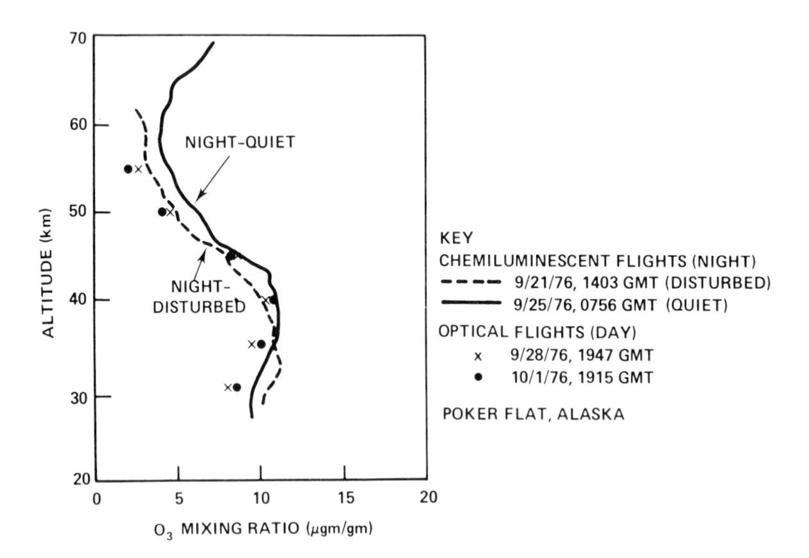

Figure 5. Examples of O_3 response to high latitude energetics during Aurorozone I. Early quiet night curve (solid) represents control night data. Post-event night response is given by the dashed curve. Daytime values are shown for reference.

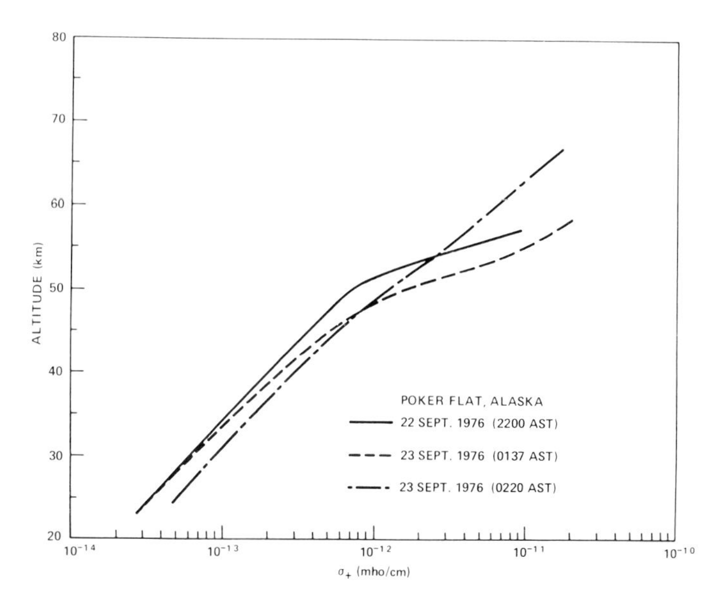

Figure 6. Conductivity profiles before, during and following the September 23, 1976 event.

The results illustrate the rapid variability of electrical parameters in space and time, and the need for simultaneous measurements of energy source functions.

The most recent results have been obtained during Aurorozone II (March 1978). As part of Aurorozone II, we developed and successfully launched a new Arcas payload (10 cm diameter, 8.2 kg). This X-ray/Gerdien payload (XRG) is depicted in Figure 7. It carries a Gerdien probe to measure ion conductivity, density, and mobility; an NaI scintillation detector to measure X-rays in four energy ranges; a guard counter to measure energetic electrons and monitor cosmic radiation; and a vertical electric field instrument incorporating antennas designed to measure atmospheric potential difference. A two-axis magnetometer is used to establish platform stability. The above instruments had been flown previously as individual payloads on meteorological class rockets, but never in a combined configuration until the development of XRG. The temporal and spatial response of the atmosphere to rapidly changing auroral X-ray phenomena make this newly developed grouping mandatory for proper interpretation of the data obtained.

A preliminary look at XRG results from Aurorozone II has already caused a restudy of Aurorozone I data, and the discovery that the precipitation of relativistic electrons was the major atmospheric energy source above 55 km during the two events (Barcus, private communication). The result for the September 23, 1976 event is shown in Figure 8, where the electron spectrum from 15 to 150 keV was determined by data obtained above the atmosphere. It was then extended to 550 keV by applying atmospheric intensity range considerations to data obtained within the

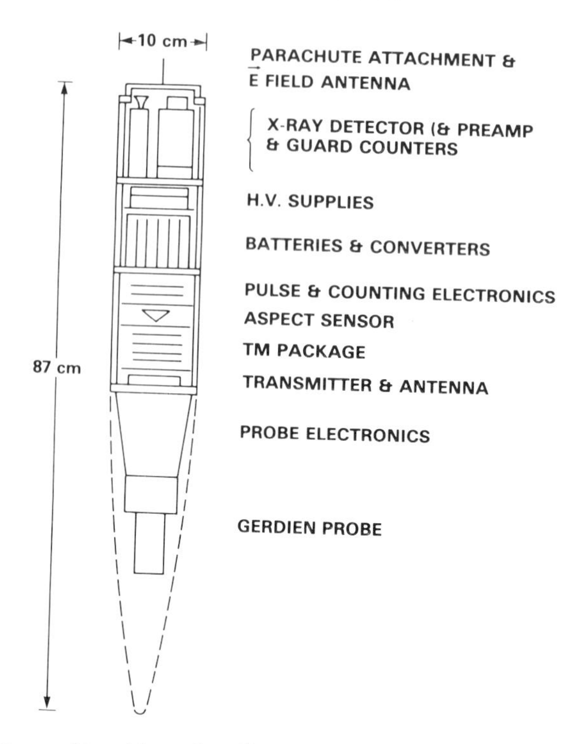

Figure 7. X-ray/Gerdien (XRG) Super Arcas payload.

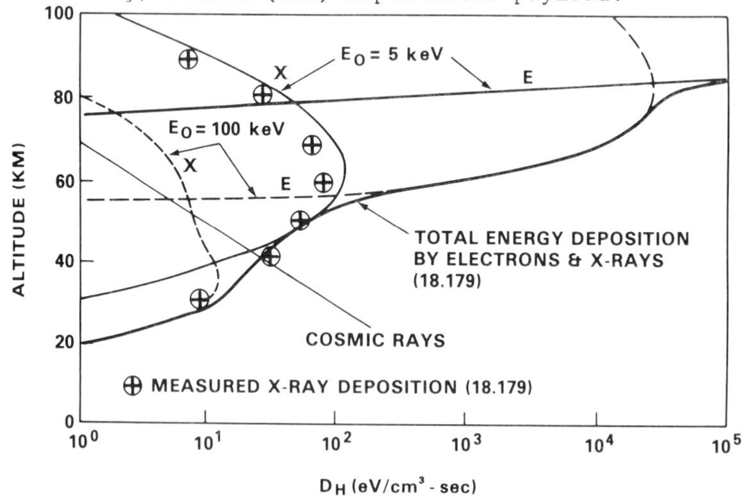

Figure 8. Energy deposition rates measured and calculated for 18.179, September 23, 1976. Measured electron (E) and calculated X-ray (X) curves are shown for distributions of 5 keV (solid) and 100 keV (dashed) folding energies. The measured X-ray deposition (E_o = 5 keV) is shown by the symbol ⊕.

atmosphere during up and downleg transits. A two-component spectrum was determined. The calculated profiles of energy deposition for both components (e-folding energies of ∿5 and ∿100 keV) resulting from direct electron impact (E) and bremsstrahlung X-rays (X) are shown in Figure 8. These are compared with the measured contribution by X-rays and with that produced by the cosmic ray background. For this event, it is clear that direct electron impact dominates above 55 km, whereas X-ray absorption is most important between 55 and 40 km. Our findings from other events also show that electron impact dominates some portion of the upper region previously thought to be under the primary influence of X-ray absorption. These radiations may also help explain the unexpected 25% O_3 depletion.

2.3 Spacelab - Atmospheric X-Ray Emission Telescope (AXET)

The rocket-balloon program is designed to trace the energetics of high latitude phenomena to stratospheric depths; it is also important to map the frequency and global extent of such energy sources. The AXET (14) is a proposed Spacelab experiment to image and measure the spatial, temporal, and spectral distributions produced in the upper atmosphere by precipitating electrons. Figure 9 illustrates the proposed concept. The sensor will operate from a remote platform situated obliquely above the atmospheric X-ray source. The height in the atmosphere where X-rays of 5 to 150 keV energies are produced by the bremsstrahlung process lies between 80 and 120 km [15]. Images (5 to 25 keV) and non-imaging information (>25 keV) are provided by spatially sensitive single-wire proportional counters combined with collimators. The sensitivity of AXET will permit spatial resolution of a 200 km scale size object from a slant range of 1500 km. Background contamination of the instrument by precipitating electrons requires that remote sensing of the source occur from an oblique location beyond the primary electron precipitation region.

Concomitant observations of relevant solar and terrestrial phenomena will allow determination of what physical relationships may exist between X-ray aurorae, solar activity, and atmospheric processes. X-rays produced by the bremsstrahlung process will normally cause their greatest atmospheric effect between 75 and 40 km, which is where the 15 to 25 keV X-rays are mainly absorbed. Observations of the auroral X-rays will also provide information on the primary precipitating belt electrons, and can be used to study the dynamics of the radiation belt processes. The X-rays carry with them information concerning effects of the electron precipitation on the atmosphere at various depths such as changes in electrical conductivity, chemical reactions, and Joule heating. In addition, they contain information concerning magnetospheric processes responsible for the electron precipitation events.

The ultimate goal would be to combine AXET with other instruments designed to observe solar, magnetospheric, ionospheric, and atmospheric processes simultaneously, to inquire into the validity of any suggested theories.

THE STATISTICAL SIGNIFICANCE OF DATA FROM SUPERPOSED EPOCH ANALYSES

B. Jamison and R. Regal

Department of Mathematics and Statistics
State University of New York at Albany
Albany, NY, U.S.A.

ABSTRACT. Superposed epoch analysis is a widely used technique to determine whether a given type of event influences a physical process which is either intrinsically random or whose measurements are perturbed by random noise. The technique is statistical, and so questions arise as to the statistical significance of the results of such an analysis. We discuss such questions, not so much in their generality but in the context of a recent application [5,6,1] of the technique to the case where the events in question are sector boundary crossings and the physical process is whatever is measured by the vorticity area index (VAI) of Olson, Roberts, and Gerety [3].

Let V_1, V_2,... be the successive values of the VAI at midnight and noon of each day beginning at midnight, Jan. 1, 1963 (the choice of time origin being made so that we can use the key days listed in Table I of [5]). Consider the list of 54 sector boundary crossings given in [5]. (For the conventions used in assigning these sector crossings to specific half-days, see [5] and [2].) For seven of these crossings, any assignment of them to a time interval as narrow as a half-day seems questionable. This leaves us with 47 sector crossings. Let $T_1,\ldots,T_{47}$ be the half-days at which the sector crossings occur. We refer to the T_i's as <u>key-half-days</u>. Let <u>d</u> be a fixed whole number. (In the application we are considering, d = 12, but it is easier to follow our discussion if we use "d" instead of "12" in the formulas.) For each k - 1,..., 47, the stretch of 2d+1 half-days from T_k-d to T_k+d is called the <u>k^{th} epoch.</u> For each such k, let V_{ki} be the value of the VAI at the <u>i^{th}</u> day of the <u>k^{th}</u> epoch, where i ranges from -d to d; that is, $V_{k,i} = V_{T_k+i}$. Now, for each i = -d, ...,d, let $\bar{V}_i$ be the average over the 47 epochs of the VAI at the <u>i^{th}</u> half-day of the epoch:

$$V_i = \frac{1}{47} \sum_{k=1}^{47} V_{ki} \qquad i = -d, \ldots, d. \tag{1}$$

B.M. McCormac and T.A. Seliga (Eds.): Solar-Terrestrial Influences on Weather and Climate. 175-179.

If this is done, it is observed that $\overline{V}_2$, the average VAI a full day after the sector crossing, is smaller than the other $\overline{V}_i$'s. Is this deviation statistically significant?

The concept of significance level as formalized in the theory of mathematical statistics requires the notion of a statistical hypothesis. (In textbook statistics, there is usually a pair of hypotheses, one "null" and one "alternative," In applications, however, the null hypothesis is usually implicit, while the alternative hypothesis is missing altogether.) Statistical significance is defined in terms of the null hypothesis only. One performs an experiment whose outcome is uncertain. (The experiment usually consists of a number of subexperiments, performed under identical circumstances, with the outcome of each subexperiment independent on the outcomes of the others.)

The outcome of the experiment is a number x; since the outcome of the experiment is uncertain, one can look at x as the value of a random variable X. A statistical hypothesis is a statement H about the distribution of X. The notion of significance level becomes applicable if, assuming H, one can figure out, for each possible x, the probability $P(|X|\geq x)$ that the experiment results in a value of X at least as extreme as x. (It is sometimes appropriate to use $P(X\leq x)$ or $P(X\geq x)$ in place of $P(|X|\geq x)$.) If, when the experiment is performed, the value of X which results is x_o, $P(|X|\geq x_o)$ (or $P(X\leq x_o)$ or $P(X\geq x_o)$) is the _significance level_.

A _null_ hypothesis H can be looked at as a translation into the language of mathematics of what presently held scientific theories predict about the outcome of the experiment. In the case at hand, the majority view among meteorologists (at least until recently) is that the passage of the Earth through sector boundaries is of so little consequence energetically as to not materially affect the VAI. Consequently, the key days should not be, so to speak, a preferred frame of reference from which to calculate (by superposed epoch analysis for example) certain gross statistical characteristics of the VAI. More precise mathematical expressions of this assumption are presupposed in the estimation of significance levels in [5], [6], and [1]. We briefly describe our way of looking at the problem. The graph of the VAI looks like oscillations superimposed on a sine wave of period about a year. Let P(t), with t measured in half-days, be the least-squares fit of such a sine wave to the VAI. It turns out that

$$P(t) = 42.00 + 12.98 \cos \omega t + 6.54 \sin \omega t, \tag{2}$$

where $\omega = 2\pi/730.5$ radians. Let $R(t) = V(t)-P(t)$. The graph of R(t) for the years 1963 to 1970 shows no obvious periodicities, and tends to oscillate around 0. This leads us to express the null hypothesis as follows: R(t) is a stationary process with mean 0. This in itself is not a sharp enough hypothesis on which to base a computation of significance level, so we also assume that R(t) is Gaussian. We have checked this assumption in a preliminary way by making plots on normal probability paper, and it does not seem unreasonable. Let σ be the standard deviation of R(t). From the data, we estimated that σ is approximately 13.23. Let $\rho(k)$ be the value for lag equal to k half-days of the autocorrelation function of R(t). Our estimate of $\rho(k)$ is displayed in Figure 1. (It was obtained by averaging estimates of $\rho(k)$ for each of the 8 yr 1963 to 1970.) In the computations needed to compute the significance level, we assumed, somewhat arbitrarily, that $\rho(k) = 0$ if $|k| \geq 60$ half-days.

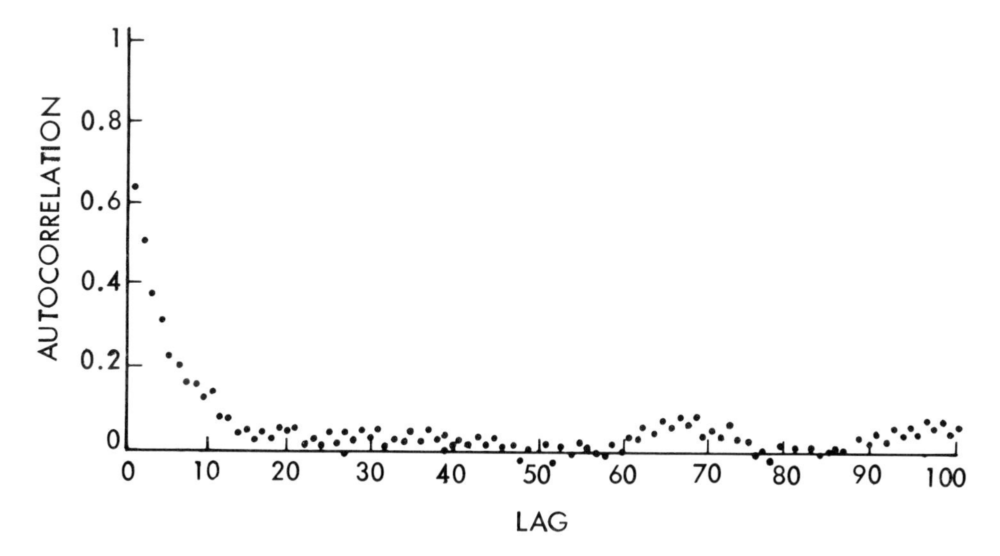

Figure 1. Average of yearly VAI Autocorrelations over years 1963-1970: LAGS 1 - 100.

Let R_{ki} be defined in terms of R(t) in the same way that V_{ki} was defined in terms of V(t). Let

$$X = \frac{1}{47} \sum_{k=1}^{47} (R_{k2} - (1/2d+1) \sum_{i=-d}^{d} R_{ki}).$$

The assumption that the mean value of R(t) is 0 implies that the mean value E(X) of X is 0. Our estimate of $\rho(k)$, and the expression for the variance of a sum in terms of the variances and covariances of the summands turns out to imply that the standard deviation $\sigma(C)$ of X is equal to about $.121\sigma$. At the conference, J. Wilcox informed us that the VAI oscillates more in the winter than in the summer, so that our preliminary estimate of σ as approximately 13 is too low. After the conference, we re-estimated σ from the values V(t) during the winter months of the time period covered, and obtained the estimate of $\sigma \cong 14$, yielding the value 1.7 for $\sigma(X)$. The observed value of X, namely -7.1, is thus 4.17 standard deviations below its assumed mean 0. The assumption that $R(t)$ is Gaussian implies that X is normally distributed. The probability that a normally distributed random variable falls as much as 4 standard deviations below its mean is 0.00003.

The interpretation of this last number as a significance level in the sense defined earlier, is, however, not valid. The data used to compute the significance level are the data which, upon superposed epoch analysis, exhibited a rather sharp minimum of $\overline{V}_i$ a full day after the key day, and it was this feature of the data which led us to select the statistic X that we did. The computation of 0.0003 as the probability that X is as negative as it turned out to be does not take into account the fact that we selected X because it - or something like it - was pronouncedly negative. (Given our assumptions, however, the probability that any of the 25 numbers

$$X_j = (1/47\Sigma_{k=1}^{47}(R_{kj}-(1/2d+1)\Sigma_{i=-d}^{d}R_{ki})$$

differs in absolute value from 0 by as much as $X=X_2$ is no larger than $2 \times 25 \times 0.00003 = 0.0015$.)
One can, however, use new data to recompute the significance level of a hypothesis which is now arrived at before looking at the data rather than after. A new supply of key days is available, and the results of [1] and [6] indicate that use of the new data would yield a result of about the same level of significance. (For another assessment of the situation, see [4].)

When we presented these results at the symposium, there were several criticisms and suggestions. The fact that $\sigma(V(t))$ is greater in the winter than in the summer shows that the assumption of stationarity for $R(t)$ is invalid. The processing of data used to obtain the value of the VAI at midnight differs from that used to obtain the value at noon. This is another source of non-stationarity for $R(t)$, but, more importantly, it affects our estimate of $\sigma=\sigma(R(t))$. (This was also pointed out by Wilcox.) The graph of the autocorrelation function of $R(t)$ suggested to one member of the audience that $R(t)$ contains a periodic trend of about 27 days that we have not removed, and the graph of $R(t)$ suggested to J. King the presence of a low-frequency periodic trend. We are grateful for these and other suggestions, and are in the process of taking them into account.

The idea of using sector boundary crossings as key days in superposed epoch analyses of various indices of terrestrial atmospheric activity is Markson's [2]. We would like to express our appreciation to Ralph Markson for suggesting that we look at the technique of superposed epoch analysis from a statistical and probability-theoretic point of view.

REFERENCES

1. C. O. Hines and I. Halevy, J. Atmos. Sci., 34, 382, 1977.
2. R. Markson, Pure and Appl. Geophys., 84, 161, 1971.
3. R. G. Olson, W. O. Roberts, and E. Gerety, Solar-Terrestrial Physics and Meteorology, Working Document II, SCOSTEP, 1977.
4. R. Shapiro, J. Atmos. Sci., 33, 865, 1976.
5. J. M. Wilcox, P. H. Scherrer, L. Svalgaard, W. O. Roberts, R. H. Olson, and R. L. Jenne, J. Atmos. Sci., 31, 581, 1974.
6. J. M. Wilcox, L. Svalgaard, and P. H. Scherrer, J. Atmos. Sci., 33, 1113, 1976.

SOLAR CYCLES AND THE WEATHER: SUCCESSFUL EXPERIMENTS IN AUTOSUGGESTION?

A. Barrie Pittock *
Laboratory of Tree-Ring Research
University of Arizona
Tucson, AZ 85721 U.S.A.

ABSTRACT. Claims relating the single and double sunspot cycles to weather and climate are critically reviewed in the light of what is known about the sun/weather/climate system. Various pitfalls in the application or lack of application of statistics to the problem are discussed and illustrated from the literature. Following a survey of the literature it is concluded that despite the great number of papers on the subject, little convincing evidence has yet been produced for real correlations between sunspot cycles and weather or climate, although evidence for correlations between weather and solar events on time scales of days appears to exist. The conclusions tend to support A. S. Monin's impression of "successful experiments in autosuggestion" and S. J. Gould's recent suggestion that unconscious manipulation of data may be a scientific norm. Some guidelines are offered as an aid to critical evaluation of our own and other scientists' work and it is suggested that progress is more likely on shorter time-scale solar-weather effects. A longer version of this paper may be found in the August 1978 issue of Reviews of Geophysics and Space Physics.

I. INTRODUCTION

The Russian meteorologist Monin [1] has stated that the presence of a connection between the Earth's weather and the fluctuations in solar activity "would be almost a tragedy for meteorology, since it would evidently mean that it would first be necessary to predict the solar activity in order to predict the weather." He dismissed the collected evidence for the influence of solar activity on the weather, saying that it"fortunately produces only an impression of successful experiments in autosuggestion."

On the other hand, according to Lesik [2] the USSR's First All-Union Conference on the Problem "Solar-Atmospheric Relationships in the Theory

*On leave from C.S.I.R.O. Division of Atmospheric Physics, Mordialloc 3195, Australia.

B.M. McCormac and T.A. Seliga (Eds.): Solar-Terrestrial Influences on Weather and Climate. 181-191.

of Climate and Weather Forecasting" held in Moscow in 1972 resolved that "...investigations...in the USSR...make it possible to assert with assurance that solar activity and other space-geophysical factors exert a substantial influence on atmospheric processes. Allowance for these is of great importance in preparing weather forecasts."

In the West, a widely circulated paper on Sun-weather relationships by King [3] marshalled a striking array of published and unpublished "evidence" which, although it contained some caveats, was all interpreted as support for such relationships. Of his 85 references (listed in a reprint circulated by SCOSTEP [4] none was cited as finding evidence against such a relationship. More objective reviewers [5,6], however, have noted many contradictions in the literature. Tucker stated that "Because of the inconclusive and sometimes contradictory evidence surveyed, a serious consideration must be whether the solar-weather relationship is likely to be a profitable hypothesis to pursue in the future." [5] He commented that "much of the data has been handled badly...(and) investigators allowed preconceived ideas to affect their judgement... Nevertheless some of the evidence cannot be dismissed..."

I have just completed a comprehensive review of the recent literature, concentrating on the possible influence of the single and double sunspot cycles (approximately 11- and 22-yr periodicities) on the weather/climate. This forces me to concur with Gani [7] who concluded in another context that "many of the arguments presented are based on poor foundations. ... There is very little statistical analysis in the climatologists' work, and some of this is either superficial or wrong."

In this summary of a much longer review [8] I will outline some of the pitfalls in the application of statistics to the problem, with illustrations from the literature. I will then summarize the principal conclusions to be drawn from the literature and comment on their implications.

2. THE CONTEXT

Weather and climate are highly variable on all time scales [9,10] and only a fraction of that variability can reasonably be ascribed to sunspot cycles [11]. For example, Pittock [12] allotted the variance in 8 yr of weekly measurements of the vertical distribution of O_3 into components due to synoptic weather systems and instrumental noise, seasonal variations, and the inter-annual and trend components. The inter-annual component only exceeded 20% of the total variance in the atmosphere above about the 30 km level, and the component due to trend (i.e., on a time-scale greater than 8 yr) reached a maximum of less than 4% of the total variance. If we are interested in an 11 or 22 yr sunspot cycle induced signal, we are forced, therefore, to study only 4% of the total variance, the other 96% being for our purposes "noise." This imposes severe limits on the statistical confidence of conclusions drawn from necessarily limited data sets.

Climatic data are also highly correlated over large distances in space, so we cannot substitute data from more stations for longer time series and assume the additional data to be independent. Instead of N stations reducing the uncertainty of any estimate by a factor of $(N-1)^{-\frac{1}{2}}$ we are limited to at most some 10 or so independent pieces of information on a global scale, so our factor at most is about 3.

The variability of the solar constant over the sunspot cycle is generally considered to be less than 1% [13,14] although accurate direct measurements of the solar constant from satellites did not commence until 1975. The latter suggest an upper limit of 0.1% may be more realistic. Physical hypotheses for an effect on the weather therefore tend to concentrate on "trigger" mechanisms of one sort or other, usually driven by either the more variable but energetically minute far UV radiation or by ionizing particles modulated by the solar magnetic field.

Nearly all existing physical hypotheses have been put forward to explain observed and supposedly significant correlations between solar indices (such as sunspot numbers) and climatological data series, but unfortunately they have not yet been used to make predictions which have been subject to independent and critical tests.

This means that most of the literature must be assessed in terms of the true statistical significance of observed cycles or correlations. It is again unfortunately a fact that statistical tests of significance, unless the evidence is unambiguous and overwhelming, are hedged about with assumptions, qualifications and uncertainties which make such purely statistically-based conclusions suspect. One has only to consider the problems posed by data selection (whether conscious or unconscious, or in space or time), autocorrelations, smoothing, and post-hoc hypotheses, to realize that a serious difficulty exists at this point.

3. STATISTICAL PROBLEMS

Different statistical methods differ greatly in their ability to identify recurrence intervals within, and to describe the properties of a given data set. Undoubtedly some sophisticated modern techniques such as the maximum entropy method [15] and the non-integer technique of power spectrum analysis [16]give very precise descriptions of given data sets. However there is an important distinction between such a <u>description</u>, and the <u>prediction</u> of the properties of another data set. Methods which give a more detailed and sensitive description of one set do not necessarily give a more reliable prediction of the properties of another (see e.g., [17]).

Climatic data series are notoriously unstable in their statistical properties, i.e., they tend to be not "well-behaved" and to be non-stationary. A classical case in point is the data series for water levels of Lake Victoria in East Africa. Brooks [18] in 1923 found a remarkable correlation of this series with sunspots over the then known record of about 20 years. No doubt some sensitive modern techniques would ascribe considerable power to a periodicity of about 11 yr in this limited data set. By 1936, however, Walker [19] had noticed the breakdown of this periodicity into what looked like the second harmonic. In 1961-64 very heavy rainfall led to a rise in lake level of about twice the amplitude of the earlier oscillations [20] with only a slow fall since [21]. There is also historical evidence of very high lake levels in the late nineteenth century [22] and of other major excursions in the paleo-record [23].

The work of Bell [24] clearly demonstrates the dangers of assuming that correlations, which might appear to be highly significant over short time spans, are stable over long time spans. Bell calculated correlation coefficients over sliding 15 yr intervals between various climatic time series and sunspots, and found fluctuations ranging from +0.5 to -0.5 over different time intervals. This largely acounts for contradictions between Wexler's map of surface pressure differences between sunspot maxima and minima [25] and that of Baur [26], and throws the real significance of these results, as well as those of Jagannathan and Bhalme [27] for Indian rainfall, into question.

This problem is partly overcome by making proper allowance for autocorrelations in each data series using the formula given by Quenouille [28] for the reduction in the number of independent observations due to the autocorrelations (e.g., see [29]). However, as the autocorrelation functions themselves do not appear to be very stable this is not a complete solution.

The *a posteriori* selection of one data set considered to have statistically significant correlations or periodicities from a much larger quantity of data not showing these properties, is another major problem in interpretation. Five data sets in each 100 independent sets might be expected to show any given correlation or periodicity at the 95% confidence level purely by chance. Such selection may be made unconsciously in the choice of area, variable, or epoch for investigation on the basis of a qualitative "feeling" that the given area is a profitable one to investigate. It may also be conscious but only implicit, e.g., Bhalme's analysis [30] of data for Rajasthan in India, which selects an area found to be "significant" by Jagannathan and Bhalme [27] in their analysis covering the whole of India. The latter paper in fact analyzed rainfall data from 105 stations, first breaking the time series at each station into six seasonal distribution parameters, and then performing power spectrum analysis on the resulting 630 sets of variables. Jagannathan and Bhalme found 26 cases in which an 11 yr periodicity was significant at the 95% confidence level. Thus 1 in 24 cases was "significant," whereas one would have expected 1 in 20 purely by chance. A similar criticism must be leveled at the surface pressure analyses of Parker [31] who finds significant effects in just the proportion of the total area which one might expect by chance.

Data selection is disturbingly evident in a superficially impressive paper by Xanthakis [32] which found apparently highly significant correlations between "mean zonal annual precipitation" series in the northern hemisphere and an index of solar activity. An examination of Xanthakis' stated data source shows that, in the 70 to 80°N zone, he omitted a complete data set from 1912 through 1924 from one station, even though his "zonal mean" from 1912 through 1921 was based on only one station. In the interval 1951-60 Xanthakis used data from only 6 or 7 stations in any given year, and omitted data from 5 other stations. Agreement between Xanthakis' curve and that using all the data is good around the central solar maximum in 1938, but becomes progressively worse towards the beginning and end of the record, as an increasingly higher proportion of the available data was omitted [8].

King's paper [3] contains a number of impressive curves which show data for highly selected data intervals, and most of which are highly smoothed (which improves the apparent correlations in a similar manner to autocorrelation). For example King's figure 3 depicts rainfall at Fortaleza in Brazil, but utilizes only the most favorable half of the data given by Markham [33]. King's curve for central England temperatures in July uses the data for 1750 through 1880, despite the availability of a series from 1659 to 1973 [34], while that for the number of polar bears caught in SW Greenland includes only the more convincing half of Vibe's curve [35]. Mason [36] and Folland [37] have pointed out that King's correlation [38] of the 11 yr solar cycle with potato yields in England for the period 1935-59 does not hold over the longer period 1890-1935.

When data do not fit simple hypotheses it is tempting to elaborate a hypothesis to fit the available data. Thus changes with time, or from station to station, in the sign of supposedly significant correlations of climatic parameters with sunspot numbers have led several authors to suggest that the effect of sunspots is reversed when some critical sunspot number is reached [32,39-41]. The validity of such elaborated hypothesis must be tested on independent data. Bell [24] has pointed out that more recent data have not borne out Lawrence's hypothesis [39].

4. THE RESULTS

A review of the literature on atmospheric pressure variations over the solar cycles [25,26,31,42-44] shows the various claimed variations to be mutually contradictory and generally not significant when allowance is made for autocorrelations and spatial selection. As Bell [24] pointed out, the results depend on which particular sunspot cycles are included on the analysis.

Comprehensive recent studies on surface temperature variations, notably that of Gerety _et al_. [45] show no significant variations over the solar cycles, with the exception of local or regional studies by Mason [36], Mock and Hibler [17], and Currie [46] using the maximum entropy method. These latter studies are of dubious predictive value.

Several comprehensive studies of precipitation data, notably those by Rodriguez-Iturbe and Yevjevich [47], Dehsara and Cehak [48] and Gerety _et al._ [45] give negative results. The exceptions are studies by Jagannathan and Bhalme [27] and Xanthakis [32] which claim significant results. The former results for India are not significant when proper allowance is made for _a posteriori_ selection of the results, while Xanthakis' result is not borne out when all the data are included. A series of papers on Southern African rainfall by Dyer _et al._ [49-52] does not appear to be based on long enough records to have any validity as a basis for prediction, particularly as evidence of earlier oscillations in other localities cited by Dyer and Tyson [52] further demonstrates the likelihood that the recently observed oscillations are ephemeral chance occurrences.

Claimed correlations between thunderstorm activity and sunspot cycles [53-58] are generally contradictory, or small and non-significant. The notable correlation of +0.88 claimed by Septer [59] and Brooks [53] for Siberia, has been found to be false [54].

Correlations of O_3 content with solar cycles remain controversial. The most comprehensive recent analysis by Angell and Korshover [60] shows global variations of O_3 with solar variations to be "not quite significant." Hill and Sheldon's analysis [61] of the Arosa total O_3 data (1932-1974), which is usually said to have a solar cycle component, found different periodicities for the O_3 and sunspot series, with a cross correlation of the two series showing "no significant phase lag relationship between them."

Claimed correlations of sunspot numbers with tropopause height appear to be based on inadequate data, or are mutually contradictory[41,62-63].

Claimed correlations of other climatic variables with sunspot cycles are equally unconvincing, with several comprehensive studies of river flows [47], tree ring series [64], and varves [9,65] giving negative results. Recent and as yet unpublished results by J. M. Mitchell, Jr. and C. W. Stockton concerning a drought area index, based on tree ring data in the western United States, may have some regional significance, but have limited (if any) predictive value. Their work is amongst the most careful and persuasive in this field.

The relevance to longer (climatic) time-scales of apparently significant short-term correlations between meteorological indices and the passage of solar magnetic sector boundaries [66-68] has yet to be established.

5. CONCLUSION AND RECOMMENDATIONS

In summary, it seems that despite a massive literature on the subject, there is at present little or no convincing evidence of statistically significant or practically useful correlations between sunspot cycles and weather on climatic time scales. The evidence to hand suggests that, if in the future more data and better analysis enable the detection of statistically significant relationships, these will account for so little of the total variance in the climatic record as to be of little practical value.

This conclusion might be obviated if a more complex hypothesis were developed capable of accounting more subtly for a greater proportion of the variance. Such a hypothesis would not be established as a viable theory, however, until it was used to make detailed predictions which were verified in independent data.

These conclusions bear out those of more objective reviewers such as Tucker [5] and Meadows [6], sharp contrast to King [3]. Indeed, the frequency with which errors, fallacies, and biases appear in the literature provides justification for the extremely critical stand taken by Monin [1] when he wrote of "successful experiments in autosuggestion", and the equally critical comments of the Russian reviewer Khromov [69].

Sir Gilbert Walker wrote in 1946: "I think it is likely that after long ages of belief in the control of our affairs by the heavenly bodies men are born with instinctive faith in the existence of periods in the weather. I lost mine when the imperative need of reliability in seasonal forecasts drove me to replace instinct by valid quantitative criteria applied to the results given by standard methods" [70].

Perhaps the explanation is broader than that, and more relevant to the institutional framework in which science is conducted. Pressures exist, both in the USSR and in "Western" countries, for individual scientists in all areas of research to keep up their numbers of publications and to make their work appear relevant to society at large; indeeed their personal and institutional support depends on these factors.

It is tempting, therefore, to please a relatively non-scientific and uncritical public by publishing plausible but non-rigorous or incomplete work with exaggerated claims and conclusions. This applies particularly in areas of great potential public interest and benefit such as long-range weather or climatic forecasting. It probably applies also in other areas of great social, economic and political interest, such as in the social sciences, economics, and environmental sciences.

Gould [71], in a recent thought-provoking paper which re-analyzes data by the nineteenth century physical anthropologist, Samuel George Morton, suggests that "unconscious or dimly perceived finagling, doctoring, and massaging (of data) are rampant, endemic, and unavoidable in a profession that awards status and power for clean and unambiguous discovery. This is the middle ground of unappreciated bias and more conscious manipulation in the interest of a truth passionately held but inadequately supported." Gould goes on to say that he raises this issue "with two hopes for alleviation -- first, that by acknowledging the existence of such a large middle ground, we may examine our own activity more closely; second, that we may cultivate as Morton did, the habit of presenting candidly all our information and procedure, so that others can assess what we, in our blindness, cannot."

It is in this spirit that I offer some simple qualitative groundrules which might prove useful for the guidance of authors, editors, referees, and readers of research papers in this controversial area of solar cycles and the weather. I would suggest the following:

(a) Understand the properties of the data - its errors, biases, scatter, autocorrelation, spatial coherence, frequency distribution, and stationarity;

(b) Choose statistical methods appropriate both to the properties of the data and the purpose of the analysis (e.g., description or prediction);

(c) Critically examine the statistical significance of the result, making proper allowance for spatial coherence, autocorrelations and smoothing, and data selection;

(d) Test the result on one or more independent data sets, or sub-sets of the original data;

(e) Endeavor to derive a physical hypothesis which can be tested on independent data sets, preferably at some other stage in the hypothesized chain of cause and effect;

(f) Estimate the practical significance of the result, e.g., the fraction of the relevant total variance which can be predicted or "explained;

(g) Set out the properties and limitations of the data, the statistical methods used (including data selection and smoothing), and any assumptions, reservations or doubts; and

(h) Do not over-state the statistical or practical significance of the result.

Such ground rules may seem to some to be a counsel of perfection, but they would certainly remove much of the confusion and controversy which presently surrounds the subject of solar-weather relationships.

Progress in this field may indeed still be possible, at least in terms of a better understanding of how the atmosphere works. However, claims that solar cycle-weather relationships may form a useful basis of climatic prediction must be treated at present, both by researcher and consumer, with a high degree of skepticism.

I want to stress, however, that the present very critical review has been confined to the literature on the 11- and 22-yr sunspot cycle relationships. My general impression is that while much of the literature on shorter-term solar-weather relationships might also be faulted, there are several soundly-based papers which establish credible relationships on a day-to-day time-scale, notably references [66-68]. This is in part due to the much greater relative availability of data on these shorter time-scales, which enables much more credible statistics to be derived. Indeed on these time-scales it is much easier to gather independent data sets with which to test mechanisms and predictions.

There is another reason for expecting significant progress to be more likely on these shorter time-scales. This is that the climatic system recovers from (or damps out) quite large short term disturbances in the global energy balance usually in a matter of days or weeks. For example, major snow storms sweeping across North America often leave huge areas of greatly increased albedo which reflect an appreciable fraction of the total incoming solar radiation back into space. Instead of such storms leading via the "snow-blitz" mechanism [72] (a positive feedback) to an ice age, negative feedback takes over and the climatic system quickly returns toward its average state.

In the same way, the much less energetic disturbances of solar origin, while they may trigger sizable transient atmospheric effects, may soon have those effects damped out. If such solar-induced transient effects can be understood and predicted it raises the possibility of significant improvements in short term weather forecasts, even though significant effects on longer time scales might not be established.

REFERENCES

1. A.S. Monin, Weather Forecasting as a Problem in Physics, MIT Press, Cambridge, MA, 1972.
2. B. Lesik, Appendix I in J.M. Wilcox, J. Atmos. Terrest. Physics 37, 237, 1975.
3. J.W. King, Astronaut. Aeronaut. 13, 10, 1975.
4. A.H. Shapley, H.W. Kroehl and J.H. Allen, Solar-Terrestrial Physics and Meteorology: A Working Document, Spec. Comm. for Solar-Terr. Phys., Nat. Acad. Sci., Wash. D.C., 1975.
5. G.B. Tucker, Weather, 19, 302, 1964.
6. A.J. Meadows, Nature, 256, 95, 1975.
7. J. Gani, Search, 6, 504, 1975.
8. A.B. Pittock, Rev. Geophys. and Space Phys., August 1978 (in press)
9. A.S. Monin & I.L. Vulis, Tellus, 23, 337, 1971.
10. J.E. Kutzbach and R.A. Bryson, J. Atmos. Sci, 31, 1958, 1974.
11. J.M. Mitchell, Jr., Quaternary Res. 6, 481, 1976.
12. A.B. Pittock, Quart. J. Roy. Meteor. Soc., 103, 575, 1977.
13. E.V.P. Smith & D.M. Gottleib, Space Sci. Rev., 16, 771, 1974.
14. O.R. White (ed.), The Solar Output and Its Variation, Colo. Assoc. Uni. Press, Boulder, 1977.
15. J.P. Burg, Geophysics, 37, 375, 1972.
16. P.T. Schickedanz and E.G. Bowen, J. Appl. Meteorol., 16, 359, 1977
17. S.J. Mock and W.D. Hibler III, Nature, 261, 484, 1976.
18. C.E.P. Brooks, Geophys. Mem. London 2 (20), 337, 1923.

19. G.T. Walker, Quart. J. Roy. Meteor. Soc., 62, 451, 1936.
20. H.H. Lamb, Geogr. J., 132, 183, 1966.
21. H. Rodhe and H. Virji, Mon. Weather Rev., 104, 307, 1976.
22. E.G. Ravenstein, Geogr. J., 18, 403, 1901.
23. R.W. Fairbridge, Quaternary Res., 6, 529, 1976.
24. G.J. Bell, Weather, 32, 26, 1977.
25. H. Wexler, Tellus, 8, 480, 1956.
26. F. Baur, Physikalische - Statistiche Regeln als Grundlagen fur Wetter und Witterungs Vorhersagen, p. 125, Akad. Verlag, Frankfurt am Main, W. Germany, 1958.
27. P. Jagannathan and H.N. Bhalme, Mon. Weather Rev., 101, 691, 1973
28. M.H. Quenouille, Associated Measurements, Butterworth, London, 1952
29. A.J. Troup, Geofis Pura Appl., 51, 184, 1962.
30. H.N. Bhalme, Indian J. Meteorol. Hydrol. Geophys., 26, 57, 1975.
31. B.N. Parker, Meteorol. Mag. 105, 33, 1976.
32. J. Xanthakis, in Solar Activity and Related Interplanetary and Terrestrial Phenomena, ed. by J. Xanthakis, Springer, N.Y., 1973.
33. C.G. Markham, J. Applied Meteorol., 13, 196, 1974.
34. G. Manley, Quart. J. Roy. Meteorol. Soc. 100, 389, 1974.
35. C. Vibe, The Danish Zoological Investigations in Greenland, Bd. 170, Nr. 5, C.A. Reitzels Forlag, Copenhagen, 1967.
36. B.J. Mason, Quart. J. Roy. Meteorol. Soc. 102, 473, 1976.
37. C.K. Folland, Weather, 32, 336, 1977.
38. J.W. King, Nature, 245, 443, 1973.
39. E.N. Lawrence, Weather, 20, 334, 1965.
40. T. Suda, Geophys. Mag. (Tokyo), 31, 515, 1963.
41. T. Suda, Geophys. Mag. (Tokyo), 37, 361, 1976.
42. H.C. Willett, Ann. N.Y. Acad. Sci, 95, 89, 1961.
43. I.V. Maksimov and B.A. Slepcov - Sevlevic, Dokl. Akad. Nauk SSSR, 201, 339, 1971.
44. M.K. Miles, Meteorol. Mag., 103, 93, 1974.
45. E.J. Gerety, J.M. Wallace, and C.S. Zerefos, J. Atmos. Sci., 34, 673, 1977.
46. R.G. Currie, J. Geophys. Res., 79, 5657, 1974.
47. I. Rodriguez-Iturbe and V. Yevjevich, Hydrol. Papers, 26, Col. State Univ., Fort Collins, 1968.
48. M. Dehsara and K. Cehak, Arch. Meteorol. Geophys. Bioklim. Ser. B, 18, 269, 1970.
49. P.D. Tyson, Ann. Assoc. Amer. Geogr., 61, 711, 1971.
50. P.D. Tyson, T.G.J. Dyer, and M.N. Mametse, Quart. J. Roy. Meteorol. Soc., 101, 817, 1975.
51. T.G.J. Dyer, S. Afr. J. Sci., 71, 369, 1975.
52. T.G.J. Dyer and P.D. Tyson, J. Appl. Meteorol., 16, 145, 1977.
53. C.E.P. Brooks, Quart. J. Roy. Meteorol. Soc., 60, 153, 1934.
54. Z.P. Kleimenova, Meteorol. Gidrol., no. 8, 64, 1967.
55. G.P. Pavlova, Tr. Gl. Geofiz. Observ. Leningrad, no. 242, 118, 1969.
56. R. Markson, Pure Appl. Geophys., 84, 161, 1971.
57. H.J.Fischer and R. Muhleisen, Meteorol. Rundsch., 25, 6, 1972.
58. M.F. Stringfellow, Nature, 249, 332, 1974.
59. E. Septer, Meteorol. Z., 43, 229, 1926.

60. J. K. Angell and J. Korshover, Mon. Weather Rev., 104, 63, 1976.
61. W.J. Hill and P.N. Sheldon, Geophys. Res. Letters, 2, 541, 1975.
62. S.I. Rasool, Geofis. Pura Appl. 48, 93, 1961.
63. H.P. Cole, J. Atmos. Sci., 32, 998, 1975.
64. V.C. LaMarche, Jr., and H.C. Fritts, Tree Ring Bull. 32, 19, 1972.
65. R.A. Bryson and J.A. Dutton, Ann. N.Y. Acad. Sci., 95, 580, 1961.
66. J.M. Wilcox, J. Atmos. Terr. Phys., 37, 237, 1975.
67. J.M. Wilcox, L. Svalgaard and P.H. Scherrer, J. Atmos. Sci., 33, 1113, 1976.
68. C.O. Hines and I. Halevy, J. Atmos. Sci., 34, 382, 1977.
69. S.P. Khromov, Meteorol. Gidrol., no. 9, 90, 1973.
70. G.T. Walker, Quart. J. Roy. Meteorol. Soc., 72, 282, 1946.
71. S.J. Gould, Science, 200, 503, 1978.
72. H. Flohn, Quaternary Res., 4, 385, 1974.

PLANETARY MOTION, SUNSPOTS AND CLIMATE

H. T. Mörth
Climatic Research Unit, University of East Anglia,
Norwich, NR47TJ, United Kingdom

L. Schlamminger
F. S. Astronomical Observatory, Erlangen
Federal Republic of Germany

ABSTRACT. Past attempts to link sunspots with the gravitational attraction by planets or with the orbital motion of the Sun have not been successful. In this paper the changes in orbital angular momenta of planets due to gravitational perturbation are considered. Ninety-eight percent of the angular momentum of the solar system is contained in the orbital motion of the four giant planets Jupiter, Saturn, Uranus, and Neptune. The potential relevance of the motion of these giant planets to the sunspot variations is investigated. The perturbation forces between two planets vary periodically over their mutual synodic period and show symmetry about heliocentric conjunction and opposition. Certain pairs and pair groupings in the giant planets have highly commensurable mean motions. Their synodic half periods correspond to the principal sunspot number frequencies. In particular, the 10 yr period appears to be associated with the relative motion of the pair Jupiter - Saturn, the 90 yr period with that of Uranus and Neptune, and the 11 yr period with the motion of the pair Jupiter - Saturn relative to the pair Uranus - Neptune. The physical link between planetary motion and sunspots is seen in the outward transfer of angular momentum from the Sun to the fringe of the solar system. We assume the existence of a basic energy flow from the Sun outward effected by accelerations of outer planets through gravitational perturbation by inner planets rhythmically modulated by the orbital configurations of all the planets. The transmission of gravitational torque in the solar system is assumed to cause changes in the global and local vorticity patterns of low inertia materials such as the solar photosphere and the terrestrial atmosphere; this provides a mechanism of control on the terrestrial atmospheric circulation and climate by extraterrestrial forces either directly or through modulation of solar activity.

The possibility of planetary motions playing a part in the changing appearance of the Sun's photosphere has been considered from the time of

B.M. McCormac and T.A. Seliga (Eds.): Solar-Terrestrial Influences on Weather and Climate. 193-207.

the first telescopic sunspot observations by Scheiner and Galileo at the beginning of the 17th century. Many papers have been written on this topic, a selection of references being appended. Some investigators tried to explain the sunspot periodicities by the gravitational pull exerted by planets on the surface of the Sun. Problems arise in this approach, as the "tide-raising" planets, Mercury, Venus, Earth, and Jupiter produce periodicities which do not match those in the sunspots. It was pointed out by Smythe and Eddy [1], that the tide-raising force, even in the case of a rare 9-planet conjunction, is only 10^{-12} of the force of solar gravity, and it is unlikely that planetary tides are a controlling factor of the motions on the Sun's surface. Others considered the solar orbital motion about the barycenter of the solar system. The changes of angular momentum in the Sun's motion indeed show periods of 22 and 179 yr. However, the gravitational motion of the Sun is regular and does not lend itself to explaining the considerable amplitude variations and phase jumps on the sunspot number curve.

In our approach, we do not consider the tidal effects of gravitational attractions, but rather their role in transferring orbital angular momenta in the solar system. Figure 1, comparing masses, attractions on the Sun, and orbital angular momenta of planets, shows that the bulk of the angular momentum in the solar system is contained in the four giant planets, Jupiter, Saturn, Uranus, and Neptune. We therefore concentrated on analyzing the motion of these four planets, and on identifying the variations which appear significant in relation to sunspots.

If i and o are two planets with orbital periods P_i and P_o, then P_i is always smaller than P_o. Relative positions of two planets, such as conjunction or position, or any specific separation, repeat over the mutual synodic period $P_{io} = P_i \times P_o / (P_o - P_i.)$ The synodic periods of the combinations of pairs of giant planets are as follows:

Pair	Symbol	Synodic Period
Jupiter - Saturn	5 - 6	19.85 a
Saturn - Uranus	6 - 7	45.36 a
Uranus - Neptune	7 - 8	171.39 a
		(a = tropical years)

A study of the distribution of planetary and satellite orbital motion shows that the ratio of the mean orbital periods of certain pairs in the solar system is close to the ratio of two small integers. This "commensurability" is usually expressed by the term $\varepsilon = P_i / P_o - I/J$, where I and J are small positive integers. In defining "small" we must take an arbitrary decision. In our calculations we limited the size of I, J to be smaller than or equal to 9. Unless this restriction is imposed, one can always find two integers whose ratio approaches P_i/P_o to any degree of accuracy. We computed the values of ε for all planet pairs in the solar system. Pairs with relatively high commensurability ($|\varepsilon|$ less than 0.01) are:

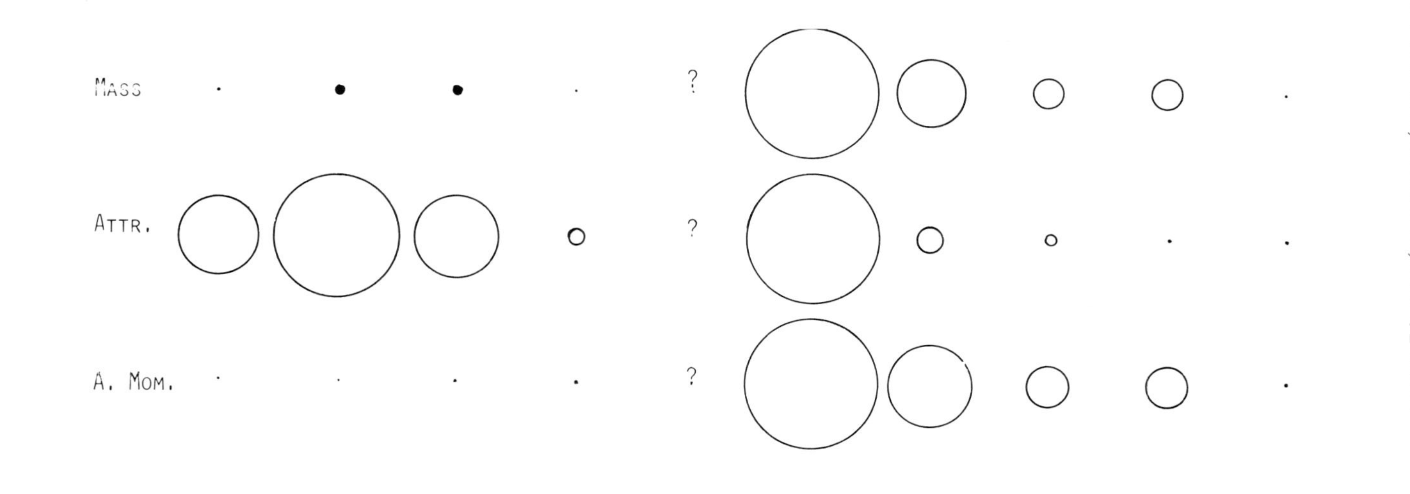

	MERCURY	VENUS	EARTH	MARS	ASTEROIDS	JUPITER	SATURN	URANUS	NEPTUNE	PLUT
No.	0	1	2	3	4	5	6	7	8	0
Mass	0.055	0.817	1.000	0.108	?	318.4	95.2	15.0	17.2	0.8
Attr.	0.95	2.16	1.00	0.03	?	2.26	0.11	<0.01	<0.01	<0.0
A. Mom.	0.03	0.69	1.00	0.13	?	726.6	293.7	65.7	94.8	5.0

Figure 1. Some comparative properties of the solar system

Pair	Symbol	I/J	ε
Mercury - Venus	0 - 1	2:5	-0.0085
Mercury - Earth	0 - 2	1:4	-0.0092
Mercury - Mars	0 - 3	1:8	+0.0030
Venus - Earth	1 - 2	5:8	+0.0098
Venus - Mars	1 - 3	1:3	-0.0063
Venus - Ceres	1 - 4	1:8	+0.0086
Mars - Ceres	3 - 4	2:5	+0.0085
Mars - Jupiter	3 - 5	1:6	-0.0081
Jupiter - Saturn	5 - 6	2:5	+0.0026
Jupiter - Uranus	5 - 7	1:7	-0.0016
Saturn - Pluto	6 - 9	1:9	+0.0074
Uranus - Neptune	7 - 8	1:2	+0.0098
Uranus - Pluto	7 - 9	1:3	-0.0049
Neptune - Pluto	8 - 9	2:3	+0.0033

Roy and Ovenden [2] have shown that commensurability of mean motions in the solar system is more frequent than may be expected from chance.

We now "pair off" neighboring planets. The pairing is based on the criterion that, overall, the commensurabilities in the pairs should be as high as possible. The two possible groupings and their respective ε values are as follows:

Pairs	0 - 1	2 - 3	4 - 5	6 - 7	8 - 9	
ε	8.5	23.9	11.8	17.3	3.3	$\times 10^{-3}$

Pairs	0	1 - 2	3 - 4	5 - 6	7 - 8	9	
ε		9.8	8.5	2.6	9.8		$\times 10^{-3}$

The second grouping is therefore our choice. The synodic periods and ε-values for these pairs are:

Pair	Synodic Period	I/J	ε
1 - 2	1.60 a	5:8	9.8×10^{-3}
3 - 4	3.18	2:5	8.5
5 - 6	19.86	2:5	2.6
7 - 8	171.39	1:2	9.8

Next, each pair is considered to be an individual system, and paired off with the neighboring system:

12 - 34 34 - 56 56 - 78

Their synodic periods and values are:

Pair	Synodic Period	I/J	ε
12 - 34	3.22 a	1:2	2.7×10^{-3}
34 - 56	3.79	1:6	6.5×10^{-3}
56 - 78	22.46	1:9	4.8×10^{-3}

Finally, the quadruples 1234 and 5678 are paired:

1234 - 5678

The quadruple commensurability is found to be almost perfect:

Pair	Synodic Period	I/J	ε
1234 - 5678	3.75	1:7	0.3×10^{-3}

We see in these results evidence of a strict control of the relative motion of planets in the solar system, probably exerted by gravitational coupling.

We now look at a potential coupling mechanism between two planets. Their mean motion is consistent with the model of a system of two point masses moving under the influence of a central attracting force of a third, much larger mass (the Sun). Each planet disturbs the motion of the other, not so much by its attraction on it as by the difference in the attractions it exerts on the Sun and on the other planet. The size of these perturbations grows with the mass of the disturbing planet and varies periodically with the relative positions of the two planets. The perturbations force can be resolved into two components; a radial component pointing to the Sun and a transversal component, normal to the radial (and almost tangential as eccentricities of the planetary orbits are relatively small). The action of the transversal component produces accelerations and decelerations in the orbital path velocities of the planets over the synodic period. The direction of these velocity changes is schematically illustrated in Figure 2. Figure 2a shows the disturbing effect from the outer planet on the inner planet. The inner planet is retarded from heliocentric conjunction to the point equidistant from the Sun and from the outer planet, then accelerated to opposition; thence retarded again to the second point of equidistance and, finally, again accelerated to conjunction. Inner planets disturb outer planets in two different ways. If the distance Sun - outer planet is greater or equal to twice the distance Sun - inner planet (Figure 2b), there are only two neutral points of the tangential component of the disturbance force: the outer planet is accelerated from conjunction to opposition and retarded from opposition to conjunction. If the distance Sun - outer planet is less than twice the distance Sun - inner planet (Figure 2c), four neutral points of the orbit-accelerating forces exist. The outer planet is retarded from conjunction to equidistance, accelerated

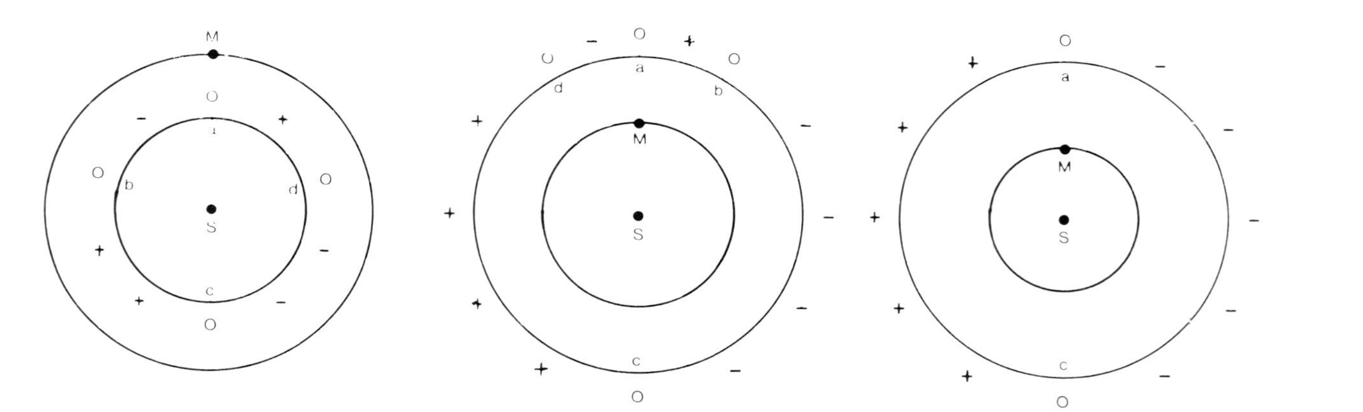

Acceleration and deceleration of perturbed outer planet mentioned in text refer to relative (clockwise) orbiting. Signs in this diagram represent changes in velocity of absolute (anticlockwise) orbiting.

INNER PLANET PERTURBED	OUTER PLANET PERTURBED	OUTER PLANET PERTURBED
	$2R_I < R_O$	$2R_I \geqq R_O$

S = SUN
M = DISTURBING PLANET

RELATIVE POSITIONS OF PLANETS:

a = Heliocentric conjunction
b = First point of equidistance
c = Heliocentric opposition
d = Second point of equidistance

DISTURBED PLANET EXPERIENCES:

0 = No acceleration
+ = Acceleration
- = Deceleration

Figure 2. Variation of transversal component of perturbation force over synodic period (schematic).

from there to opposition; thence retarded to the second point of equidistance and, finally, accelerated to conjunction. A classical example is provided by the pair Jupiter - Saturn. Approaching their conjunction, Jupiter is ahead and Saturn behind the positions computed from Kepler's Law.

The variations of the transversal component of the perturbation force, as expressed by the rate of change of angular momentum of the disturbed planet, is shown in Figure 3 for the planet pairs Jupiter - Saturn, Uranus - Neptune, and Saturn - Uranus. These diagrams illustrate the symmetry of perturbations in pairs of neighboring planets over a complete relative orbit. Heliocentric conjunction and opposition divide the synodic periods into two equal portions (synodic half-periods) over which the perturbation forces are equal and opposite in direction.

Schlamminger [3] compared the principal peak frequencies derived from power spectrum analysis of sunspot numbers with the periods arising from planetary motions, and found that the three peaks containing most of the power (at 10, 11, and about 90 yr) correspond to the half synodic periods of planet groupings in the four giant planets:

Spectral Peak	Synodic Half-Period	Pair	ε
9.7 - 9.9 a	9.93 a	5 - 6	2.6×10^{-3}
10.9 - 11.2	11.23	56 - 78	4.8
89.6 - 110	85.70	7 - 8	9.8

This point is illustrated by Figure 4, which shows two maximum entropy spectral analyses and a correlogram published by Cohen and Lintz [4]. The discrepancy in the values for the long period may be expected to arise from the relative shortness of the sunspot record.

The synodic half-period in a pair of planets corresponds to the time which elapses from heliocentric conjunction to opposition, or from opposition to conjunction. The period 9.93 a applies to the relative motions of Jupiter and Saturn; the period 11.23 a to that of the pairs Jupiter - Saturn and Uranus - Neptune, and the period 85.7 a to that of Uranus and Neptune. It is difficult to visualize geometrically the synodic period of the system of two planet pairs, such as 56 - 78. The individual synodic periods in each pair are repetition periods of geometric configurations, the time intervals that elapse between two successive occurrences of the same heliocentric longitude separation angle between the two planets. The synodic period of the pair-pair configuration is the time that elapses between two successive occurrences of the same longitude separation in each planet pair. For the pair 56 - 78 this period is 22.46 yr. If the value zero is taken to represent a "pseudo conjunction" of the two pairs (not of the four individual planets", then the value 180° would correspond to a "pseudo opposition"; the time interval between them corresponds to the synodic half-period (11.23 a) of the pair-pair group.

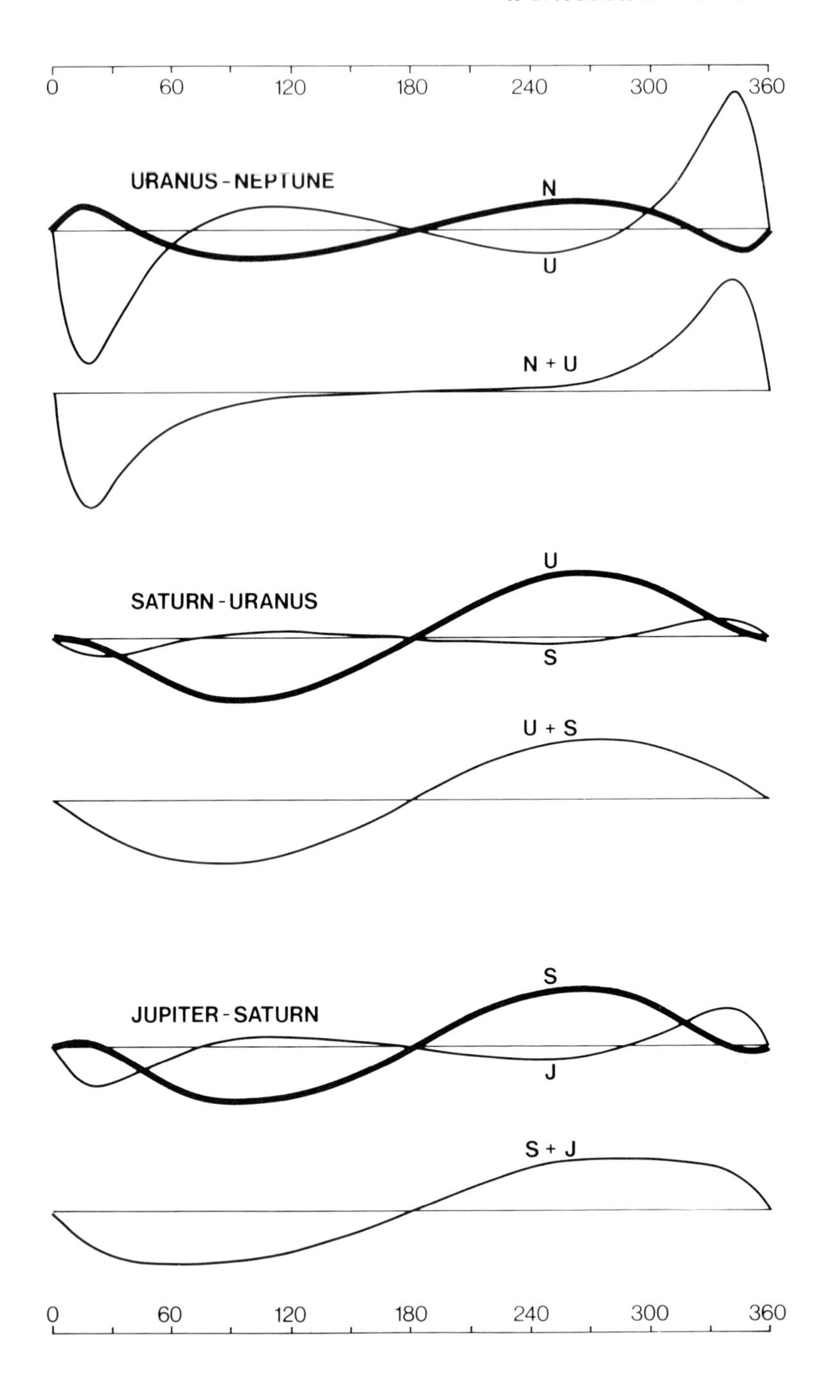

Figure 3. Rate of change of orbital angular momentum by perturbation in giant planets.

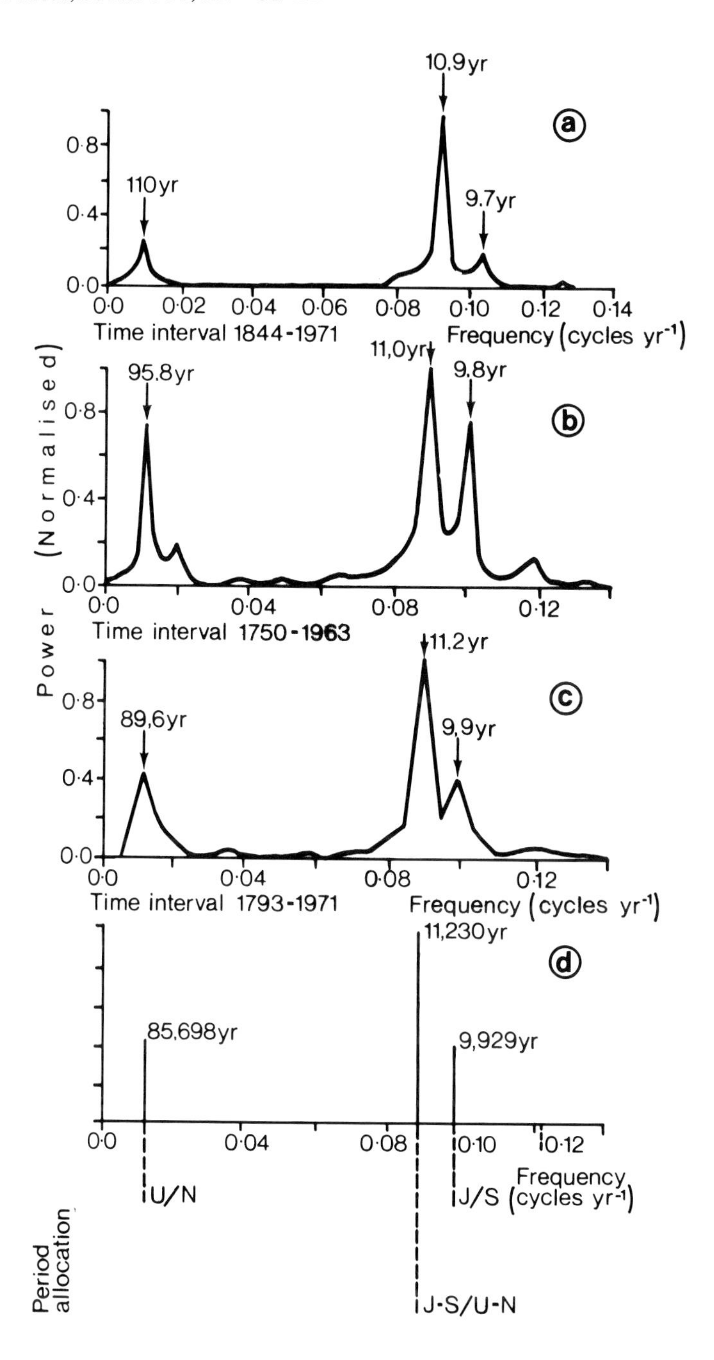

Figure 4. Power spectra (A,B), correlogram (c) of relative sunspot numbers and synodic half periods of giant planets (D).

Having shown compatibility of sunspot frequencies with frequencies in relative motion of the four giant planets, we now compare sunspot numbers with angular separation in Figure 5. It appears that the major peaks and some of the lesser fluctuations in the sunspot numbers correspond to either of two planetary parameters:

(1) The predominant association over the period of record from 1730 is between sunspot number maxima and 90 or 270° separations in the 56 - 78 pair-pair configuration;

(2) During two relatively short periods around 1780 and 1950 (approx. 170 yr apart), peaks in the sunspot number curve occur shortly after 90 and 270° separations of Jupiter and Saturn.

The long period peak amplitude variation appears to be associated with the angular separation of Uranus and Neptune; greatest amplitudes at angular separation 90°, smallest amplitudes near conjunction and opposition. This point is further illustrated by Figure 6, which compares the variations in sunspot maxima amplitudes with a simple function of longitude separation angles in the pairs.

For the latter part of the record period, magnetic observations of sunspots are available. These show alternations in magnetic polarity from one peak to the next and suggest that the sunspot cycle is really 22 and not 11 yr long. Schlamminger [3] pointed out that the changes in the resultant planetary perturbation forces may well initiate a polarity change in the solar magnetic field which leads to an oscillation of magnetic polarity of the sunspots with a period of 22.46 yr.

An angular separation of about 90° produces a maximum of perturbation of the outer planet by the inner planet in each pair. Therefore the variations of sunspots appear to relate primarily to the rhythm of _outward_ directed perturbation forces in the solar system and we hypothesize that sunspots arise as a result of outward directed energy fluxes from the Sun, modulated by the relative positions of the planets. Ninety-degree separation in planet pairs allows maximum outward flux, while 0° (180°) separation blocks it.

Sukurai [5] reports that low sunspot numbers are associated with relatively high solar equatorial rotation speed. It is conceivable that during periods of blocked outward energy flux the angular momentum of the outer equatorial solar layers builds up. According to Eddy _et al._ [6], the solar equatorial rotation was on average faster by 4% or more during the Maunder minimum than now observed.

One of the principal objections to the influence of planets on the sunspot cycle has been based on the observation than no sunspot cycle was discernible during the Maunder minimum while the planets still orbited as regularly as ever. This argument can be refuted on two counts. Firstly, Gleissberg [7] has shown that, although very diminished, the sunspot cycle was still discernible throughout the Maunder minimum. Secondly, if planetary motion is only a modulating influence on a solar energy emission which reached a low during the Maunder minimum, the continued orbiting of the planets during this period is no longer an obstacle.

Within the framework of this presentation, it is not possible to go into detailed aspects of solar-terrestrial weather relationships for which a vast amount of literature is accumulating. We merely wish to point out the possible relevance of applying this planetary sunspot theory to the general problem of weather and climate variation.

We assume that periodically changing fluxes of angular momentum result from the action of planetary perturbations in the solar system. The Earth, as any other planet, can participate in the transmission and temporary storage of angular momentum exchange differentials in a rhythm that is determined by the relative position of neighboring planets. This may cause variations in the orbital velocity and spin of the planet, in the orbital velocity or spin of satellites and material rings (Saturn, Uranus), or in the circumpolar and local vorticity of fluid motion in oceans and atmosphere. The latter are likely to respond first because of their relatively small inertia.

Besides the synodic period, two other periodicities arise from the relative orbiting of two planets when their mean motions are nearly commensurate. For example, 13 times the period of Venus being nearly equal to eight times that of the Earth, near coincidence occurs every fifth conjunction in the same part of the orbits producing a resonance effect with a period of 8.1 yr. Since the coincidence is not perfect, successive resonance positions drift slowly around the orbits, the resonance drift period being 240 yr for Venus-Earth, nearly 2,700 yr for Jupiter-Saturn.

Periods arising from the Earth's interaction with neighboring planets are

	Synodic	Resonance	Resonance drift
Venus - Earth	19.2 mo	8.1 yr	240 yr
Earth - Mars	25.6 mo	no	no

some of which have been observed in climatological and geophysical data.

Because of the inclinations between planetary orbits there arise small components of the disturbance forces at right angles to the orbital planes. These produce long-period variations in orbital elements of the planets, known as Milankovitch periods in the case of the Earth. Thus arises the possibility that the Milankovitch variations influence climate by a gravitational mechanism.

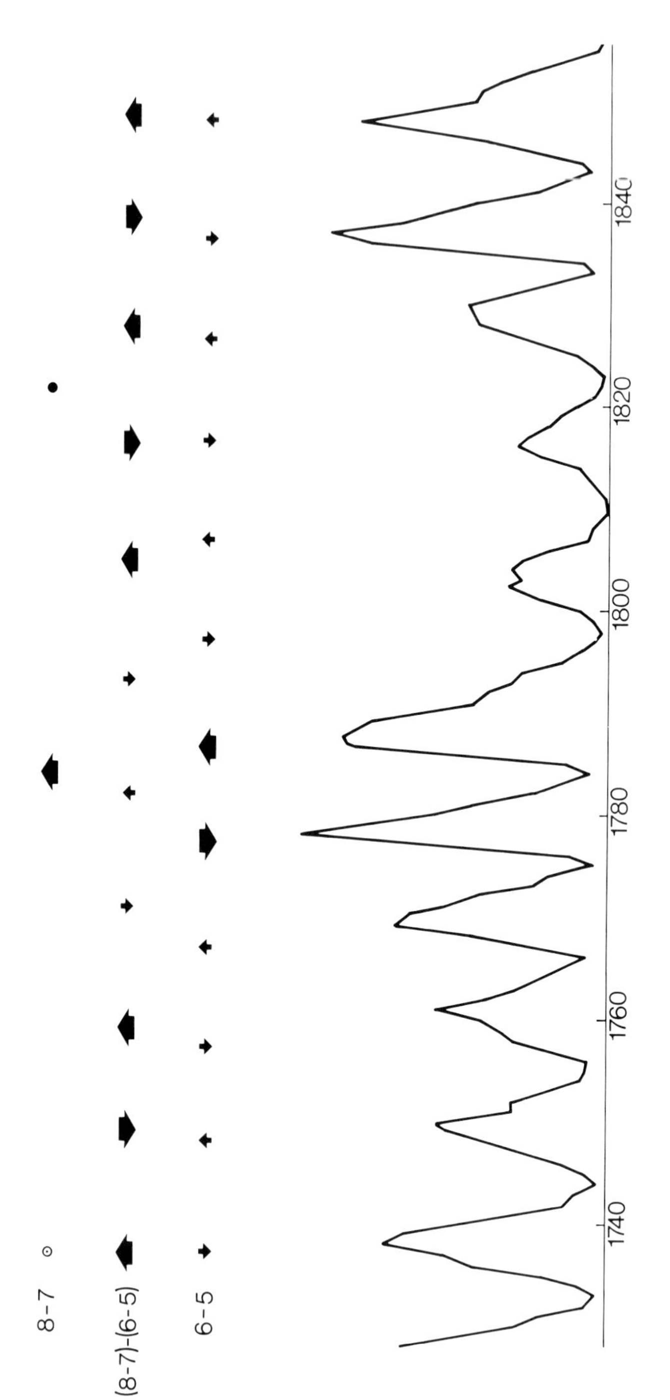

Figure 5. Angular separation of pairs of giant planets and relative sunspot numbers
Period 1730 - 1977.

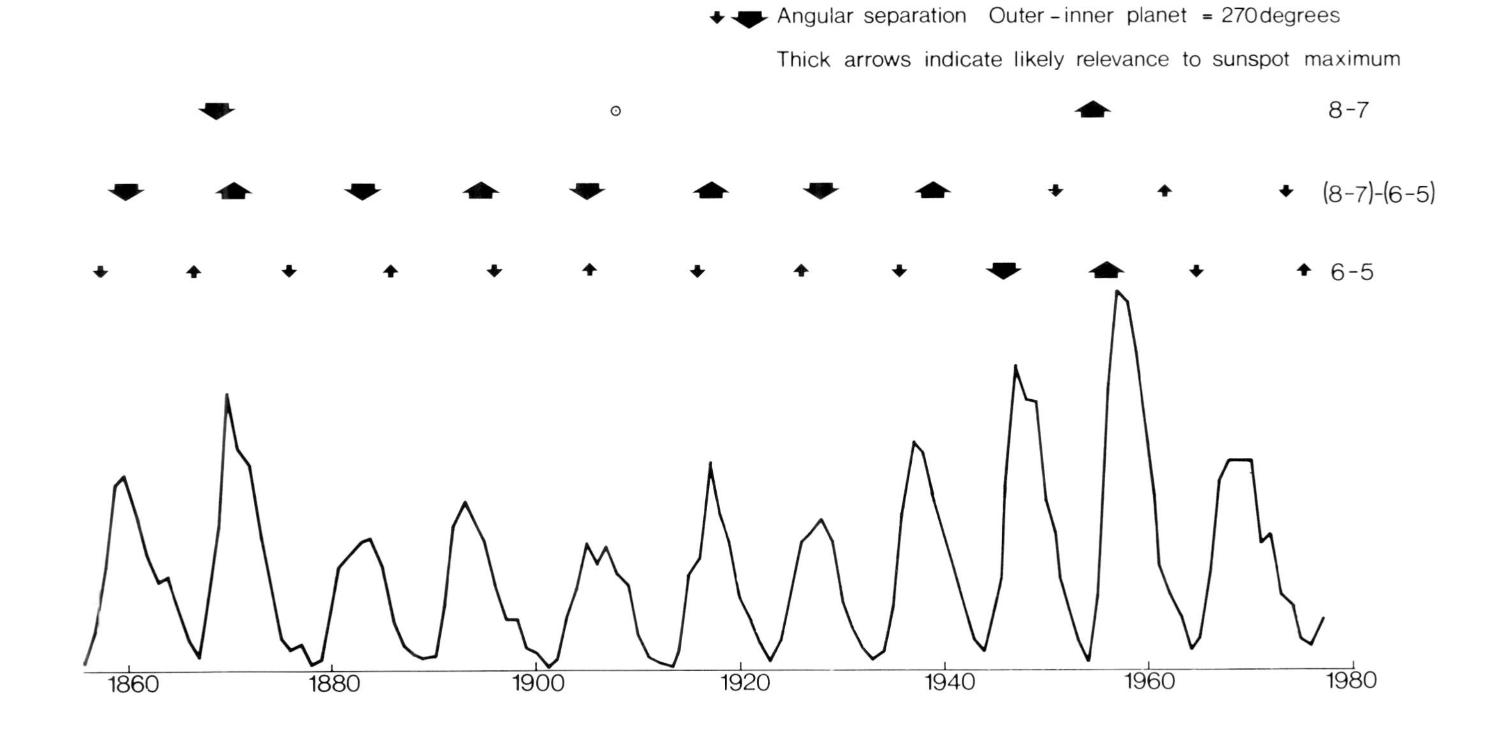

Figure 5. Continued

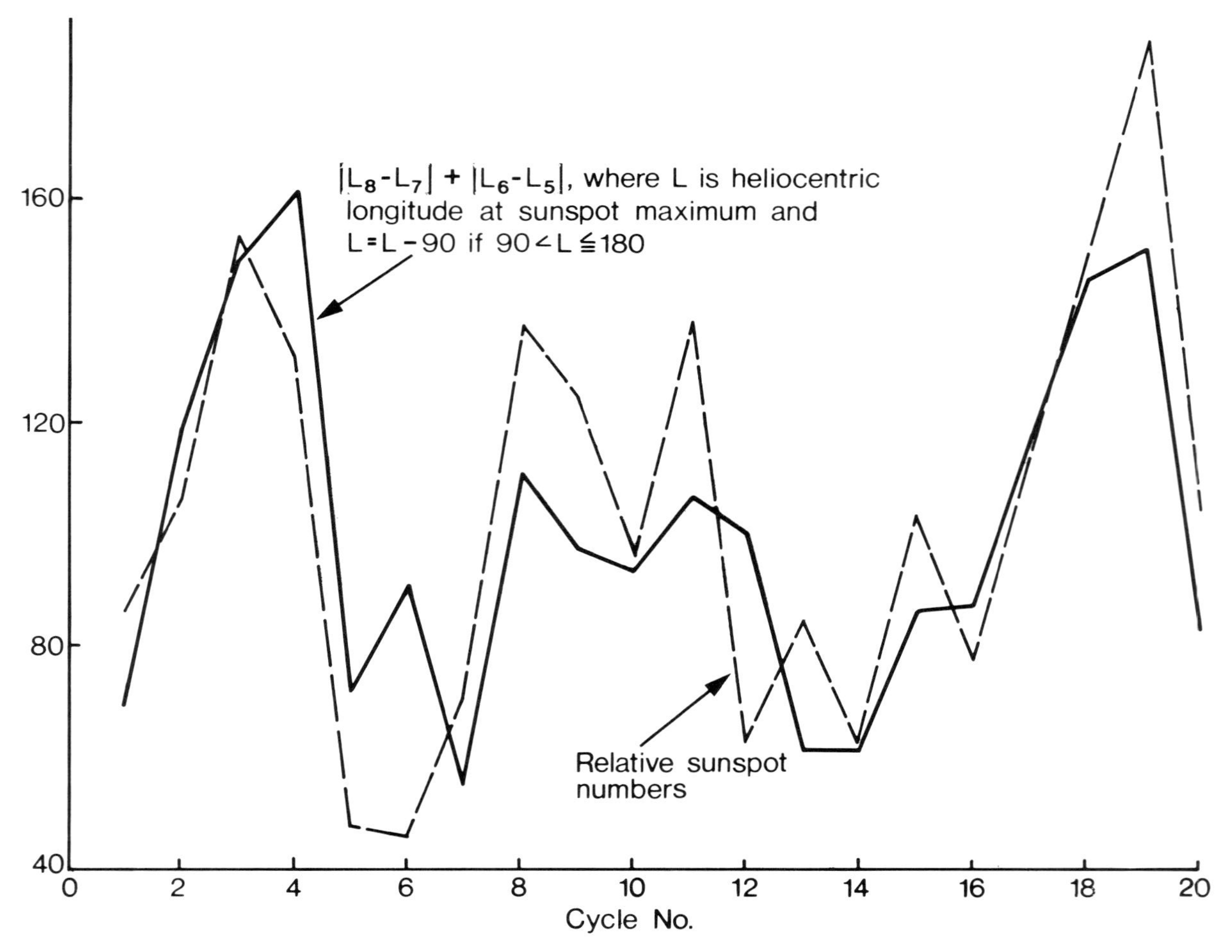

Figure 6. Relevance of angular separation in planet pairs Jupiter-Saturn (5-6) and Uranus-Neptune (7-8) to variations of peak values at sunspot maxima.

REFERENCES

1. C. M. Smythe and J. A. Eddy, Nature, 266, 1977.
2. R. E. Roy and M. W. Ovenden, Mon. Not. Roy. Astr. Soc., 114, 232, 1954.
3. L. Schlamminger, Sunspot periods show a synchronization with the resonance periods and the perturbation periods of planets (unpublished), 1977.
4. T. J. Cohen and P. R. Lintz, Nature, 250, 398, 1974.
5. K. Sukurai, Nature, 269, 402, 1977.
6. J. A. Eddy, P. A. Gilman, and D. E. Trotter, Solar Phys., 46, 3, 1976.
7. W. Gleissberg, Sterne und Weltraum, 7-8, 229, 1977.

SOME OF THE LITERATURE ABOUT PLANETARY INFLUENCES ON VARIATIONS IN SUNSPOT NUMBERS

1859 Wolf, R,; Compt Rend. Acad. Sci. Paris; Vol. 48, p. 231.
1872 De la Rue, W., Stewart, B., Loewy, B.; Proc. Roy. Soc. London; Vol. 20. pp. 210 - 218.
1899 Birkeland, Kr.; Recherches sur les taches du soleil et leur origine; Skrift. Vidensk., Vol. 1, Christiania.
1900 Brown, E. W.; A possible explanation of the sunspot period; Monthly Notices, Royal Astron. Society, Vol. 60, p. 599 - 606.
1911 Schuster, A,; The influence of the planets on the formation of sunspots; Proc. Roy. Soc. London; Vol. 85, p. 309 - 323.
1918 Pocock, R. J.; Monthly Notices, Roy. Astr. Soc. London; Vol. 79, p. 54.
1936 Sanford, F.; Smiths. Misc. Coll., Vol. 95, 11 pp.
1937 Stetson, H. T.; Sunspots and their effects; McGraw Hill, 1937.
1946 Johnson, M.; Correlations of cycles in weather, solar activity, geomagnetic values and planetary configurations; Phillips and Van Orden, San Francisco.
1947 Clayton, H. H.; Solar cycles; Smiths. Misc. Coll., Vol. 106, No. 22.
1954 Anderson, C. N.; Geophys. Res.; Vol. 59, p. 455
1954 Nelson, J.; Transactions, Inst. Rad. Eng.; CS-2, p. 19.
1962 Suda, T.; J. Met. Soc. Japan; 40, p. 278.
1965 Jose, P.; Astron. Journ. 70, p. 193 - 200.
1965 Ward, F. I.; Astrophys. Journ., 141, p. 534.
1965 Wood, K. D., Wood, R. M.; Nature, Vol. 208, p. 129 - 131.
1966 Trellis, M.; CR Acad. Sci., B, 262, p. 312.
1967 Takahashi, K.; Rad. Res. Lab.; Vol. 14, p. 237.
1967 Bigg, E. K.; Astr. Journal; Vol. 72, p. 463 - 466.
1968 Bollinger, C. J.; Tellus XX, 3, p. 412 - 416.
1969 Nickel, G. H.; Rep. Lawrence Radiation Lab., URCL - 502264.
1970 Bureau, R. A., Craine, L. B.; Nature, 228, p. 984
1972 Wood, K. D.; Nature, 240, p. 91 - 93.
1972 Sleeper, H. P. Jr.; Planetary resonances, bistable oscillation modes and solar activity cycles; NASA Contractor Report 2035, prepared by Northrop Services Inc., Huntsville, Alabama.
1975 Okal, E., Anderson, E. L.; Nature, 255, p. 511.
1977 Smythe, C. M., Eddy, J. A.; Nature, 266, p. 435.

USE OF DIFFERENCE EQUATION METHODS FOR PREDICTING SUNSPOT NUMBERS

G. W. Brier
Atmospheric Science Department
Colorado State University
Fort Collins, Colorado 80523 U.S.A.

ABSTRACT. A new method is described for examining time series for multiple periodicities where the physics of the situation suggests forcing functions with a number of possible periods p_1, p_2, ... p where the signals may not be additive but include effects of interactions or nonlinearities. Application of the method to the prediction of the monthly values of the Wolf relative sunspot number indicates that forecasts can be made for periods up to 100 mo in advance, with mean-square-errors comparable to those previously obtained for predictions made for only one or two years ahead.

1. THEORY

Methods of Fourier analysis are commonly used when it is known that periodic excitations may be acting as forcing functions on a physical system. This paper presents an alternative approach to the analysis and prediction of time series when it is desired to test for the existence of multiple periodicities where the physics of the situation suggests a number of possible periods p_1, p_2, p_3 ... p_n that may interact in some complex way. In the procedure described here, there are no assumptions regarding the phase, amplitude or <u>wave-form</u> of response of the system to the forcing functions with periods p_1, p_2, ... p_n. However, it is assumed that these periods have been specified <u>a priori</u> on the basis of theoretical considerations or previous information. The basic idea is to use difference equations and filtering methods to provide multiple estimates of the signal contained in observation Y_t at time t, to reduce the noise below that obtained by other methods. Consider initially a periodic signal free of noise with a repetition period p_1. Then at time t the signal y_t is given by the prediction equation

$$y_t = y_{t-\ell_1}, \tag{1}$$

where ℓ_1 represents the lag which is equal to the period p_1. If there is an additional cycle of length p_2, then (1) will be in error and it is easy to show that the proper equation is

B.M. McCormac and T.A. Seliga (Eds.): Solar-Terrestrial Influences on Weather and Climate. 209-214.

$$y_t = y_{t-\ell_1} + y_{t-\ell_2} - y_{t-\ell_1-\ell_2} \tag{2}$$

where ℓ_2 is the lag equal to cycle length p_2.

In general, if there are n cycles of length p_1, p_2, ... p_n, the signal $y_t(1,2,...n)$ will be given by 2^n -1 terms on the right hand side. There are no restrictions on wave form, amplitude or phase of the cycles. Of course, the real problem comes because of noise on the system when the observation Y_t is composed of the signal y_t and an error ε_t. For example, the signals $y_{t-\ell_1}$, $y_{t-\ell_2}$ and $y_{t-\ell_1-\ell_2}$ in Equation (2) may have superposed errors ε_1, ε_2 and $\varepsilon_{1,2}$ respectively. To control this noise, we make multiple estimates of the signal y_t in the following way:

$$\hat{y}_t(n_1,n_2) = Y_{t-n_1p_1} + Y_{t-n_2p_2} - Y_{t-n_1p_1-n_2p_2}, \tag{3}$$

where n_1 and n_2 take all possible combinations of integer values permitted by the data set. The average $\hat{y}_t(n_1,n_2)$ of these estimates can be used as the prediction.

The Equations (2) and (3) are valid only if the signals from the various periodic contributions are additive. But there is no assurance that in complex physical systems the signals are additive so we consider some way around this difficulty. For example, suppose we have non-sinusoidal signals with repetition periods p_1 = 8 yr and p_2 = 12 yr. For annual values using harmonic analysis it would take at least 11 Fourier coefficients to represent these signals and more coefficients would be needed to represent any interaction terms. The period $p_{1.2}$ = 24 yr is an exact repetition period (ERP) so we can now use the prediction equations

$$\hat{y}_t\ (n) = Y_{t-n24} \tag{4}$$

regardless of whether the 8 and 12 yr signals are additive or not.

But suppose p_1 = 8 and p_2 = 12.2. These periods are no longer commensurable and 24 yr is no longer an ERP but only an approximate repetition period (ARP). At 48 yr the two periods are out of phase by twice the previous amount, and in the discussion and application that follows here we will define an ARP whenever p_1 and p_2 are out of phase by no more than 18° ($\cos 18^o = 0.95$). Thus we would consider (4) as valid only for $n \leq 2$ so far as p_1 and p_2 are concerned.

Because of the noise on the system, predictions made from equations such as (1) through (4) will have an error arising from two sources. One might have some measure of control on these, but not on the error associated with Y_t, the observation being predicted. We can write the predictions in the general form

$$\hat{y}_t = \sum_{i=1}^{m} a_i Y_{t-\ell(i)}, \quad (\Sigma a_i = 1) \tag{5}$$

where the $Y_{t-\ell(i)}$ are the m previous values of Y at lag $\ell(i)$ selected for use in the prediction equation. If the variance of the signal is s^2, the variance of the observations is $\sigma_Y^2 = s^2 + \sigma^2$ where σ^2 is the error variance ε^2. When $s^2 = 0$ the mean square error (MSE) expected will be given by

$$E(\text{MSE}) = \sigma_Y^2 (1 + K), \tag{6}$$

where

$$K = \sum_{i=1}^{m} a_i^2 . \tag{7}$$

We are not especially concerned here with the assumption $\overline{\varepsilon_i \varepsilon_j} = 0$ since our predictions will deal with large time lags ($\ell = |i-j| >> 50$) where it is unlikely that ε_i and ε_j are correlated unless they contain a signal of some kind which has not yet been removed.

For a sufficiently long series of Y_t with variance σ_Y^2, the actual MSE and correlation between Y and y can be calculated and compared with expectation. If the ratio of the MSE to E(MSE) is sufficiently less than unity, we assume that we have evidence of a signal s and examine the residuals $e_t = (Y_t - \hat{y}_t)$ for evidence of further predictable signal. Spectrum analysis or auto correlation studies might suggest further deterministic periods to be tested or possibly an autoregressive or ARIMA process.

2. APPLICATION OF DIFFERENCE EQUATION PREDICTION AND FILTERING (DEPAF).

The application presented here is an investigation of the predictability of the monthly sunspot numbers from January 1749 to December 1977. The DEPAF analysis presented here is based upon the consideration of 9 basic frequencies associated with the planetary periods of the solar system. Their choice in no way depends upon an empirical analysis of the sunspot series (SS) and, in fact, if they have no relationship to SS, the application of DEPAF to the series could not result in useful and valid prediction equations. Emphasis in the analysis is not on these periods $p_1, p_2, \ldots p_9$ but in certain repetition periods involving combinations of these frequencies. Since these periods are not commensurable, a repetition period selected will be what we have defined as an ARP in the previous section. The ARP's used in the analyses here have been selected on an *a priori* basis without regard to considerations of the observed solar activity.

The ARP's used here will be designated as p_1', p_2', p_3', ... etc. A convenient way for examining the nine planetary periods for ARP possibilities is to examine the function

$$\overline{C}_t = 1/\Sigma w_i^2 \cdot \sum_{i=1}^{9} w_i \cos 2\pi f_i t \qquad (8)$$

for small increments of t beginning with t = 0 and ending, say with t = 300 yr. At least three reasonable sets of weights w_i can be considered such as $w_i(1) = 1/9$, $w_i(2) = M_i R_i^{-2}$ and $w_i(3) = M_i R_i^{-3}$ where M_i is the mass of planet i and R_i is its mean distance from the sun. Since the value of $\overline{C}_t = 1$ when t = 0, inspection of the graph of $\overline{C}_t$ for high values would indicate when a number of the cycles are back in phase. Figure 1 shows the graph of $\overline{C}_t$ for the first 14 yr with weights proportional to $M R^{-3}$ (the tidal force). Notice the highest peak that occurs at 11.08 yr. The curves based on weights $w_i(1)$ and $w_i(2)$ show approximately the same form, with the highest peak always at 11.08 yr. The curves in Figure 1 are, of course, essentially the autocorrelation function.

When curves such as that shown in Figure 1 are extended to year 200 there are only three additional major peaks that exceed the peak at 11.08 yr when weighting functions $w_i(2)$ and $w_i(3)$ are used. These peaks are at 24.07, 59.03, and 83.1 yr. For the case of equal weight ($w_i = 1/9$) there is a peak at 179.0 yr which is not approached in magnitude anywhere between 11.08 and 179.0 yr. The ARP's to be used for the difference equation predictions were found to be $p_1' = 133$ mo, $p_2' = 96$ mo, $p_3' = 289$ mo, $p_4' = 708$ mo, $p_5' = 997$ mo, and $p_6' = 2148$ mo. There are no ARP's shorter than 8 yr. Prediction equations using combinations of p' all showed predictive skill, a typical example being

$$\hat{y}_t = Y_{t-133} + 1/2Y_{t-96} + 1/2Y_{t-192} - 1/2Y_{t-229} - 1/2Y_{t-325} \qquad (9)$$

which had a correlation of 0.56 between Y and $\hat{y}$. Further improvement (r = 0.63) was obtained with equation (10), not shown here, which had 27 terms ($\Sigma a_i^2 = 2.854$) and used p_1' through p_5'. Spectrum analysis of the residual errors showed 50% of the variation due to a narrow band around a period of 10 yr. Since Figure 1 suggested the possibility of an ARP near that period, further examination disclosed an ARP at 119 months which had previously been missed. Preliminary correlation and spectrum analysis suggested that an autoregressive or ARIMA model could improve the forecast for periods of several months up to a year or two in advance.

In addition, a spectrum analysis (Figure 2) made on the 2628 months of sunspot data from 1749 to 1967 inclusive showed peaks corresponding to the ARP's around 10, 11, 59, and 83 yr. This interesting coincidence suggests that further investigation is needed on the controversial planet-sunspot relationships.

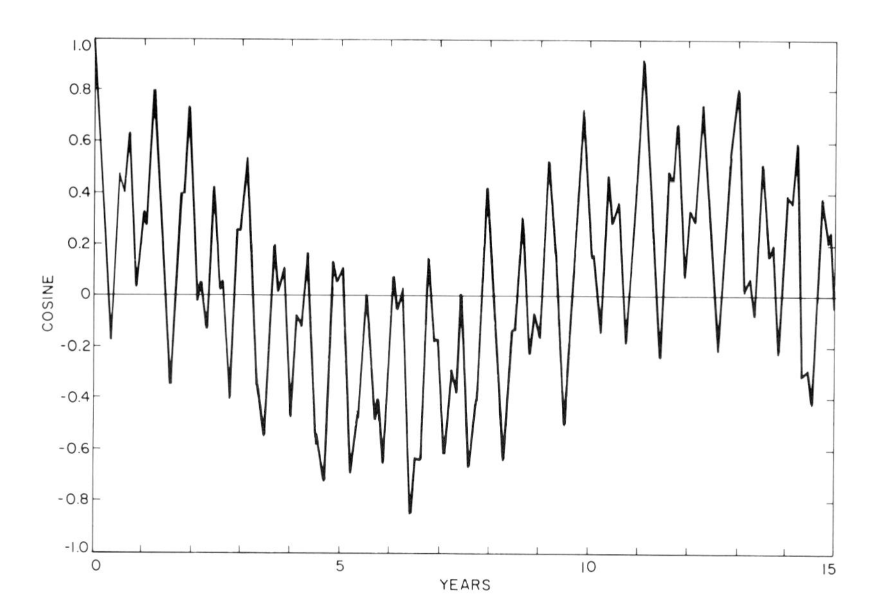

Figure 1. Weighted average of nine cosine curves based on planetary periods. The weights used are proportional to M R^{-3} where M is the mass and R is the mean distance from the sun.

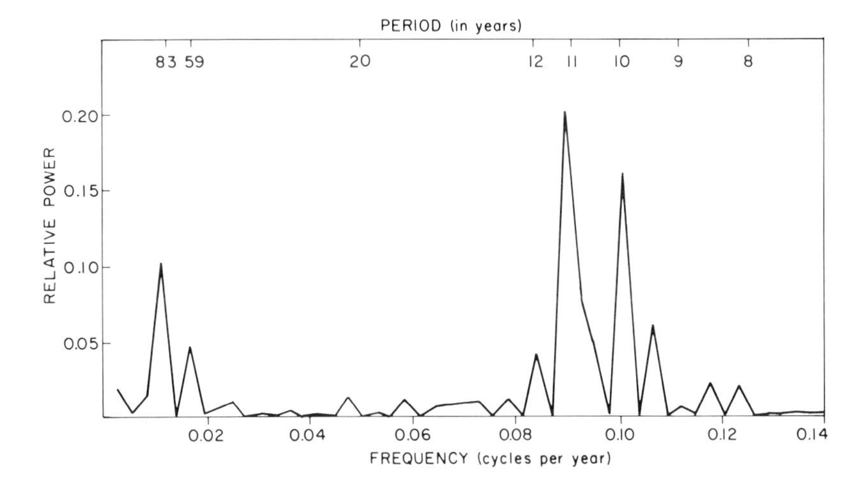

Figure 2. Cosine transform (unsmoothed) of 2148 autocorrelations of 2628 monthly sunspot numbers.

ACKNOWLEDGMENTS

This research was supported in part by the National Oceanic and Atmospheric Administration under Grant No. 04-7-022-44007 and the National Science Foundation under Grant ATM76-18860.

ATMOSPHERIC ELECTRICITY AND THE SUN-WEATHER PROBLEM

Ralph Markson
Building W-91
Massachusetts Institute of Technology
Cambridge, MA 02139 U.S.A.

ABSTRACT. The principal problem of Sun-weather research remains; there is no obvious physical mechanism to explain the findings. This paper discusses the possible role of atmospheric electricity in providing a feasible mechanism. It is suggested that solar controlled conductivity variations in the lower stratosphere over thunderstorms control current flow in the global circuit; this may ultimately influence cloud physical processes and thus atmospheric energy. The proposed mechanism and a method for testing it have resulted from a sequence of related studies and an experimental investigation of the variation of ionospheric potential. Some previous findings relating solar variability to atmospheric electricity are reviewed. Verification of the global circuit concept and the introduction of the solar sector parameter in Sun-weather research are discussed. A planned critical experiment is outlined to test the electrical part of the proposed mechanism by conducting atmospheric electrical balloon soundings at least once a day for a year or more. This would result in relating spatial and temporal variations in ionospheric potential and stratospheric conductivity to controlling solar factors. A "geoelectric index" representative of global thunderstorm currents could thus be established to complement the "geomagnetic index" which has proven invaluable in solar-terrestrial research.

1. INTRODUCTION

This paper discusses the possible role of atmospheric electricity in Sun-weather relationships, and suggests an electrical mechanism linking solar variation to weather parameters. Because the energy input to the atmosphere is essentially constant (the solar constant varies by less than 1% on a time scale up to at least decades), and because the meteorological parameters that do change with solar variability are confined to the upper atmosphere, which is very weakly coupled to the lower atmosphere, I have been led to consider atmospheric electricity as the agent linking solar variation and weather. The only solar controlled variable that does change appreciably in the lower stratosphere and troposphere is the ionizing radiation. This radiation affects the conductivity of

B.M. McCormac and T.A. Seliga (Eds.): Solar-Terrestrial Influences on Weather and Climate. 215-232.

the atmosphere, thereby controlling the currents flowing in the global circuit, and hence also the electric field intensity. Since ambient electric fields and ionizing radiation may be important in controlling cloud electrification and rainfall, a sequence of events is envisioned whereby solar activity may control the redistribution of energy already present within the atmosphere.

Because most solar-terrestrial parameters seemed to be controlled to some extent by the Earth's position in the extended solar magnetic field [1], and because we did not know what physical mechanisms were involved, I decided to use solar sector boundary crossings as timing marks in superposed epoch analyses of thunderstorm variation. The first attempt at this is seen in Figure 1.

At this time (1969) an interval of only 14 mo containing both the thunderstorm and sector crossing data was available. In this study [2] there appears to be a relationship between thunderstorm variation and the Earth's position in the solar magnetic sectors, with thunderstorms maximizing on the first day of negative sectors (US thunderstorm data only, for details see Lethbridge [3]). It is necessary to appreciate the importance of normalizing the sectors to unit length (by "stretching" or "shrinking" them, since sectors vary in length from 4 to over 20 days), thereby calculating the thunderstorm index as a function of position in a sector. If this procedure is not adopted, i.e., if the thunderstorm index at increments of one day is used, then there is no effect. This suggests that phase differences between sectors of different lengths cause signals not near to the key day (boundary crossing) to cancel each other. If so, the implication is that this solar signal varies as a function of the Earth's geometric position within a solar sector rather than the signal being at a fixed interval of time from a sector boundary crossing. For example, high speed solar wind streams occur near the middle of solar sectors [4].

Subsequently the solar sector approach was used by Wilcox et al.[5] to find a relationship between sector crossings and the vorticity area index (VAI) developed by Roberts and Olson [6]. A signal in the VAI was found within about 1 day of the sector crossing, but this may be a result of the use of unnormalized sector lengths as mentioned above. We expect to repeat the VAI analysis using unit length sectors.

2. THE CLASSICAL PICTURE OF ATMOSPHERIC ELECTRICITY

At this stage it is necessary to outline the classical picture of atmospheric electricity and to consider briefly some of the reported findings concerning the influence of solar activity on the parameters of atmospheric electricity. Figure 2 shows the well known universal time ("unitary") variation in the electric field measured over the oceans on which the classical picture of atmospheric electricity is based. Experimentally it has been found that this electric field is maintained by the flow of a conduction current from the ionosphere (where the conductivity is high

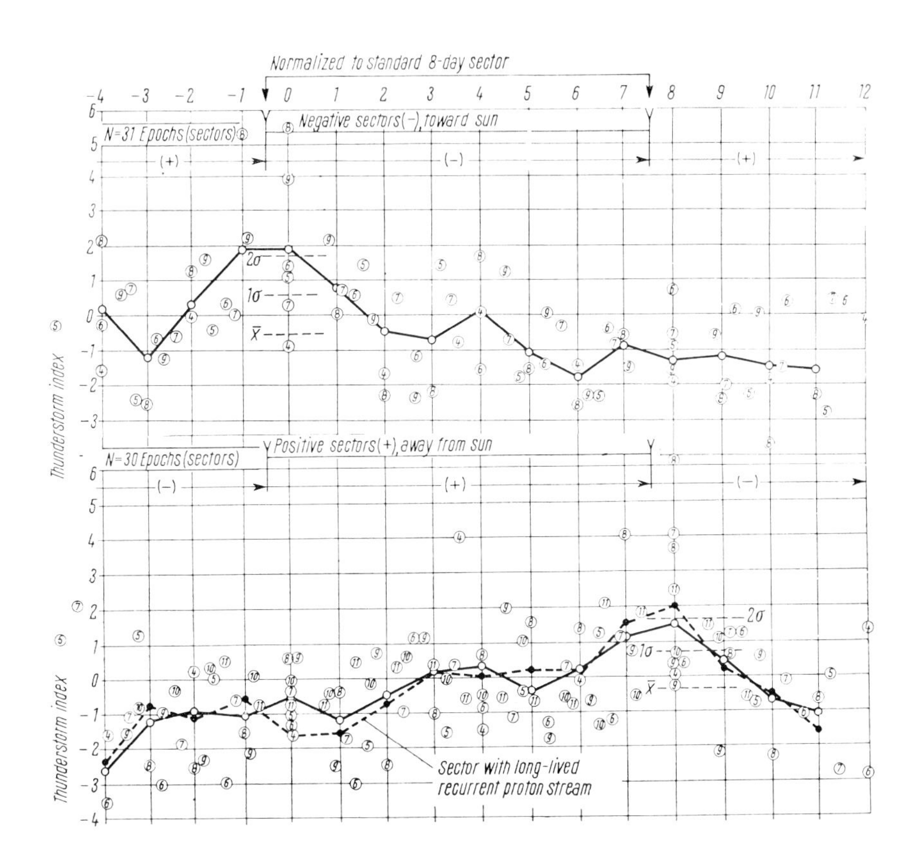

Figure 1. Thunderstorms as a function of the Earth's position in a solar sector. Transitions to adjacent sectors of opposite polarity seen at days 0 and 8. The curves are drawn through the locus of points closest to each daily increment of time. Numbers in points give days in sector being normalized to unit length, i.e., each point is the average for all sectors of that length at that increment of the sector's length [2].

enough for it to be regarded as an equipotential surface) through the weakly conducting lower atmosphere to the Earth. To maintain this current flow it is necessary that electrical generators be active in the circuit and, following the work of Wilson [7] in 1916, it is now generally accepted that the sum of global thunderstorms constitutes the generator. This theory is given considerable support by the strong similarity evident in the curves of the universal time variation in thunderstorm activity and in electric field strength shown in Figure 2. In addition, aircraft [8,9] and balloon [10] measurements over thunderstorms have confirmed that positive current of the proper magnitude flows upward over these storms in accordance with the global circuit model.

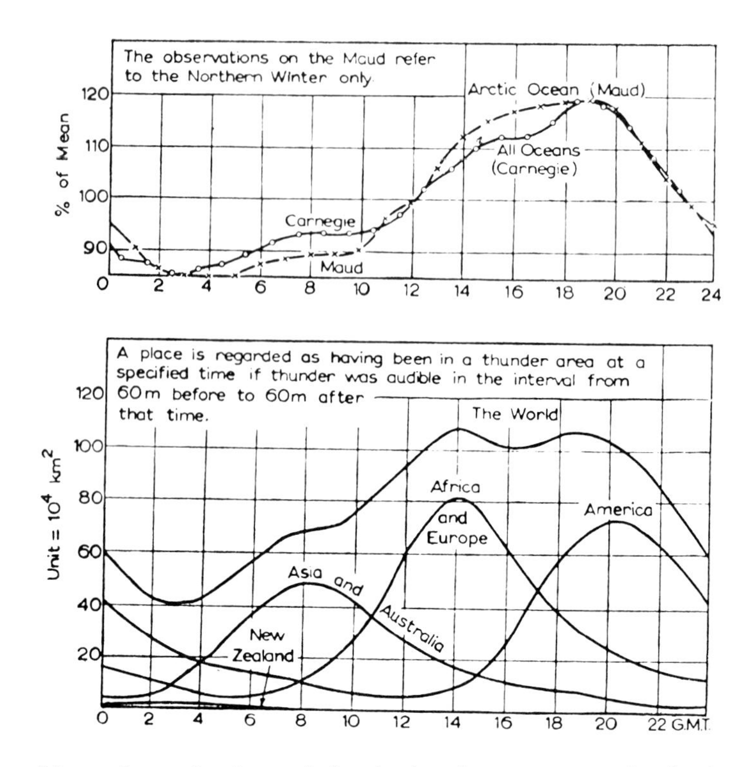

Figure 2. Diurnal variation of land thunder areas and of electric field intensity over the oceans [3].

The global circuit of atmospheric electricity is summarized in Figure 3 where it is seen that currents flow upwards from thunderstorms to the conducting ionosphere. Current between the conducting Earth and the thunderstorm flows because of ionic conduction, point discharge, charge carried on precipitation and lightning. The return branch of the circuit is the downward current between ionosphere and Earth in the fair-weather regions of the globe remote from thunderstorms. It will be noticed in Figure 3 that the upper part of both the generator and return, or "load", branches of the circuit are readily accessible to incoming ionizing radiation. The relative importance of each region in controlling current flow in the global circuit will be discussed shortly.

The first extensive review of possible solar modulation of atmospheric electricity was made by Bauer [11] who found positive and negative correlations between electric field intensity and the 11-yr sunspot cycle. Cooper [12] was evidently the first to report atmospheric electric variations related to solar flares. More recent and convincing work along these lines is summarized in Figures 4 and 5. In Figure 4, Olson [13] has shown the variation of air-Earth conduction current density (measured from balloons) with the solar cycle. He divided his data into two sets, one for the hours near the maximum in the unitary variation of

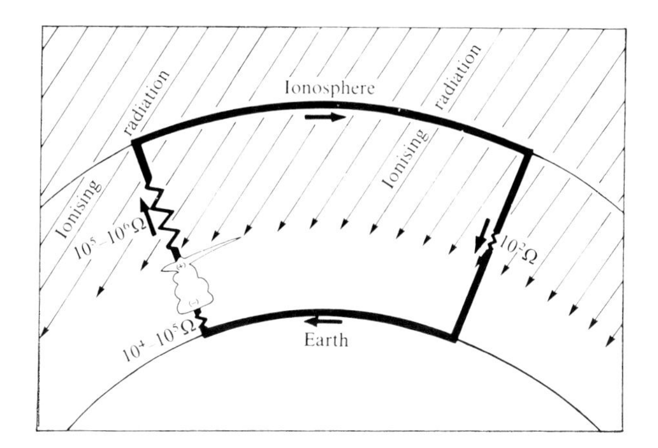

Figure 3. The atmospheric electrical global circuit. Large arrows indicate flow of positive charge. Estimated resistances of circuit elements are given. The thunderstorm depicted represents the global electrical generator, i.e., the totality of all thunderstorms [18].

the field strength, the other near the minimum hours. The scatter plots we have added show that there is negative correlation between air-Earth current density and sunspot number for both sets of data. (The correlation coefficients are 0.67 ($p<0.05$) and 0.74 ($p<0.01$) for the maxima and minima, respectively.) This suggests solar modulation of galactic cosmic radiation as responsible for the variation. In Figure 5 the increase in air-Earth current density shortly after the occurrence of H_α flares found by Reiter [14,15] and by Cobb [16] is shown. Here, to obtain a larger data set, we have combined Reiter's data over two periods into a single histogram (Figure 5(a)). The results of Cobb's report (Figure 5(b)) show a trend similar to Reiter's. These results indicate a rapid response of atmospheric electricity to solar flares and suggest solar ionizing radiation as the causal agent.

3. EXTENSION OF THE SOLAR SECTOR/THUNDERSTORM ANALYSIS

In 1973, when the Goddard Solar-Weather Symposium was being planned, W. O. Roberts invited me to give a paper under the condition that I discuss physical mechanisms, the feeling being that there were sufficient statistical associations between Sun and weather. This led to consideration of how variable solar activity might be influencing the atmospheric electrical global circuit. It is well known by workers in atmospheric electricity that conductivity changes caused by ionizing radiation variations do not occur sufficiently low in the atmosphere to significantly affect fair-weather electric fields. However, the thought occurred that there was one element in the global circuit that contained an appreciable part of the circuit resistance and was sufficiently high that it would be influenced by solar modulated ionizing radiation; this

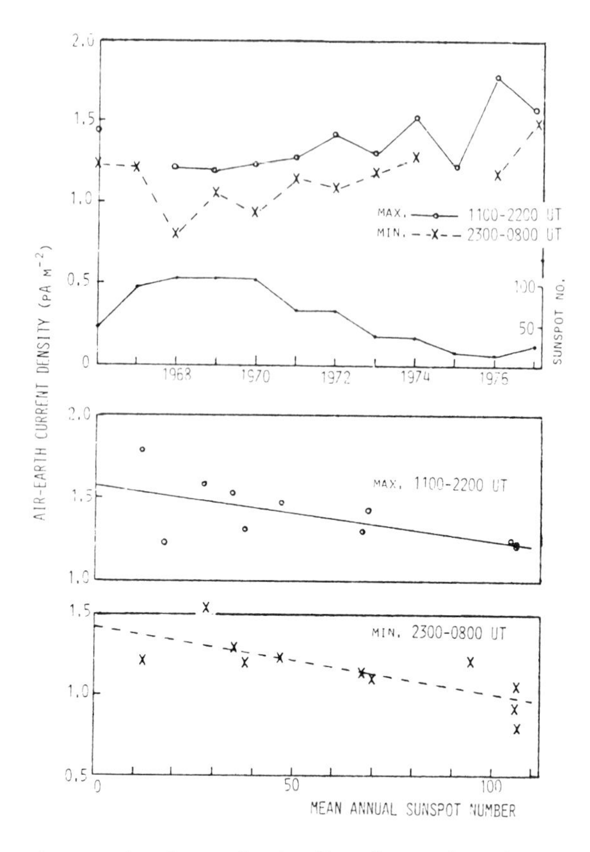

Figure 4. The variation of air-Earth conduction current density through the period of a solar cycle. Balloon measurements of Olson [13]. These data are divided into two sets, one corresponding to the hours of maximum electric field and the other the hours of minimum electric field of the "unitary" diurnal variation. Scatter diagrams showing the inverse correlation between current density and sunspot numbers have been made for each data set.

was the columnar resistance over the global thunderstorm generator. This idea formed the basis of the Goddard Symposium paper [17]. Recently the concept has been expanded and discussed in greater detail [18].

As a result of the independent findings of Svalgaard [19] and Mansurov [20] that solar sector crossings could be inferred from terrestrial geomagnetic records, it became possible to expand the data base of my original solar sector/thunderstorm analysis. The original study had 61 epochs from late 1964 through 1965 while the new one had 179 epochs covering most of the solar cycle from 1957 through 1965. It now proved possible to divide the data into two independent sets, one including sunspot minimum (1962 through 1965) and the other sunspot maximum (1957 through 1961). The results are shown here as Figure 6, and it can be

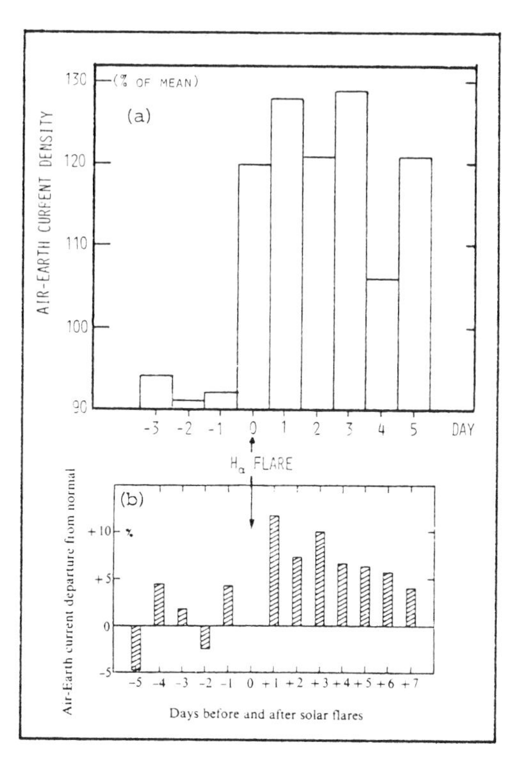

Figure 5. Variation from the mean of air-Earth current density before and after solar flares due to Reiter [14,15] (a) and Cobb [16] (b). See text.

seen that the same trend is apparent in both sets of data as was shown by the earlier, less extensive, analysis. The importance of normalizing the sectors to unit length is shown in Figure 6(c) where the signal in the averaged thunderstorm index is obliterated if the sector lengths are not normalized. These results were reported at the IUGG meeting in Grenoble, France in 1974 [21].

Recently Lethbridge, who provided the original thunderstorm index, has updated the data for the period 1966 through 1976. In addition to this I have gone back to near the beginning of the continuous thunderstorm record in 1947. Thus three independent data sets (n=754) have been assembled covering US thunderstorms for the decades 1947-1956, 1957-1966, and 1967-1976. These sectors have been examined as before, but with the unit sector length broken into 40 equal length subdivisions and a thunderstorm index plotted at each increment through the sector. The index consists of the sum of all thunderstorm days (US Weather Bureau records for the eastern 2/3 of the United States) minus the 21-day running mean centered on that day. In this index no filtering has been done with the daily thunderstorm sums, while cube roots of this quantity were taken in the original index of Lethbridge [3]. Thus the index is a residual

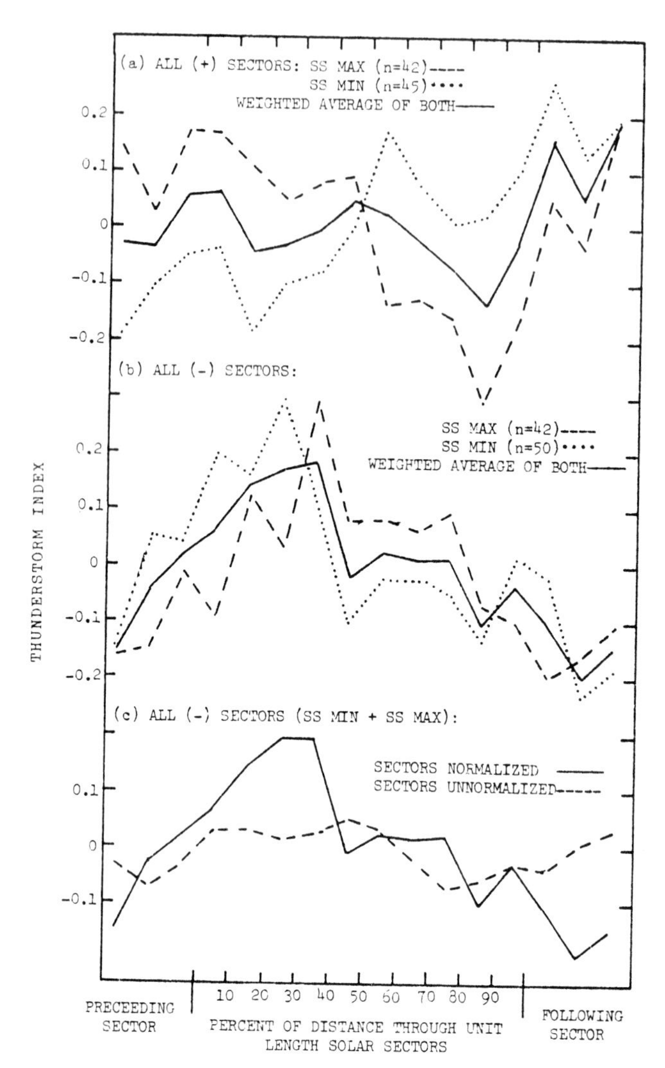

Figure 6. Variation of US thunderstorms with the Earth's position within a solar sector. In (b) the characteristic (-) sector decrease in thunderstorm number through the sector is seen. In (c) the effect of not normalizing the sectors to unit length illustrates the absence of a signal [21].

with seasonal and year-to-year variations removed. Figure 7 depicts the results of this latest study, and again the same trend is seen as in the earlier studies. It is difficult to test the significance of these results as the number of degrees of freedom is not known, because of using a data point several times in the partitioning process. It is not as high as 40 (the number of sets corresponding to the sector subdivisions) minus 2 (number of variables), or 38. However, for the record

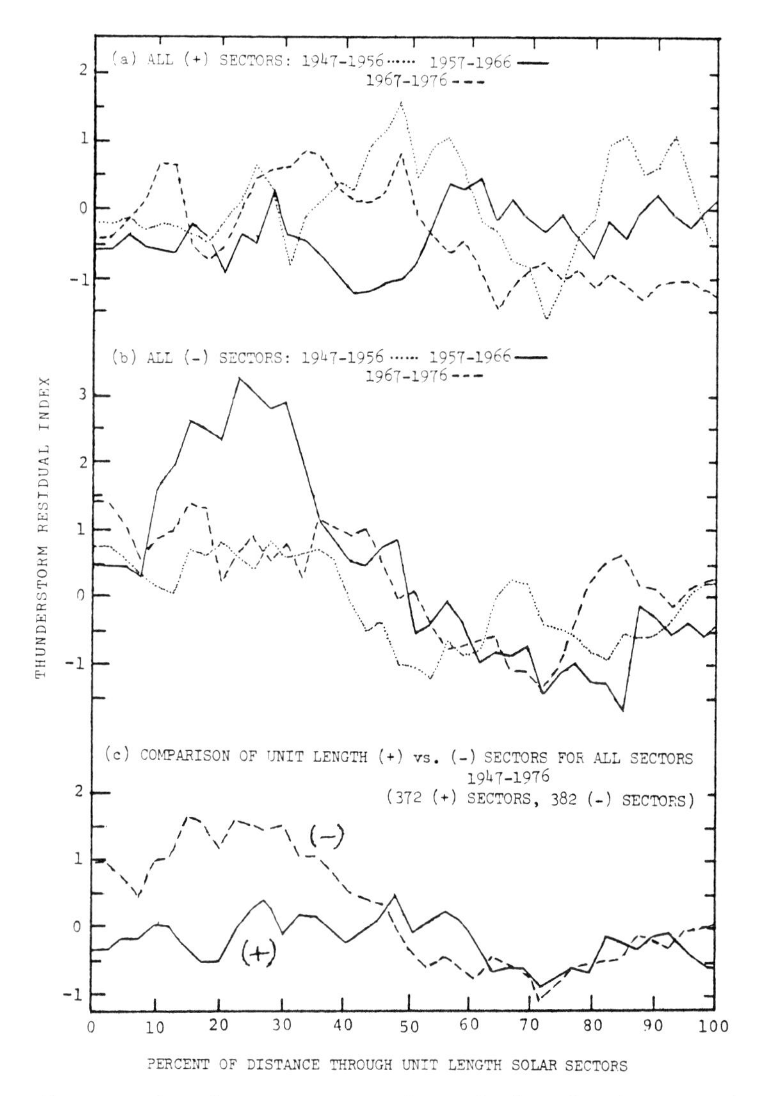

Figure 7. New results from an extension of the thunderstorm/solar sector analysis. Again note the characteristic variation of thunderstorm activity through the sector for (-) sectors.

the correlation coefficients are as follows:

(1947-1956) vs (1957-1966), r = 0.66,
(1947-1956) vs (1967-1976), r = 0.48,
(1957-1966) vs (1967-1976), r = 0.54.

The general trend exists in each independent data set for the negative sectors. The fact that three separate correlations are relatively high adds to confidence in the statistical significance. When these negative sectors are further subdivided into summer and winter half years (not shown) the same characteristic variation exists in each of the six curves. In contrast, there appears to be no systematic variation through the positive sectors, nor are they correlated.

Silverman [22] has found similar results by analyzing electric field variations at Arctic stations through solar sectors. He finds (a) that the average electric field intensity is greater during negative sectors than during positive sectors, and (b) that there is an increase in field intensity in the first part of a sector and a decrease in the latter part.

4. MEASURING THE VARIATION OF IONOSPHERIC POTENTIAL

In order to investigate the global circuit hypothesis including the validity of the ionosphere as an equal potential surface assumption, and possible solar influences on the global circuit, a program was conducted to develop a method for continuous recording of the temporal variation of ionospheric potential from an aircraft at constant altitude well above the exchange layer [23]. Anderson [24] had earlier demonstrated the feasibility of this approach with measurements every 15 minutes during 2 diurnal cycles. Figure 8 shows the results of estimates of ionospheric potential obtained during my program; when a 2-h running mean is plotted from these data (Figure 9) the unitary variation defined by the Carnegie curve (see Figure 2) is clearly depicted. It was found that in the proper low noise regions airborne measurements allow the temporal variation of ionospheric potential to be followed. In general, this cannot be accomplished from the ground because of local influences within the exchange layer. Even in favorable locations, such as the Arctic and Antarctic, usually 7 to 10 days' data are needed for statistical averaging before the unitary variation becomes evident [25]. The ability to record the variation of ionospheric potential in real time is demonstrated in Figure 10, which shows the agreement between simultaneous airborne measurements made at locations 7000 km apart [23]. While the results shown in Figures 8, 9, and 10 provide important support for the classical picture of atmospheric electricity, they also point the way to an experimental test of the proposed mechanism (to be discussed).

5. THE PROPOSED ATMOSPHERIC ELECTRICAL MECHANISM

This mechanism, explained in Reference 18, is summarized in Figures 11, 12, and 13. In Figure 11 we see some direct short-term responses due to a solar flare. Holzworth and Mozer [26] have shown that solar controlled ionizing radiation variations can have large effects on conductivity down to at least 15 km where an order-of-magnitude increase occurred. While this was a very large flare in certain respects, numerous other flares have had a greater density of high energy particles [27] which penetrate deepest into the atmosphere. (The short-term response of atmospheric electricity to solar flares was shown in Figure 5.) On the longer (solar cycle) time scale, there are also considerable changes

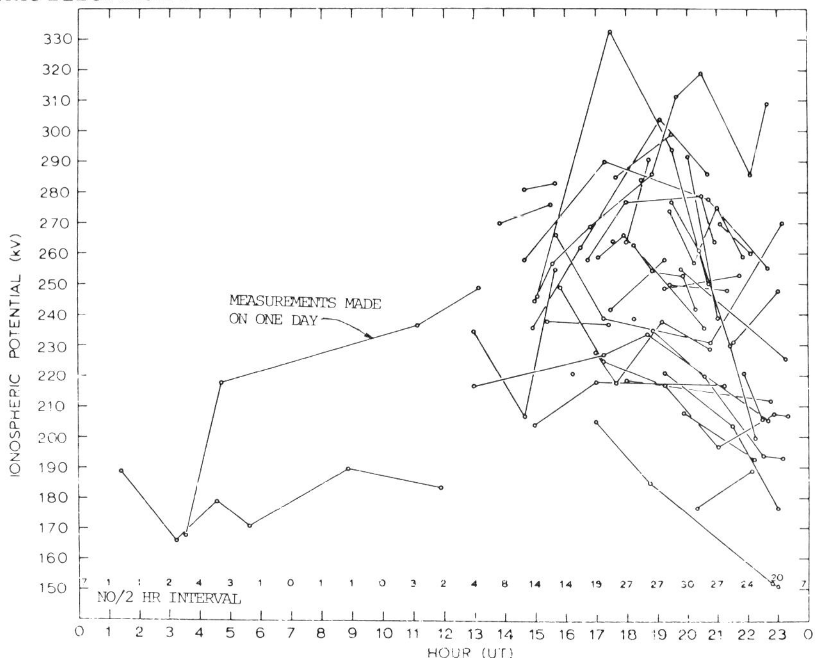

Figure 8. Summary of the ionospheric potential estimates obtained during an airborne investigation (1971-1972). Straight lines connect data points obtained on the same day. Note the "unitary" variation present every day with multiple measurements [2].

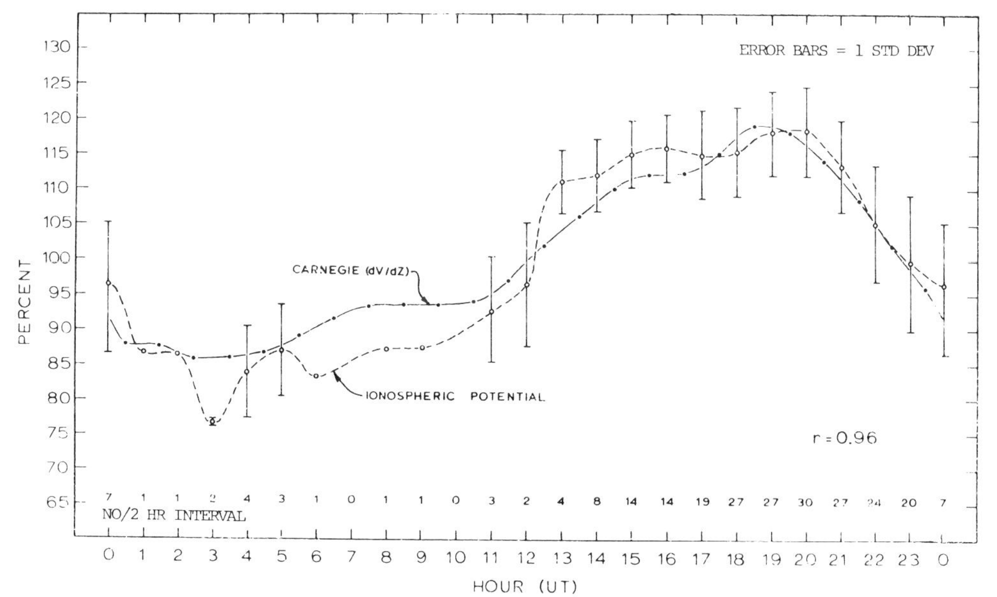

Figure 9. Normalized 2-h running mean obtained from the data of Figure 8. The number of soundings per 2-h interval is shown at the bottom. The normalized "unitary" electric field variation of the Carnegie, shown in Figure 2, is reproduced here for comparison. The correlation coefficient between the two curves is 0.96 [23].

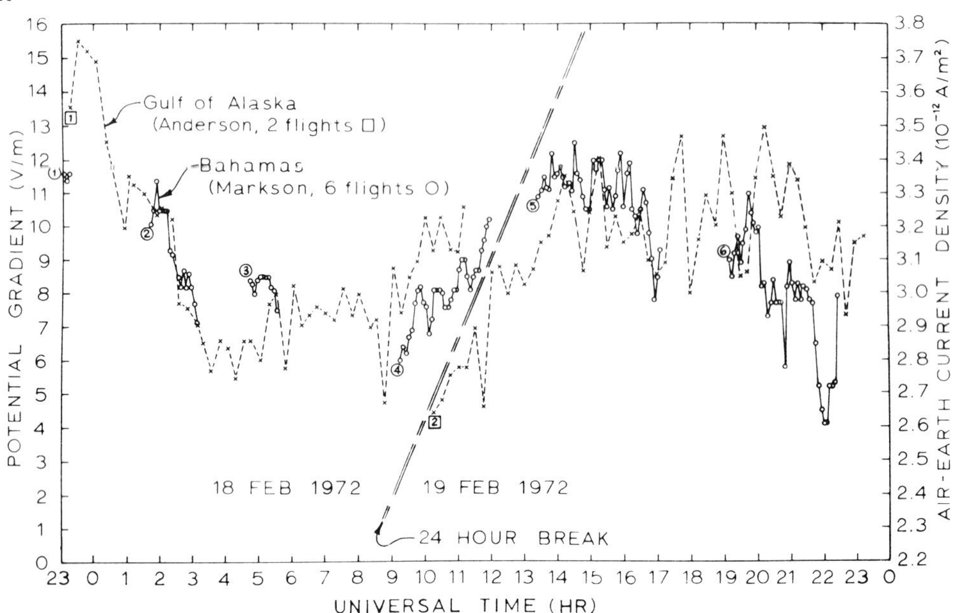

Figure 10. Comparison of constant altitude measurements of the vertical electric field in the Bahamas with simultaneous measurements of air-Earth current density over the Gulf of Alaska. Five minute averages are plotted for the Bahamas data, one data point was obtained every 15 min for the Alaskan record [23].

in conductivity in the lower stratosphere caused by solar modulation of galactic cosmic radiation. The typical variations in ionization rate at 20 km as a function of latitude are illustrated by Figure 12 [28]. (The corresponding atmospheric electrical response was suggested in Figure 4.) Additional evidence from balloon experiments by Winckler [29] show that ionization in the 20 to 30 km altitude range at mid-latitude typically increases by 10 to 100 times during Forbush decreases in galactic cosmic radiation measured on the Earth. These aspects of the proposed mechanism have been expanded on recently [30].

Figure 13 illustrates that the upper parts of both the generator and load branches of the global circuit are accessible to ionizing radiation (of both solar and galactic origin). Variations in conductivity will modulate the effective resistance of these two branches of the circuit and, as shown in Figure 13, if the resistance over the thunderstorm generator decreases by, say 50%, the current flowing from the generator would increase by 45%. If the load resistance remained constant, the ionospheric potential (and fair-weather electric field intensity) would also increase by 45%; in fact, the increase would be 40% because of the small (5%) decrease in return path resistance. If electric field intensities in fair-weather parts of the world were enhanced, this could increase the electrification of developing cumulus clouds, assuming one or more of the numerous induction theories of cloud electrification were

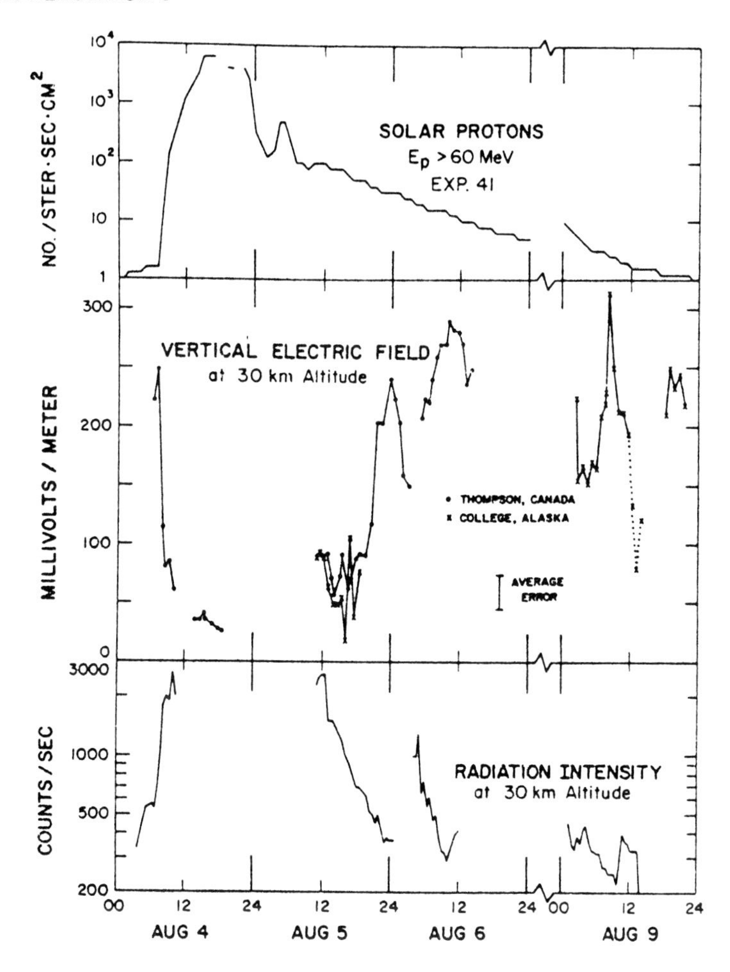

Figure 11. Comparison of simultaneous measurements of solar protons from a satellite vs vertical electric field and ionization rate from a balloon. A solar flare occurred at 0617 UT on 4 August 1972 [26].

valid [31]. In addition, an increase in conductivity in the vicinity of the tops of cumulonimbus clouds would amplify electrification according to the convective charging mechanism of Grenet [32] and Vonnegut [33]. Cloud physical effects could result. Thunderstorm intensity electric fields in clouds may influence phase changes of water and droplet coalescence forming heavy rainfall and adding latent heat to the atmosphere [34]. This could influence mesoscale and synoptic scale meteorology as well as the general circulation as schematically suggested in Figure 14.

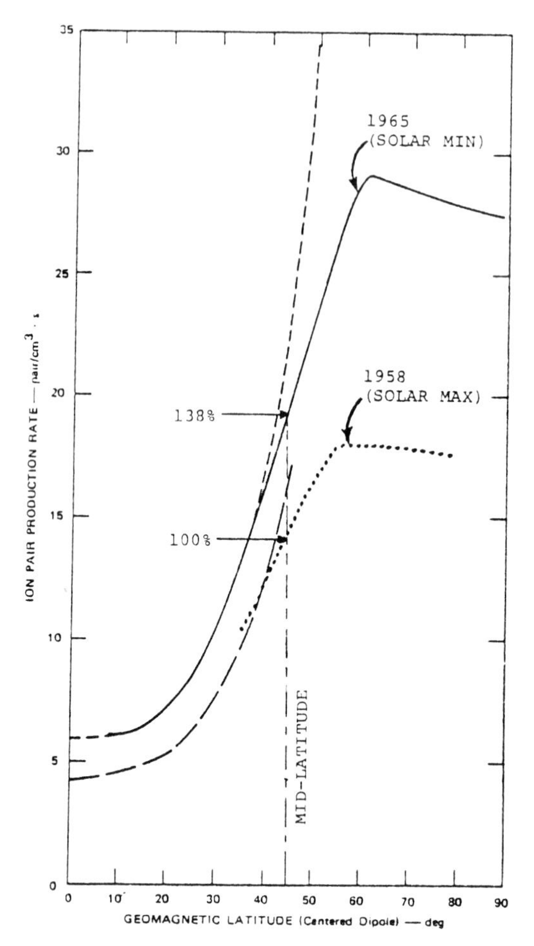

Figure 12. Variation of ion production rate at 20 km as a function of geomagnetic latitude and solar activity. Note the midlatitude variation referred to in the text [28].

6. TESTING THE PROPOSED MECHANISM

The unique features of the atmospheric electrical global circuit provide an advantageous regime for testing the atmospheric electrical part of the proposed mechanism. A measurement of ionospheric potential at only one location by one instrument is sufficient to characterize the electrification of the Earth's atmosphere over its entire surface (except for the polar cap regions where ionospheric/magnetospheric generators can make a significant local contribution). Such a single measurement can be compared to the procedure required to determine the global pressure or vorticity fields, these require hundreds of simultaneous balloon soundings which are routinely conducted every 12 h. By adding just one more radiosonde that measures the vertical electric field, conductivity and ion production rate, we would acquire a whole new family of parameters to

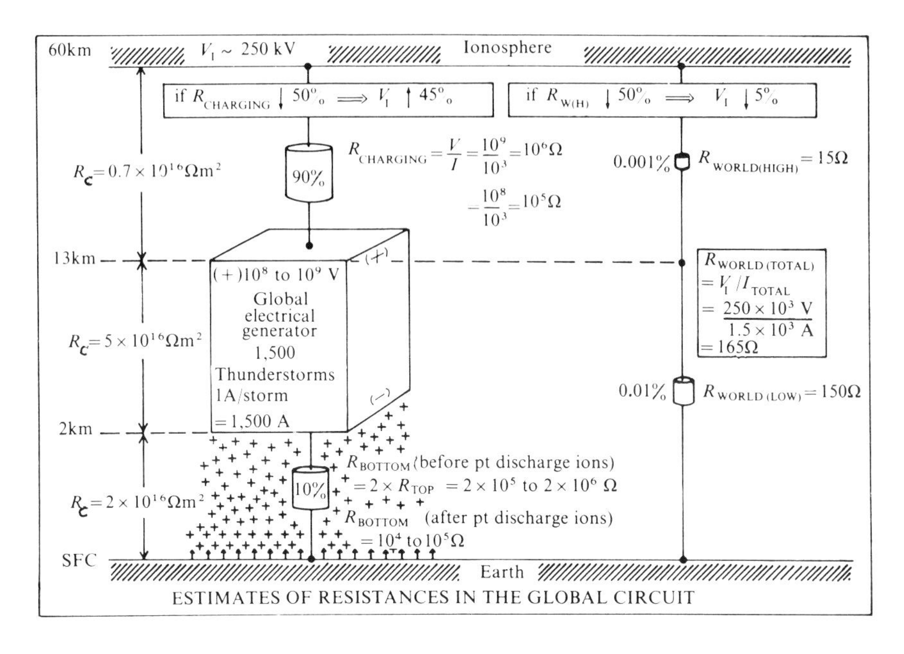

Figure 13. Details of the global circuit indicating how element resistances were derived and the distribution of resistance through the circuit [18]. Columnar resistances are not resistances *per se*; they must be divided by the applicable horizontal cross-sectional area to derive the resistance from top to bottom of an air column.)

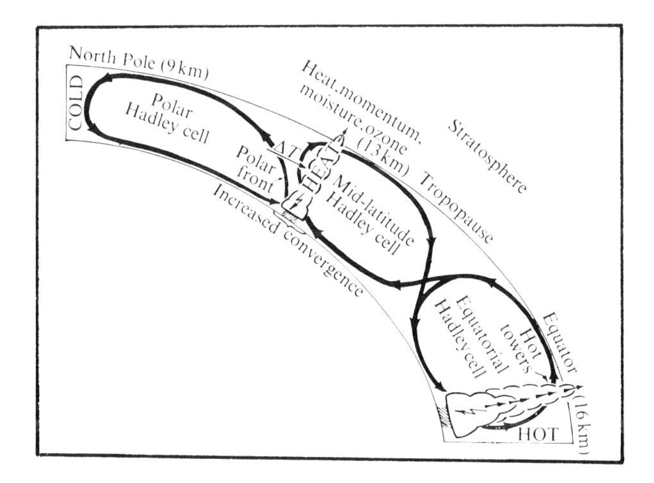

Figure 14. Longitudinal cross section of the atmosphere illustrating how solar modulation of cloud physical processes might influence the general circulation [18].

study relative to the Sun-weather problem. Such measurements could provide the temporal variation of ionospheric potential and stratospheric conductivity for comparison with the solar-terrestrial phenomena which are routinely monitored, such as solar flares, proton events, sector crossings, and solar wind characteristics. One objective would be to identify the particles or interplanetary magnetic field conditions which may cause effects in the global circuit. It is suggested that a minimum of one atmospheric electrical sounding per day be conducted at the same time (near the minimum in the unitary cycle, 0300-0400 UT) to minimize the influence of this variation. Soundings would be made more frequently (e.g., every 2 h) from just before predicted periods of solar activity and thereafter for several days to allow for unknown particle transit times. These soundings should be made from particularly suitable locations such as Mauna Loa in Hawaii, a mountain peak generally above the noise filled exchange layer, in the middle of the Pacific Ocean. Thus the balloons would not drift over mountainous or polluted regions which influence the air-Earth conduction current and thus introduce error into the balloon soundings of ionospheric potential [23]. It would be desirable to obtain an unbroken time series for an extended period, e.g., 1 yr or longer, to facilitate statistical correlations. Also, there are about 30 solar sector crossings per year which is about the minimum number required for reliable statistical analysis. The ionospheric potential variation would constitute a "geoelectric index" to complement the "geomagnetic index." It would be used in much the same fashion to relate atmospheric electrical variations to solar controlled parameters. Provision would be made for soundings at several widely separated sites (a) to allow for unfavorable weather at one site, and (b) through simultaneous soundings to examine the magnitude of ionospheric potential differences across the polar cap region and from the auroral zone to lower latitudes as a function of solar and geomagnetic activity.

Another approach to monitoring ionospheric potential is a technique utilizing a balloon or kite tethered by a conductor to a high voltage d.c. power supply [35]. While more complicated and logistically difficult than the simple radiosonde balloon, this would offer the possibility for continuous monitoring of ionospheric potential and the other variations. Such time resolution might be important in identifying direct effects due to solar particles or sector crossings.

The possible cloud physical effects resulting from solar activity and atmospheric electrical variations could be investigated through separate experiments. One would be satellite microwave observations of the spatial and temporal variations of liquid water and water vapor in the atmosphere [36]. The variation of global lightning (as a measure of thunderstorm activity) can best be monitored optically from satellites [37-39].

Through simultaneous observations of the Sun, the solar wind, the interplanetary magnetic field, particles entering the atmosphere, atmospheric electrical responses, and cloud physical parameters, we might be able to identify the causal chain by which variable solar activity may be modulating weather.

ACKNOWLEDGMENTS

Numerous individuals have provided information and stimulating discussion at various points in the work summarized here. I wish in particular to thank H. R. Anderson, R. V. Anderson, E. A. Bering, H. Dolezalek, R. H. Holzworth, M. D. Lethbridge, M. S. Muir, R. H. Olson, W. O. Roberts, R. G. Roble, R. Shapiro, S. Silverman, B. Vonnegut, and J. M. Wilcox. Support for this research has been provided by J. Hughes, Atmospheric Science Program, Office of Naval Research, under Contract N00014-74-C-0336.

REFERENCES

1. J. M. Wilcox, Space Sci. Rev., 8, 258, 1968.
2. R. J. Markson, Symp. Elect. Processes and Problems in the Stratosphere and Mesosphere, IAGA Meeting, Madrid, 1969, published in Pure Appl. Geophys., 84, 161, 1971.
3. M. D. Lethbridge, J. Geophys. Res., 75, 5149, 1970.
4. A. J. Hundhausen, Coronal holes and high speed wind streams, ed. by J. B. Zirker, Colo. Assoc. Univ. Press, 1977. Also see other articles in this volume.
5. J. M. Wilcox, P. H. Scherer, L. Svalgaard, W. O. Roberts, and R. H. Olson, Science, 180, 185, 1973.
6. W. O. Roberts and R. H. Olson, Rev. Geophys. Space Phys., 11, 731, 1973.
7. C. T. R. Wilson, Proc. Roy. Soc. A, 92, 555, 1916.
8. O. H. Gish and G. R. Wait, J. Geophys. Res., 55, 473, 1950.
9. B. Vonnegut, C. B. Moore, R. P. Espinola and H. H. Blau, J. Atmos. Sci., 23, 764, 1966.
10. C. G. Stergis, G. C. Rein and T. Kangas, J. Atmos. Terr. Phys., 11, 83, 1957.
11. L. A. Bauer, Terr. Magn. Atmos. Elec., 29, 23 and 161, 1924.
12. F. L. Cooper, Am. J. Sci., 240, 584, 1942.
13. D. E. Olson, Symp. Influence of Solar Activity and Geomagnetic Changes on Weather and Climate, IAGA/IAMAP Joint Ass'y, Seattle, 1977.
14. R. Reiter, Pure Appl. Geophys., 72, 259, 1969.
15. R. Reiter, Pure Appl. Geophys., 86, 142, 1971.
16. W. E. Cobb, Month. Weather Rev., 95, 905, 1967.
17. R. J. Markson, Possible Relationships Between Solar Activity and Meteorological Phenomena, NASA Rept. SP-366, ed. by W. R. Bandeen and S. P. Maran, Govt. Printing Office, Washington, D.C., 1975
18. R. J. Markson, Nature, 273, 103, 1978.
19. L. Svalgaard, J. Geophys. Res., 77, 4027, 1972.
20. S. M. Mansurov, Geomagn. Aeron., 9, 622, 1969.
21. R. J. Markson, Symp. High Atmosphere and Space Problems of Atmospheric Electricity, Gen. Sci. Ass'y, IUGG, Grenoble, 1975.

22. S. Silverman, Symp. High Atmosphere and Space Problems of Atmospheric Electricity, Gen. Sci. Ass'y, IUGG, Grenoble, 1974.
23. R. J. Markson, J. Geophys. Res., 81, 1980, 1976.
24. R. V. Anderson, J. Geophys. Res., 74, 1697, 1969.
25. H. Kasemir, Pure Appl. Geophys., 100, 75, 1972.
26. R. H. Holzworth and F. S. Mozer, J. Geophys. Res., in press 1978.
27. J. A. Lockwood, W. R. Webber, and L. Hsieh, J. Geophys. Res., 79, 4149, 1974.
28. R. D. Hake, E. T. Pierce, and W. Viezee, Stratospheric Electricity, Stanford Research Institute, Menlo Park, Calif., 1973.
29. J. R. Winckler, J. Geophys. Res., 65, 1331, 1960.
30. Correspondence between L. C. Hale and R. J. Markson, Nature, in press 1978.
31. J. A. Chalmers, Atmospheric Electricity, 2nd ed., Pergamon, London, 1967.
32. G. Grenet, Ann. Geophys., 3, 306, 1947.
33. B. Vonnegut, Proc. conf. atmos. elect., Paper 42, AFCRL, Bedford, Mass. 1955.
34. For further discussion of cloud physical effects and references see [18].
35. B. Vonnegut, R. J. Markson, and C. B. Moore, J. Geophys. Res., 78, 4526, 1973.
36. D. H. Staelin, P. W. Rosenkranz, F. T. Barath, E. J. Johnston, and J. W. Waters, Science, 197, 991, 1977.
37. J. A. Vorpahl, J. G. Sparrow, and E. P. Ney, Science, 169, 860, 1977.
38. J. G. Sparrow and E. P. Ney, Nature, 232, 540, 1971.
39. B. N. Turman and B. C. Edgar, Spring meeting of Amer. Geophys. Union, Miami, Fla., 1978.

ELECTRICAL COUPLING BETWEEN THE UPPER AND LOWER ATMOSPHERE

R. G. Roble
National Center for Atmospheric Research*
Boulder, CO 80307 USA
P. B. Hays
Space Physics Research Lab
University of Michigan
Ann Arbor, MI 48105 USA

ABSTRACT. A quasistatic model of global atmospheric electricity is used to examine the electrical coupling between the upper and lower atmosphere.

1. INTRODUCTION

Within recent years it has been suggested that a solar-terrestrial coupling occurs through variations in atmospheric electricity that in turn affect thunderstorm development and output [1-5]. Measurements of ground currents and potential gradients have shown atmospheric electrical responses to solar flares [6-9], geomagnetic and auroral activity [6,9-11], solar sector boundary crossings, [12,13], and solar cycle variations [14,15]. These solar- and upper atmosphere-induced variations are superimposed upon complex electrical variations associated with meteorological and anthropogenic processes in the lower atmosphere [15].

To study the electrical processes in the lower atmosphere and the coupling between solar- and upper atmosphere-induced variations upon the global electrical circuit, we have constructed a quasistatic model of global atmospheric electricity [16]. In this paper, we review the essential features of the model and discuss the results obtained thus far on solar-terrestrial electrical coupling processes.

2. NUMERICAL MODEL

The numerical model of global atmospheric electricity that is used to examine the electrical coupling between the Earth's upper and lower atmosphere has been described in detail by Hays and Roble [16] and Roble and Hays [17]. A schematic diagram of the global quasistatic model is shown in Figure 1. It is assumed that thunderstorms act as point dipole

*The National Center for Atmospheric Research is sponsored by the National Science Foundation.

B.M. McCormac and T.A. Seliga (Eds.): Solar-Terrestrial Influences on Weather and Climate. 233-241.

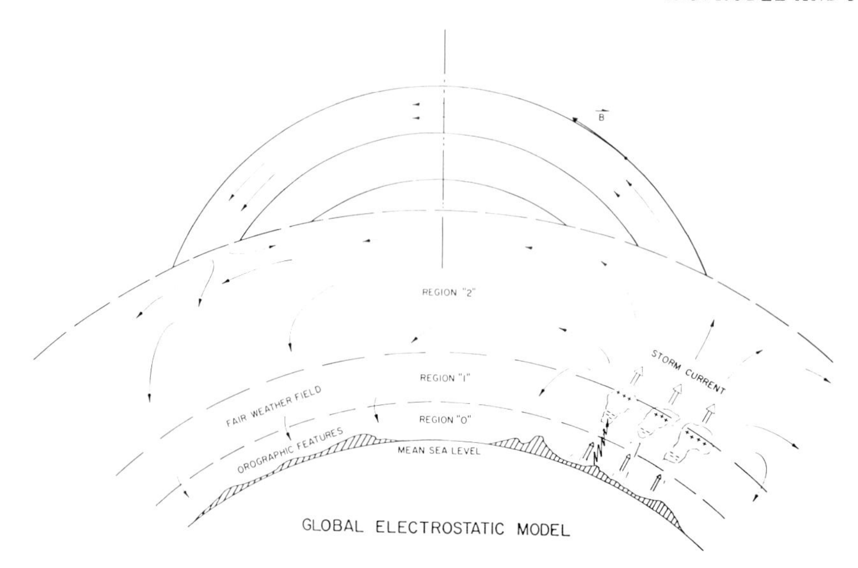

Figure 1. Schematic diagram of the global model of atmospheric electricity.

current generators, each with a positive center at the top of the cloud and a negative center a few kilometers lower than the positive center. In fair-weather regions far away from the storm centers, the distribution of the electrostatic potential above the Earth is determined by the current return from the sources to the Earth's surface. The geometry of the model is based upon an atmosphere broken into four coupled regions, as illustrated in Figure 1. Region 0 represents the lower tropospheric boundary region which has a variable conductivity in the horizontal and vertical; it also includes the Earth's orography. Region 1 represents the upper troposphere below the negative current source region within the thunderstorm and region 2 represents the stratosphere and mesosphere above the positive current source region of the thunderstorm. For regions 0, 1, and 2 it is assumed that the electrical conductivity is isotropic. Region 3 represents the ionosphere and magnetosphere where the conductivity is anisotropic and where magnetically conjugate regions are connected along geomagnetic field lines through the magnetosphere. The mathematical details of the model, the boundary conditions, and the matching conditions between various regions are described by Hays and Roble [16].

The height and latitudinal distribution of electrical conductivity used for the model calculations are shown in Figures 2a and 2b, respectively. The vertical profiles are compared to the measurements of (1) Morita et al. [18], (2) Sagalyn [19], (3) Mozer and Serlin [20], (4) Cole and Pierce [21], and (5) Tran and Polk [22]. We assume that the electrical conductivity in the upper atmosphere is maintained by cosmic ray activity which varies by about a factor of two between the equator and pole [15]. Since the conductivity is maintained primarily by small mobile ions, it varies as the square root of the cosmic ray ion production rate. Also

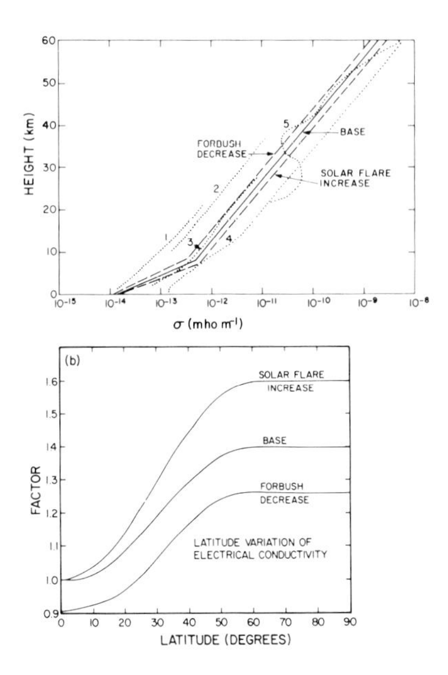

Figure 2. Model electrical conductivity variations with (a) altitude and (b) latitude. The assumed variations for a solar flare increase and subsequent Forbush decrease are also shown.

shown in Figure 3 are the assumed variations of electrical conductivity due to a sudden increase in cosmic ray ionization resulting from a solar flare and a subsequent Forbush decrease. Each thunderstorm is assumed to have a point positive current source at the top of the thundercloud and an equal negative current source at the bottom of the cloud. On a global scale there is no net current source within the model and the thunderstorms act as pumps, recirculating current from the lower to the upper atmosphere. We arbitrarily consider 2000 individual thunderstorms occurring at 1900 UT and randomly distribute the point thunderstorms in latitude and longitude in accordance with the hourly probability of thunderstorm occurrence as defined by Chrichlow et al.[24]. Northern hemisphere summer months are assumed; therefore at the 1900 UT maximum in ionospheric potential, thunderstorm activity occurs over Central America, Florida, and the Rocky Mountains, in addition to the Amazon basin and parts of Central Africa. It is assumed that at 1900 UT, 600, 100, 1000, 280, and 20 thunderstorms are randomly distributed in latitude and longitude in the vicinity of Africa, Asia, Central America, Argentina, and the Alps, respectively. The storm centers are also randomly distributed in the vertical.

3. MODEL SOLUTION

The calculated electric potential along the $\sigma = 7.3 \times 10^{-12}$ mho m^{-1} constant conductivity surface, which occurs at approximately 25 km at the

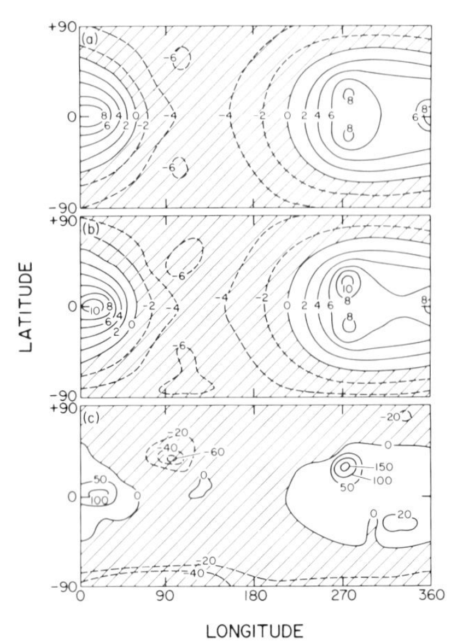

Figure 3. Contours of calculated potential difference $(\Phi-\Phi_\infty)$ along various conductivity surfaces at approximately (a) 105 km, (b) 50 km, and (c) 25 km. All values are in volts when multiplied by 10^3.

equator and slopes downward to 23 km in both polar regions, is shown in Figure 3c. The calculated fair-weather ionospheric potential is Φ_∞ = 291,000 V and the quantity $(\Phi-\Phi_\infty)$ along the constant conductivity surface is plotted. Over the thunderstorm regions of Central America, Africa, and Argentina, the value of $(\Phi-\Phi_\infty)$ is positive and electric current flows upward from the thunderstorm region to ionospheric heights. The maximum value of $(\Phi-\Phi_\infty)$ is 16,000 V whereas minimum values occur over the mountainous fair-weather region, with -5700 V over both Tibet and Antarctica. The calculated value of $(\Phi-\Phi_\infty)$ along the $\sigma = 4.74 \times 10^{-10}$ mho m^{-1} surface near 50 km in the equatorial region is shown in Figure 3b. The potential distribution has spread considerably from the concentrated regions over the thunderstorms at 25 km to a more uniform distribution at 50 km. The calculated value of $(\Phi-\Phi_\infty)$ along the $\sigma_m = 4.54 \times 10^{-6}$ mho m^{-1} conductivity surface, which is the base of the magnetosphere in region 3 of the model at roughly 105 km, is shown in Figure 3a. Φ_∞ is the global mean average along this surface. The maximum value of $(\Phi-\Phi_\infty)$, which occurs over the thunderstorms in Africa, is 975 V and the minimum value is -600 V over the Central Pacific region. For these model calculations it is assumed that the geomagnetic and geographic poles are coincident. At the geomagnetic equator the field lines are horizontal and the vertically directed electric current is restricted from spreading laterally. The maximum potential develops in this region. In Central America, the thunderstorm region is displaced from the geomagnetic equator and therefore currents flow from the storm region along the geomagnetic field line into the conjugate hemisphere. The potential perturbation then attenuates as it penetrates to lower altitudes in the conjugate hemisphere.

The calculated potential ϕ along the constant conductivity surface $\sigma_1 = 4.3 \times 10^{-13}$ mho m^{-1}, which is approximately 8 km in the equatorial regions, is shown in Figure 4a. Under each thunderstorm region the potential is strongly negative. The minimum value of several million volts occurs under the Central American thunderstorm region. In the fair-weather region away from the thunderstorms the calculated potential is positive with respect to the ground, indicating a return current to ground. The calculated potentials at 4 km and 2 km respectively are shown in Figure 4b and 4c. They are similar to the potential calculated on the 8 km surface but at these lower altitudes the distortion effects of the Earth's orographic features become evident. The potential becomes zero when a surface intersects a high mountain that protrudes above that altitude, and the potential becomes greatly distorted when the surface is near an orographic feature such as Antarctica, Greenland, or the Himalayas. On the 2 km surface the outline of continental features begin to appear.

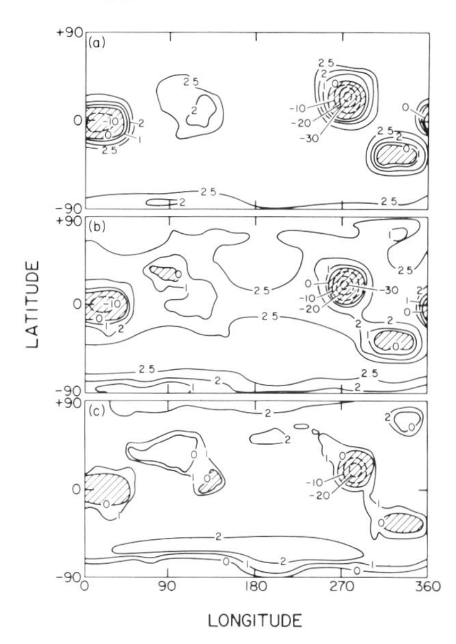

Figure 4. Contours of calculated potential ϕ along various height surfaces at approximately (a) 8 km, (b) 4 km, and (c) 2 km. All values are in volts when multiplied by 10^5.

The calculated potential gradient and the ground current along the Earth's orographic surface are contoured in Figure 5a and 5b respectively. The highest positive electric potential gradient and electric current flow occur in the Himalayas and other high mountainous regions. These occur because of the high conductivity at mountaintop level and the effect of orography in distorting the electric potential and in amplifying the electric field. The current into the ground in the fair-weather regions is balanced by a current from the ground into the thunderstorm regions, leaving no net current flow on a global scale. The

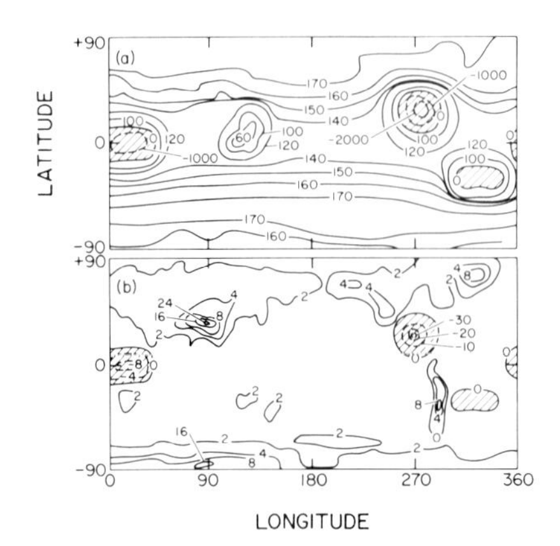

Figure 5. Contours of calculated (a) ground potential gradient V m^{-1} along the earth's orographic surface and (b) ground current, in A m^{-2} when multiplied by 10^{-11}.

electric field at sea level is 134 V m^{-1} at the equator and 173 V m^{-1} in the polar regions, which compares well with the measurements of 120 V m^{-1} over the equator and 155 V m^{-1} at 60° latitude that were made by the Carnegie expedition [15]. Table I compares the parameters of the model with the estimate of global circuit parameters made by Mühleisen [25]. There are considerable uncertainties in these global estimates, yet the comparison of parameters is encouraging.

4. VARIATIONS DURING SOLAR FLARES

During solar flares the intensity of the cosmic ray flux bombarding the Earth changes in a manner that has been discussed in detail by Forbush [23]. For our calculations we assume two phases of the conductivity variation, as shown in Figure 2. During the solar flare there is a sudden increase in the cosmic ray intensity at ground level, which generally lasts from one-half to one hour. The intensity at the ground can increase 5 to 50% during this period depending upon the magnitude and energy spectrum of the particle distribution. This increase is latitudinally dependent, being largest in the polar regions and smallest in the equatorial region. Following the sudden increase in conductivity the Forbush effect sets in and a world-wide decrease in cosmic ray intensity occurs that can last from one to several days.

The major calculated global electrical properties for the solar flare increase and the Forbush decrease, as well as the base 1900 UT case, are given in Table II for comparison. The calculated global electrical resistance decreases from 273 to 247 ohms during the solar flare (cosmic ray intensity increase) and increases to 303 ohms during the Forbush decrease. The calculated ionospheric potential decreases from 291,000 V to 244,000 V during the cosmic ray increase but then increases to

323,300 V during the subsequent Forbush decrease. The total global electric current decreases from 1065 A to 988 A, then increases to 1063 A during the same sequence. The ground electric field at both the equator and the pole is also modulated, decreasing first and later increasing. These calculations indicate that changes in electrical conductivity due to solar flares are capable of altering the electric circuit on a global scale. There is much uncertainty in the actual global representation of conductivity during solar flares and we stress that these calculations were made to be illustrative of a possible solar-terrestrial influence on the electrical structure of the atmosphere. This study is not a critical evaluation of a specific event.

Table I. Comparison of Mühleisen (1977) estimate of global circuit parameters with model calculations at 1900 UT.

	Mühleisen [14]	Model
Number of thunderstorms	1500 - 1800	2000 (specified)
Mean value of current intensity over one thunderstorm cell	0.5 - 1.0 A	0.51 A
Global current	800 - 1800 A	1065A
Ionospheric potential	180 - 400 kV (range) 278 kV (mean)	291 kV
Columnar resistance	$1.3 \times 10^{17}\ \Omega\ m^2$	$1.39 \times 10^{17} \Omega\ m^2$
Total resistance	230 Ω	273

5. GEOMAGNETIC SUBSTORMS

In the Earth's polar region there exists a two-cell magnetospheric plasma convection flow that is associated with a horizontal potential pattern at ionospheric heights. In the polar cap there is a potential drop from the dawn to the dusk side that varies from 50 to 250 kV depending upon geomagnetic activity. Equatorward of the polar cap boundary at approximately 75° magnetic latitude the potential is assumed to recover linearly across the auroral belt to the global average ionospheric potential. As discussed by Park [4] the large horizontal scale components of this pattern map effectively to the ground and cause a variation of the potential gradient at the ground. In our calculations we consider the effects of the Earth's orography and the tilted geographic and geomagnetic poles. Acting alone, the positive and negative potential perturbations at high latitudes to the ionospheric potential at the upper boundary cause a downward and upward current respectively to flow to the ground. Because the model requires that the divergence of the current must be equal to zero, any imbalance in the downward and upward

Table II. Calculated electrical properties of the atmosphere for the 1900 UT base case and the solar flare variations.

	Ionospheric-Potential (V)	Total Current (A)	GROUND POTENTIAL GRADIENT Equator (Vm^{-1})	Pole (Vm^{-1})	Global Resistance (ohms)
Base Case (1900 UT)	291,000	1065	134	173	273
Solar Flare Increase	244,000	988	113	166	247
Forbush Decrease	323,300	1063	148	192	303

current system must be compensated by the global circuit, causing a change in the ionospheric potential. Such an imbalance may occur when the positive/negative current is aligned over Antarctica, where the surface conductivity on the high mountain plateau is large, and when the negative/positive current is aligned over a cloud-covered ocean, where the surface conductivity is low. For certain large geomagnetic substorms with the upper boundary potential pattern properly aligned over the continent and ocean, the current imbalance may require as much as a 5-10% change in the ionospheric potential maintained primarily by thunderstorms. The interaction of other upper atmospheric generators with the lower atmosphere is discussed by Roble and Hays [17].

REFERENCES

1. R. Markson, Pure Appl. Geophys., 84, 161, 1971
2. R. Markson, Nature, 273, 103, 1978.
3. J. R. Herman and R. A. Goldberg, J. Atmos. Terr. Phys. 40, 121, 1978.
4. C. G. Park, J. Geophys. Res., 81, 168, 1976.
5. D. Sartor, Problems of Atmospheric and Space Electricity, S. C. Coroniti, Ed., Elsevier Pub. Co., Amsterdam, 307, 1965.
6. W. E. Cobb, Mon. Weather Rev., 95, 905, 1967
7. R. Reiter, Pure Appl. Geophys., 72, 259, 1969.
8. R. Reiter, Pure Appl. Geophys., 86, 142, 1971
9. R. Reiter, Pure Appl. Geophys., 94, 218, 1972.
10. D. E. Olson, Pure Appl. Geophys., 84, 118, 1971.
11. T. V. Lobodin and N. A. Paramonov, Pure Appl. Geophys., 100, 167, 1972.
12. C. G. Park, Geophys. Res. Lett. 3, 475, 1976.
13. R. Reiter, Electrical Processes in Atmospheres, H. Dolezalek and R. Reiter, Eds., Steinkopff Verlag, Darmstadt, 759, 1977.
14. R. P. Mühleisen, Electrical Processes in Atmospheres, H. Dolezalek and R. Reiter, Eds., Steinkopff Verlag, Darmstadt, 450, 1977.
15. H. Israel, Atmospheric Electricity, Vol. 1 and 2, Natl. Tech. Info. Service, Springfield, VA, 1971.

16. P. B. Hays and R. G. Roble, J. Geophys. Res., submitted, 1978.
17. R. G. Roble and P. B. Hays, J. Geophys. Res., submitted, 1978.
18. Y. Morita, H. Ishirawa, and M. Kanada, J. Geophys. Res., 76, 3431, 1971.
19. R. C. Sagalyn, U. S. Air Force Handbook of Geophysics, McMillan, 1960.
20. F. S. Mozer and R. Serlin, J. Geophys. Res., 74, 4739, 1969.
21. R. K. Cole, Jr. and E. T. Pierce, J. Geophys. Res., 70, 2735, 1965.
22. A. Tran and C. Polk, report to Air Force Cambridge Labs, on contract F-19628-70-C-0090, U. of Rhode Island, Dept. of Electrical Engineering, August 1972.
23. S. E. Forbush, Handbuch der Physik, Geophysik III, J. Bartels, Ed., Springer-Verlag, New York, 1966.
24. W. Q. Chrichlow, R. C. Davis, R. C. Disney, and M. W. Clark, Office of telecommunications/ITS research report 12, Boulder, CO, April 1971.

INFLUENCES OF SOLAR ACTIVITY ON THE ELECTRIC POTENTIAL BETWEEN THE IONOSPHERE AND THE EARTH

Reinhold Reiter
Institute for Atmospheric Environmental Research
of the Fraunhofer Society
D-8100 Garmisch-Partenkirchen
West Germany

ABSTRACT. Since 1954, we recorded the electric field E, air earth current density I, and additional electric and meteorological parameters on a high mountain top station (3 km). In 1960 we were able to show that E and I rise simultaneously and significantly 2 to 4 days after solar flares, but only if recorded on genuine fair weather days above the mixing layer. A carefully elaborated sequence of data is available from 1964 to 1975, i.e., a period of 12 yr. It shows that 1 to 2 days after sector structure boundary passages of the IMF, type -/+, E and I rise quite significantly by 20 to 30% (fair weather, above the mixing layer). During passages of the type +/-, the reaction of E and I is somewhat weaker (15%); the maximum is found precisely on the day of the passage. The in phase variation of E and I above the mixing layer is representative of the electric potential of the ionosphere.

This paper gives a survey of the investigations by our Institute which is relevant to the subject. In 1954 two neighboring mountain peaks recording stations were installed for measuring the electric field E and air Earth current density I as well as additional parameters of atmospheric electricity, meteorology, and aerosol physics. Figure 1 shows the first result [1] of solar terrestrial investigations. It is based on nearly 5 yr of observations performed during increasing and maximum solar activity. Because of the high flare frequency at that time we chose only days on which at least 2 flares occurred with importance ≥ 2. Further it was necessary that genuine fair weather prevailed, that there were no active generators near the station causing electric charges (whirling-up of dust or ice crystals), that the station was above the mixing layer, i.e., that 50 to 70% of the columnar resistance was below the station level [2], and, consequently, the local conductivity of the air was nearly constant. If all these requirements were met then it is possible to obtain globally representative data of E and I. Apart from measurements at high mountain stations, measurements performed in polar zones [3], on oceanic islands [4], and on ships (e.g. the famous Carnegie series) are suitable for this purpose.

B.M. McCormac and T.A. Seliga (Eds.): Solar-Terrestrial Influences on Weather and Climate. 243-251.

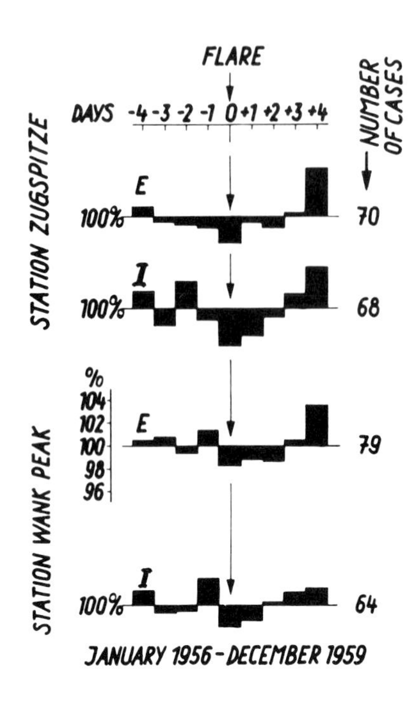

Figure 1. Oldest result of a superimposed epoch analysis of electric field E and air earth current density I around solar H_α-flares. The data were acquired during genuine fair weather days from 1954 - 1959 at the high mountain observatory Zugspitze at 3 km altitude which is above the mixing layer (see [1]).

Figure 1 shows (after [5]) that E and I increase at both stations from the flare day on until the +4 day. Minima or maxima of 4 independent data series of 9 groups each (-4 to +4 day) falling accidentally on the same group (day) means only 0.015% probability. Hence, the result is significant and shows that the electric potential of the ionosphere increases after solar flares. The study was repeated later and led qualitatively to the same result yet with the following differences: The maximum was observed on the second or third day after the flare and the increase of E and I amounted to 30 to 60% (significance level indicated by broken lines). The main difference was the more conspicuous increase of E and I resulting from using only "active" flares between 20°W and 20°E of heliographic coordinates. The former results, depicted in Figures 1 and 2, are in reasonable agreement with those of analogous measurements made by Cobb [4] at Mauna Loa. Inspired by the investigations of Wilcox et al. [6] we finally based our further evaluation of E and I data (homogeneous data from more than 12 yr) on the sector structure boundary passages of the interplanetary magnetic field (Figure 3). The diagrams a and b give the variations of E and I over the total period during -/+ and +/- passages. The type of variation around the key days within the significance range (vertical bars) is congruent

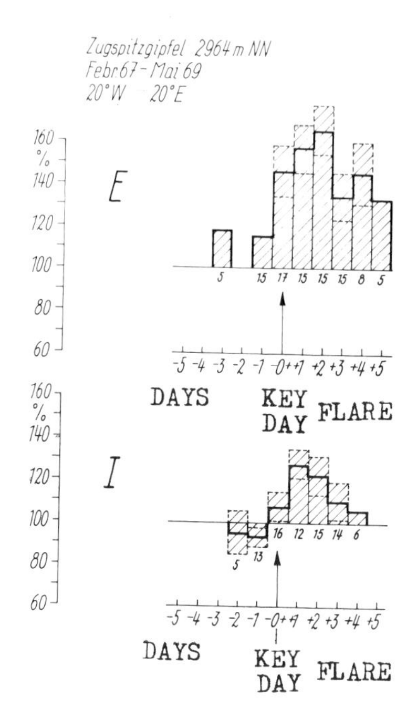

Figure 2. Same as Figure 1, yet the result of a 1967/69 study, solar flares being confined as to their heliographic position. Broken lines denote the significance level. 80 key days (after [5]).

for E and I: Maximum on +1 day after -/+, or precisely on the key day at +/- polarity changes. If the data are restricted to the period of maximum solar activity (77 (80) key days instead of 170) we obtain Figures 3c and 3d. Despite the reduction of the number of cases the amplitude and significance materially increases at -/+ polarity changes (20%). Here we again have used exclusively fair weather days with the station remaining above the mixing layer. Due to the parallelism of E and I this result shows that obviously the electric potential of the ionosphere is influenced by the solar event. (Park [5] found in 17 cases in the Antarctic a decrease of E after polarity changes; since I has not been measured we lack an essential, globally representative criterion. Moreover it is possible that deviating atmospheric-electric conditions prevail near the Pole.) At the present time no definite decision can be made on the causes of these correlations. There is one reasonable explanation and to date there are no arguments against it.

The electric potential of the ionosphere is predominantly maintained by the sum of all thunderstorms on the Earth. Markson [7] found rather early a correlation between sector structure passages and northern United States thunderstorm frequency: After -/+ passages the thunderstorm index slightly increases whereas a minimum is found before +/-

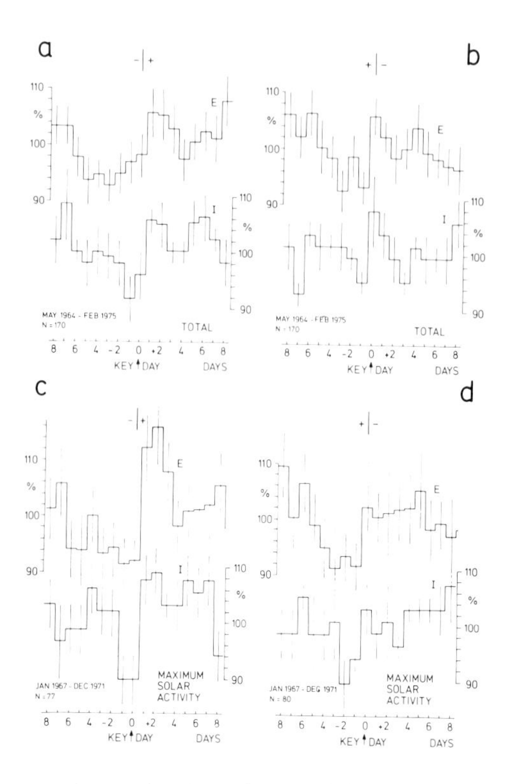

Figure 3. Epoch analysis of E and I around days with sector structure passages of different polarity (+/-, -/+): a and b: total data of 12 yr; c and d: period of max. solar activity 1967 - 1971.

passages. This finding is not at variance with Figures 3c and 3d although such a comparison remains questionable because the data stem from different periods and the thunderstorm detection only holds for a small part of the Earth surface. Nevertheless our earlier investigations have shown that even in an area of only 1000 km diameter the thunderstorm frequency is influenced by solar events. This can be seen in Figure 4 (key days analogous to Figure 2) from the increase in the sferics pulse frequency from the flare day on until +2 day (see Figure 2). This result, which agrees well with an analogous study of Bossolasco [8], is not caused by a solar influence on the radiowave propagation because the reception zone of sferics is limited to maximum of 500 km radius.

The difficulty to directly confirm that first the number, duration, and electric output of thunderstorms on the Earth are increased by solar action and thus, second, the ionospheric potential lies in the fact that thunderstorms, or even more important cloud to surface discharges, are not continuously detected on a global scale. We know the general thunderstorm distribution from the classical era of climatology (Figure 5

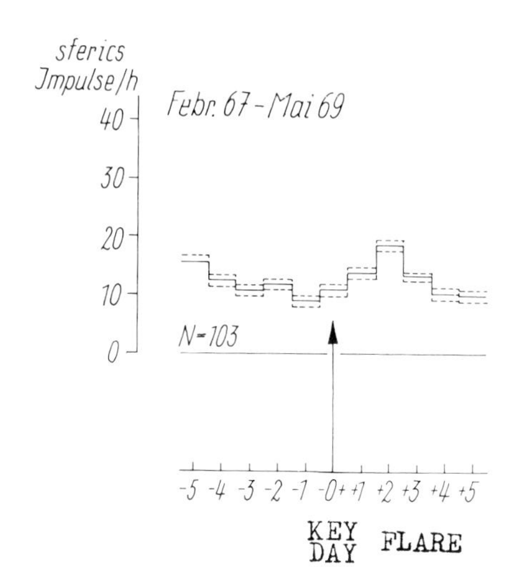

Figure 4. Superimposed epoch analysis of sferics pulses around solar flares, otherwise analogous to Figure 2. 103 key days.

after [9]), and that the most frequent thunderstorms occur in the tropical zones above land. Satellite detection of thunderstorms could confirm this pattern (Figure 6); however, a refined resolution as to place, time, and intensity of thunderstorms around the globe is missing.

We tried to make some progress in this direction by analyzing the atmospheric-electric daily variations (E, I, and L = local conductivity) during extremely favorable weather conditions at the station Zugspitze. Figure 7 gives the time in UT during which thunderstorm development above the continents is most likely to be vigorous (14.00 LT after [9]). At first we averaged the daily course (dotted line) of E, I, and L on those fair weather days prior to and during which no flares have been observed (Figure 8 - just one single example of the analysis, for details see [10]). Then, we averaged the daily variations of E, I, and L immediately after a violent flare (solid curves in Figure 8) during genuine fair weather days. The interval between these two periods should not exceed a few months. In the cases analyzed we found, with congruent and time-constant L-values, not only much higher daily means of E and I after solar flares, but also corresponding peaks in the daily course of E and I which can be correlated as to time with the strongest solar irradiation of various tropical zones (c.f. Figure 7 with Figure 8).

The noted E, I pulsations extending over several days after flares are conspicuous whereas we obtain on flare free fair weather days the pure, global "Carnegie daily variation" (E,I - dotted line in Figure 8).

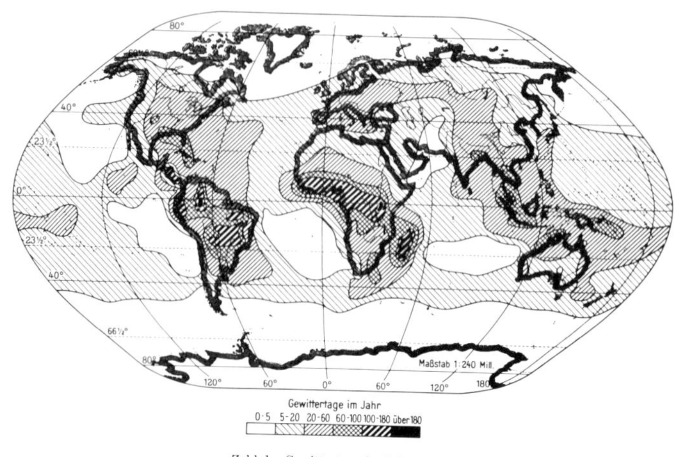

Figure 5. Climatology of thunderstorm frequency on the earth.

Figure 6. Individual case of thunderstorm distribution on the earth detected by satellite.

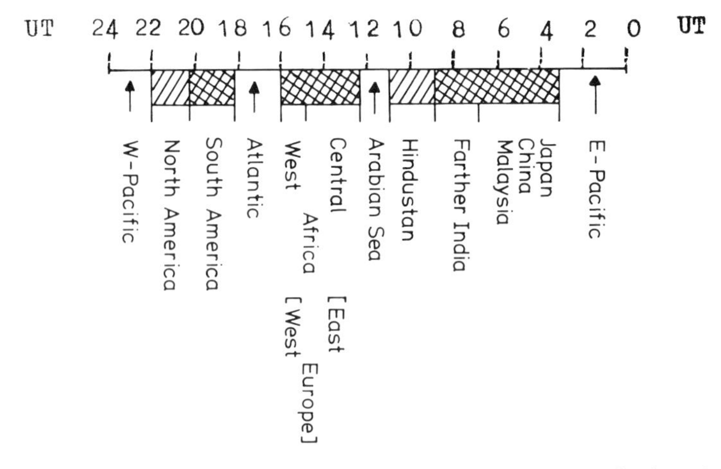

Figure 7. Data in UT with the greatest probability of thunderstorm development above the most important parts of the continents (14.00 local time).

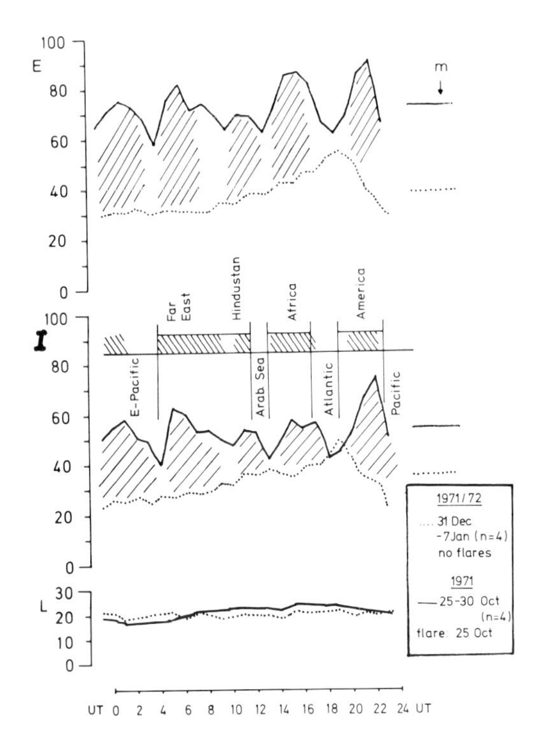

Figure 8. Daily variation of E and I on genuine fair weather days above the mixing layer at the high mountain station (3 km): the one in case when no flares occurred (dotted line) - here one finds the usual "Carnegie" daily variation; the other on days after solar flares (solid line) - the mean values (m) for E and I (right margin) are clearly increased and one finds peaks if the thunderstorm frequency above the afore stated zones was high (c.f. Figure 7).

Although this observation cannot be regarded as a definite proof it is a plausible indication of solar influence on the ionospheric potential via the global thunderstorm frequency. Other investigations (Figure 9) also show that we may assume a meteorological link, e.g., the control of the stratospheric-tropospheric exchange, studied on the tropospheric Be7 content (see other paper of the author) or the vorticity area index as a result of sector structure passages. It was also possible to show that the reliability of weather forecasts [11] in agreement with [12] is controlled by solar events. So it is apparent from Figure 10 that when a stratospheric intrusion is triggered by solar events (days with maximum Be7 concentration in the troposphere) the frequency of bad forecasts rises (Figure 10a) whilst that of the good ones lowers (Figure 10b).

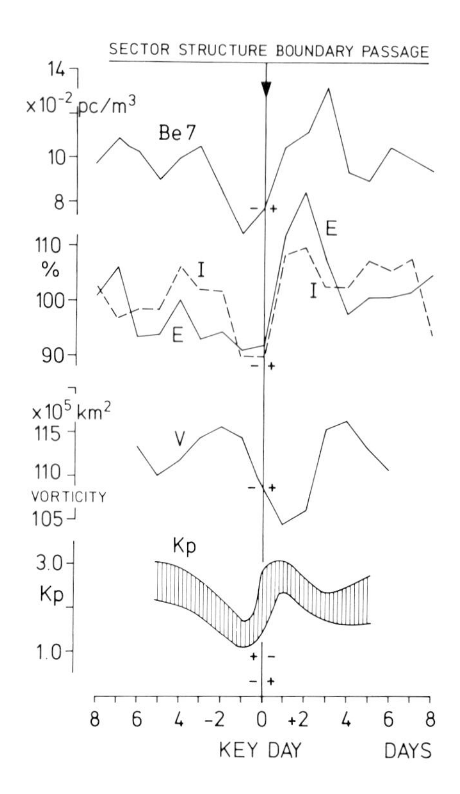

Figure 9. Comparative representation F (after [14]) of superimposed epoch analyses around sector structure passages of the following variables (from top to bottom): Be7-concentration in the troposphere as a measure of the intensity of the stratospheric-tropospheric exchange; E and I; vorticity area index V after [6]; geomagnetic activity Kp after [6].

Multiple publications (see e.g. [13] show that solar events control the state of the lower stratosphere. It is therefore reasonable to postulate solar control of the electric potential [14] of the ionosphere as a mechanism for the meteorological link and make this the subject of further studies.

REFERENCES

1. R. Reiter, Final Report, Contract AF 61 (052)-55, 1960.
2. R. Reiter, Felder, Ströme, Aerosole, Darmstadt, 1964.
3. C. G. Park, Geophys. Res. Letters 3, 475, 1976.
4. W. E. Cobb, Mon. Weath. Rev. 95, 905, 1967.
5. R. Reiter, PAGEOPH 72, 259-267, 1969.
6. J. M. Wilcox, P. H. Scherrer, L. Svalgaard, W. O. Roberts, R. Olson, and R. L. Jenne, J. Atmos. Sci. 31, 581, 1974.

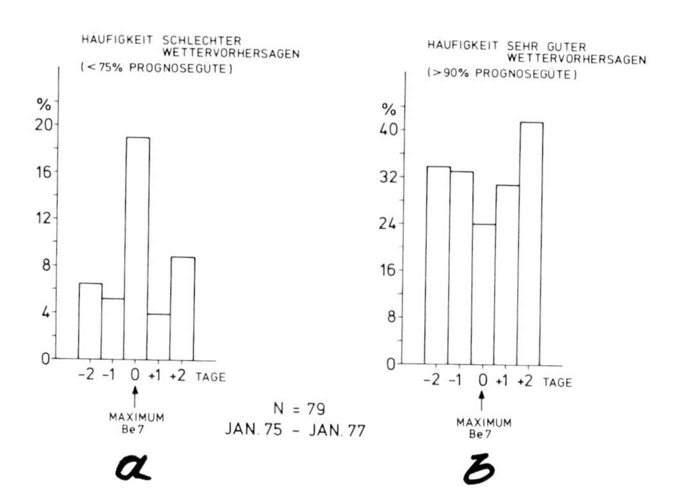

Figure 10. Epoch analysis of the reliability of weather forecasts (a - frequency of bad, b - frequency of good forecasts) around days with maximum concentration of Be7 in the troposphere, i.e., the increased frequency of stratospheric intrusions.

7. R. Markson, PAGEOPH 84, 161, 1971.
8. M. Bossolasco, I. Dagnino, A. Elena, and G. Flocchini, 1st Univ. Navale di Napoli Meteorol. e Oceanogr., 213, 1972.
9. J. Blüthgen, Allgemeine Klimageographie, Berlin, 1964.
10. R. Reiter, Riv. Italiana di Geofisica XXIII, 193, 1974.
11. R. Reiter, IAMAP General Assembly, Seattle, 1977.
12. M. F. Larsen and M. C. Kelley: Geophys. Res. Letters, 4, 337, 1977.
13. J. W. King, Aeronautics and Astronautics, 13, 10, 1975.
14. R. Reiter, J. Atm. and Terr. Phys., 39, 95, 1977.

THUNDERSTORM FREQUENCY AND SOLAR SECTOR BOUNDARIES

Mae D. Lethbridge

Department of Meteorology
The Pennsylvania State University
University Park, PA 16802 U.S.A.

ABSTRACT. With solar sector boundary dates as key days in the superposed epoch method and with thunderstorm frequency over latitude band in Northern United States as the variable, a maximum in thunderstorm frequency was found one day after boundary passage in three independent sets of data. This appears to be statistically significant on the 1% level.

There has been considerable interest the past few years in the interplanetary solar sector structure and, in particular, the passage of solar sector boundaries during which time there are variations in the velocity of the solar wind, its density, the magnitude of the geomagnetic indices, etc.

A possible relationship between solar sector boundary passage and thunderstorm frequency was first suggested by Markson in 1969. He found an increase in thunderstorm frequency during boundary passage when the polarity of the solar field changed from + to - [1]. Subsequent to this, Wilcox et al. [2] showed a decrease in the vorticity area index (VAI), a measure of synoptic scale circulation intensity, one day after passage of the sector boundaries during the winter months.

In the present study we have investigated the statistical relationship between thunderstorm frequency and the solar sector boundaries. An index of thunderstorm frequencies was computed with thunderstorm data compiled from an area of the United States extending from the Atlantic Coast to 102°W longitude and from 30°N to 45°N latitude. There were 102 stations and, initially, 11 years of data — 1966 to 1976. The data were compiled for the entire area and for three latitude bands, 30° to 35°, 35° to 40°, and 40° to 45°. The variable was the cube root of the thunderstorm frequency for each day of the 11 yr in each latitude band and for the entire area. The means of the daily cube roots during the 11 yr were smoothed and subtracted from the cube root of each day to remove any seasonal effect. This gave an index which ranged from approximately -3 to +3.

B.M. McCormac and T.A. Seliga (Eds.): Solar-Terrestrial Influences on Weather and Climate. 253-257.

To ascertain if passage of the solar sector boundaries had an influence on thunderstorm generation, the sector boundary dates given by Svalgaard [3] were used as key days in the superposed epoch method with the thunderstorm index "THDA" as the variable. At first, boundary dates for the entire year over the 11 yr period were used as the key days, totaling 330. Nothing of any significance resulted; the curves seemed to be random.

In the work of Wilcox et al. [2], a relationship could be found between solar sector boundaries and the VAI only for the winter months. Consequently, an analysis was made for the key days occurring from November through March. There was no maximum of any significance for the total area, but for the 40° to 45°N latitude band there was a maximum peak at +1 day which appeared possibly significant. The data were then divided into key days when the polarity changed from + to - and when it changed from - to +.

For both of these independent sets of data, thunderstorm activity appeared to be relatively high one day following sector passage (see Figure 1). However, these peaks are not particularly significant statistically. Other peaks in Figure 1, though not for the same lag in both samples, are almost as large.

The results presented in Figure 1 were used to suggest the working hypothesis, namely that between latitudes 40° to 45°, thunderstorms in the U.S.A. are most likely one day following a sector passage during which the polarity changes from + to -.

Figure 2 shows a test of the hypothesis, based on two additional sets of data, each about 10 yr long. The figure shows that there is a peak one day after sector passage during both periods. Separately, these two peaks are not particularly significant in the statistical sense. But we shall see that they appear to be significant in combination.

The probability of such peaks occurring by chance can be judged by two simple tests. Given the seven days for which data are shown in Figure 2, the probability of a peak occurring by change one day after sector crossing is 1/7. The probability of peaks occurring by chance in two independent samples at the same lag is $1/49 \sim 2\%$. Based on this simple nonparametric test, we then conclude that the hypothesis is significant on the 2% level.

A more accurate test requires normal distribution of means of thunderstorm indices, as well as independence of successive values of the means. Both these conditions are well satisfied.

The mean thunderstorm index for a one-day lag during 1947-56 lies $0.86\sigma_m$ above the general mean (σ_m is the standard deviation of the means). The probability of exceeding $0.86\sigma_m$ by chance is 20.5%. During the 1957-65 period, the corresponding mean lies above the grand mean by $1.8\sigma_m$ with an exceedance probability of 3.6%. Since these two periods

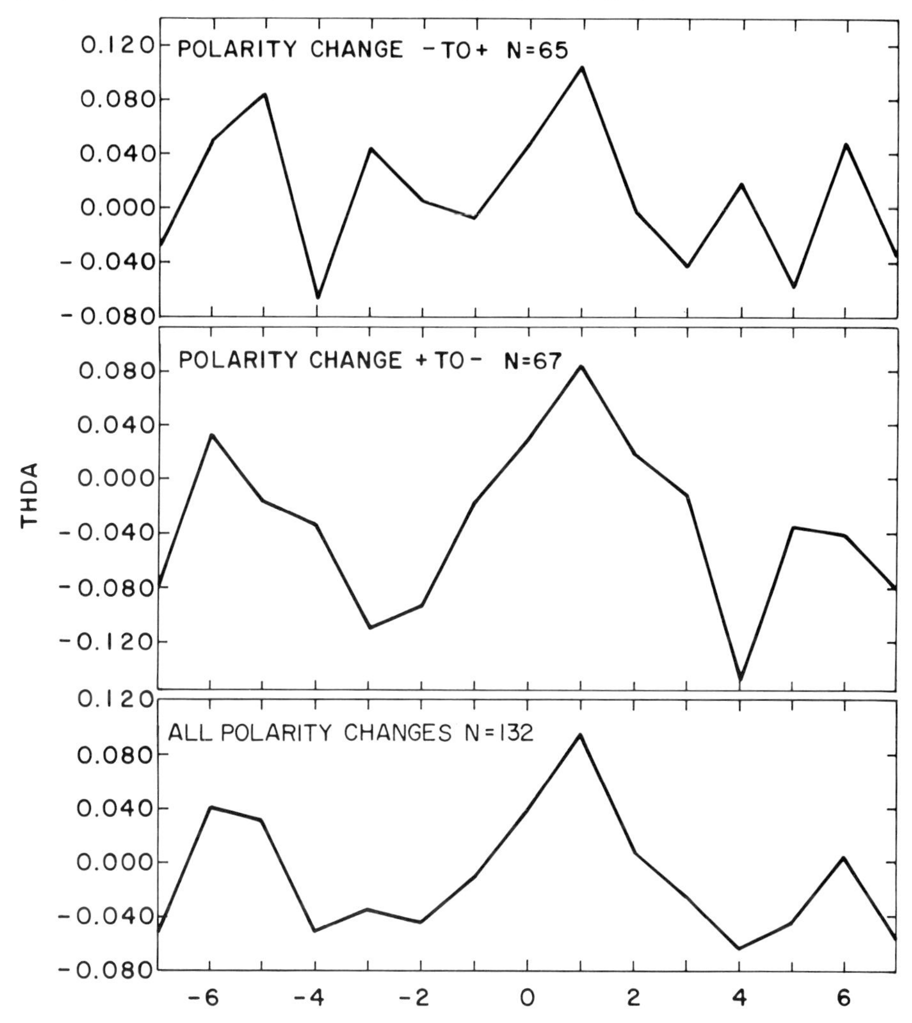

Figure 1. Key Day - Solar Sector Boundary - 1966-1976, November - March, 40° - 45°N latitude.

are independent, the combined probability of the observed peaks for both periods to exceed their representative values by chance is 0.205 x 0.036 = 0.0074 = 0.74%. Thus, the observed occurrence of peaks in thunderstorm activity is significant at better than the 1% level.

Since the same result, namely, a maximum at +1 day was found independently in the two test samples, there is likely to be a relationship between the solar sector structure and thunderstorm frequencies in winter, with latitude bands approximately 40° to 45°, when the solar structure changes from + to - polarity.

The restrictions may possibly be attributed to the following facts:

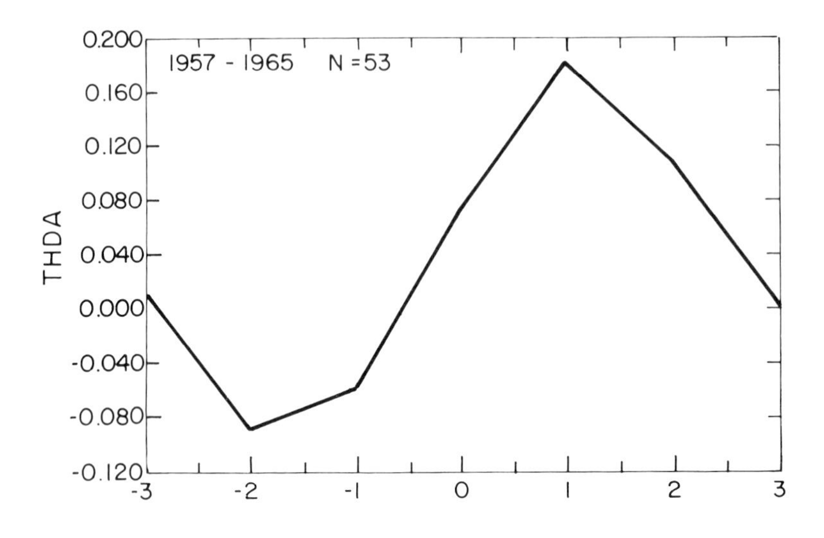

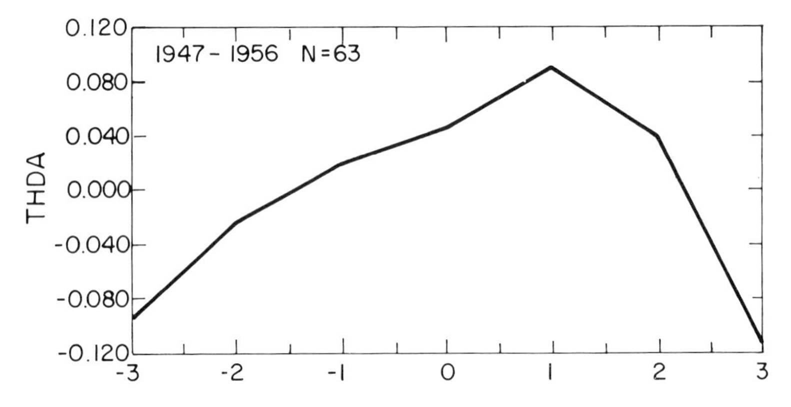

Figure 2. Key Day - Solar Sector Boundary - November - March
Polarity change + to -; 39° - 45° N. latitude.

(1) in winter the angle between the solar wind and the Earth's magnetic dipole is different from the angle in the summer; (2) access of energetic particles to the Earth's polar regions varies according to the polarity; and (3) the northernmost latitude band analyzed is closest to the area of high magnetic latitudes which are affected most strongly by solar particles.

It would be desirable to repeat the above analyses with thunderstorm data from latitudes higher than 45° and, also, with data that would cover all longitudes, especially in the southern hemisphere. The hypothesis would suggest that thunderstorm frequencies would be relatively high when the polarity change there was - to + rather than at + to - as found in the northern hemisphere.

1. R. Markson, Pageoph., 84, 161, 1971.
2. J. Wilcox, P. H. Scherrer, L. Svalgaard, W. O. Roberts, and R. H. Olson, Science, 180, 185, 1973.
3. L. Svalgaard, Interplanetary Sector Structure 1947-1975, SUIPR Rep. No. 648, Inst. for Plasma Res., Stanford Univ., 76 pp., 1976.

THE ROLE OF ATMOSPHERIC ELECTRICITY IN SUN-WEATHER RELATIONSHIPS

M. S. Muir

Physics Dept., Natal University, Durban, 4001, South Africa*

ABSTRACT. A mechanism is examined whereby atmospheric electrical conductivity may control the development of thunderstorms in certain regions. The variation in thunderstorm numbers involves large enough energy changes to account for certain changes in tropospheric circulation.

1. INTRODUCTION

Correlations between solar activity and certain weather parameters are well known. In recent attempts to explain these it has been suggested [1,2] that solar modulation of atmospheric electricity might be the underlying mechanism. Markson [1] has shown that it is possible to account for the observed increase in the electric field following proton flares [3] and solar magnetic sector boundary crossing [4] without need for any change in the thunderstorm generator itself. The observed change in the number of thunderstorms following solar flares reported by Bossolasco et al. [5] and possible solar cycle variations in thunderstorm numbers [6,7] have occasioned two possible explanations by Markson [1]. These involve a direct local effect due to increased ionizing radiations over large cumulus clouds and an indirect global effect due to increased electric fields near developing cumulus clouds.

2. THUNDERSTORMS AND ATMOSPHERIC ELECTRICAL CONDUCTIVITY

It is postulated here that the electrical conductivity of the atmosphere may also form part of the indirect world-wide effect of solar activity on thunderstorm development. It is well known that there is an inverse correlation between ionization (and hence conductivity) and sunspot number. Since the change in ionization with solar cycle varies with geomagnetic latitude [8] it may be expected that the change in conductivity will also vary with latitude, changes being larger toward the poles.

*Present address: Airborne Research Associates, 46 Kendal Common Rd., Weston, MA 02193, USA.

B.M. McCormac and T.A. Seliga (Eds.): Solar-Terrestrial Influences on Weather and Climate. 259-262.

It is suggested here that thunderstorm development (i.e., the build-up of the storm to electrical breakdown potential) depends on prevailing electrical conditions, being enhanced under conditions of low conductivity. Thus, assuming constant solar and thermodynamic conditions, thunderstorms should develop more readily at lower than at higher latitudes (see Figure 1). At a fixed latitude thunderstorm development should be enhanced during sunspot maxima when conductivity is at a minimum. On a shorter time scale a flare or sector boundary crossing which results in a short term decrease in galactic cosmic radiation (such as a Forbush decrease in the case of a flare) will mean that conductivity will be decreased and thunderstorm development enhanced. The change in conductivity produced by ionizing radiation of solar origin is largely restricted to heights above thundercloud tops and this is important in controlling the current flow between cloud and electrosphere, as pointed out by Markson [1]. It might be expected that a given solar flare will have a maximum effect on conductivity at solar maxima when the ambient conductivity is at its minimum value. This effect is apparent in the results of Bossolasco et al. [5] shown in Figure 2. It should be noted that these figures for thunderstorm incidence may be enhanced by a solar cycle variation in VLF propagation conditions, shown by Nestorov [9] to be at their best at solar maxima.

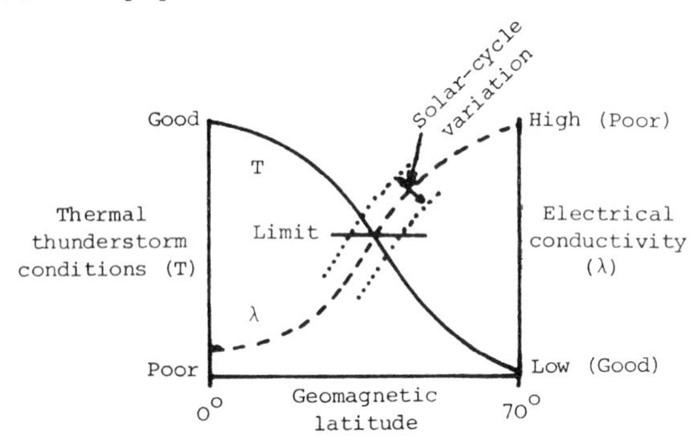

Figure 1. Diagrammatic representation of the relationship between thermodynamic and electric conditions suggested as being necessary for thunderstorm development

In the foregoing the prevailing thermodynamic conditions have been considered constant. In some geographic regions, however, thermal conditions are such that thunderstorm development is highly probable irrespective of conductivity. In other regions the variation of conductivity is such that even at its highest level it is low enough for thunderstorm development to occur. However, when thermal conditions are such that thunderstorm development is possible only if the prevailing conductivity is low enough, a solar cycle signal will be apparent with a positive correlation between thunderstorm number and sunspot number. In this way the correlation between these two parameters found by Stringfellow [6] and the noncorrelation found by Lethbridge [7], shown in Figure 3, may be explained. The suggested situation regarding thermal and electrical conditions is summarized in Figure 1. It will be appreciated that the

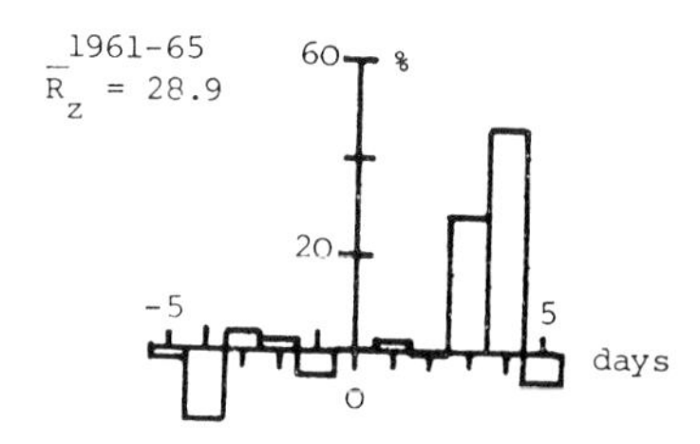

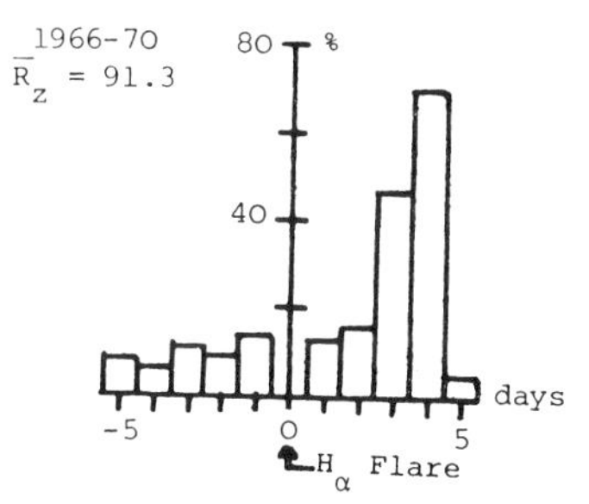

Figure 2. The variation of thunderstorm incidence with the occurrence of a solar flare. After Bossolasco et al [5].

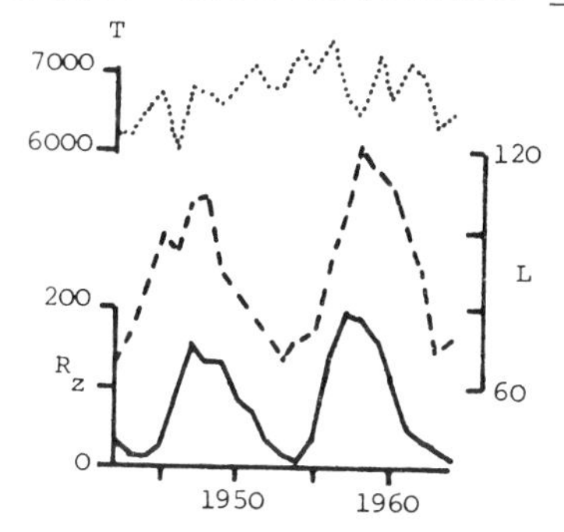

Figure 3. The variation in sunspot number (R_z, solid curve) compared with the variation in thunderstorms in Britain (lightning index L, dashed curve, after Stringfellow [6]) and in North America (T, dotted curve, after Lethbridge [7]).

real situation is a very complex one, and inspection of the configuration of the real geomagnetic field [10] and the global distribution of thunderstorm activity [11] will readily show this.

3. ENERGY CONSIDERATIONS

Willis [12] has calculated the energy involved in changes in vorticity above the 300 mb level to be of the order of 10^{17} J. Braham [13] has estimated that the energy supplied to the atmosphere by a single thunderstorm is of the order of 10^{15} J, while a lower value of 10^{14} J was arrived at by Koppany [14]. According to Chalmers [15] there are probably about 3000 thunderstorms active at any instant on a global scale. This implies an input of energy to the upper levels of the atmosphere of between 3×10^{17} and 3×10^{18} J. Thus a change of between 3 and 30% in the number of active thunderstorms as a result of solar modulation could account for the energy change in the Northern Hemisphere low pressure system. If the supply of energy to such large scale weather events can be satisfactorily accounted for, some of the smaller scale phenomena should be more easily explained. For instance, the correlation between mean sunspot number and sea-surface temperature found by Muir [16] was accounted for by Colebrook [17] in terms of direct heat exchange with the atmosphere and stems from changes within the atmospheric circulation over the whole North Atlantic.

The mechanism proposed here is speculative but such speculation has the merit of defining certain areas which could profitably be investigated. In this case these include the role of the background conductivity in the development of the thunderstorm, the influence of thunderstorms on atmospheric circulation and the phase relationships between the various correlating parameters.

REFERENCES

1. R. Markson, Nature, 273, 103, 1978.
2. M. S. Muir, S. Afr. J. Phys., in press 1978.
3. R. Reiter, Pure Appl. Geophys., 86, 142, 1971.
4. R. Reiter, Naturwiss., 63, 192, 1976.
5. M. Bossolasco, I. Dagnino, A. Elena and G. Flocchini, Instituto Universitario Navale di Napoli, Meteorol. Oceanog., 1, 213, 1972.
6. M. F. Stringfellow, Nature, 249, 332, 1974.
7. M. D. Lethbridge, Quoted by R. Markson, Pure Appl. Geophys., 84, 161, 1971.
8. R. D. Hake, E. T. Pierce, and W. Viezee, Stratospheric Electricity p. 79, Stanford Research Institute, Menlo Park, California, 1973.
9. G. T. Nestorov, J. Atmos. Terrest. Phys., 39, 741, 1977.
10. J. C. Cain and S. J. Hendricks, NASA Technical Note, D-4527, 1968.
11. H. Israel, Atmospheric electricity vol. II Israel Program for Scientific Translations, Jerusalem, Figure 340, 1973.
12. D. Willis, J. Atmos. Terrest. Phys., 38, 685, 1976.
13. R. R. Braham, J. Meteorol., 9, 227, 1952.
14. G. Koppany, Meteorol. Mag., 104, 302, 1975.
15. J. A. Chalmers, Atmospheric electricity 2nd ed., Pergamon Press, Oxford, p 307, 1967.
16. M. S. Muir, Nature, 266, 475, 1977.
17. J. M. Colebrook, Nature, 266, 476, 1977.

ACKNOWLEDGMENT is made to the South African National Council for Aeronomy, Geomagnetism and Space Science for financial assistance.

POSSIBLE MECHANISMS OF SOLAR ACTIVITY - WEATHER EFFECTS INVOLVING PLANETARY WAVES

H. Volland
Radioastronomical Institute, University of Bonn
5300 Bonn, W. Germany

ABSTRACT. Several types of short term and long term effects of solar activity on the weather have been recently discussed in the literature in relation to sun-spots, sector structure of the interplanetary magnetic field, and solar flares. In this paper, I want to investigate the possibility that ultra-long waves may be involved in the coupling process between active areas on the Sun and weather changes within the troposphere. The most likely coupling link in this respect is enhanced electromagnetic radiation from active longitudes on the Sun which penetrates at least down to stratospheric altitudes. That radiation may change the stratospheric wind and temperature profile. The ultra-long-wave geopotential field in the troposphere is sensitive to these changes so that an initial middle atmospheric effect may be detectable in the planetary wave propagation within the troposphere. Or, the enhanced solar radiation absorbed within the stratosphere may change the albedo so that the solar radiation reaching the ground may generate directly tropospheric planetary waves. The expected heat input and planetary wave structures and magnitudes are estimated for various time scales, in particular for waves with periods of 27 days.

1. INTRODUCTION

Several types of short term effects [14,24] and long term effects [6,18] of solar activity on the weather have been recently discussed in the literature in relation to sunspots, sector structure of the interplanetary magnetic field, and solar flares. To date, there is no general agreement about the reality of these effects because the apparently very small effects are difficult to reproduce. Moreover, plausible mechanisms which link events on the Sun, the interplanetary medium, or the upper atmosphere with the troposphere are missing. Any direct energetic coupling from the upper into the lower atmosphere is highly improbable for energetic reasons [25]. Therefore, Herman and Goldberg [12] and Markson [17] hypothesize that the solar activity may influence the electric conductivity of the lower and middle atmosphere and thus may indirectly be involved in the triggering of thunderstorms. Such mechanism should have short time scales and thus might explain short term effects of the order of days.

B.M. McCormac and T.A. Seliga (Eds.): Solar-Terrestrial Influences on Weather and Climate. 263-274.

I want to review another mechanism which may be responsible for long term effects with time scales of one month or longer in which planetary waves are involved. It is well known that solar EUV- and UV radiation varies with the 11 yr solar cycle. Moreover, long-living active longitudes on the Sun corotate with the Sun and are related to the M-regions which are responsible for recurrent geomagnetic activity [19]. Small variations of the solar constant associated with the active longitudes are not implausible. In fact, Foukal et al.[8] claim that a 27-day variation in the solar constant exists on the ground with amplitudes of the order 0.1%. They attribute this variation to changes in the O_3 concentration induced by variations of the solar UV facular radiation. Solar cycle variations in the O_3 content have been reported by Paetzold [20], and model calculations of the effect of solar UV variations on the stratospheric structure and composition show that the O_3 column sum may increase 14% from solar minimum to solar maximum, while the temperature may increase by 12 K [3].

A 27-day periodicity in the solar radiation reaching the ground may generate forced planetary waves within the troposphere [22]. Eleven yr cycle variations of the stratospheric structure and composition may influence ultra-long-wave propagation within the troposphere [1,9].

In this paper I will estimate the heat input due to a small 27 day variation in the solar constant and the corresponding response of the atmosphere. Furthermore, I want to study the influence of the stratospheric structure on planetary wave propagation.

2. HEAT INPUT DUE TO A 27 DAY PERIODICITY OF THE SOLAR CONSTANT

The atmosphere and the Earth's surface absorb the solar short wave radiation and reradiate in the infrared. During periodic changes of that radiation of the order of one month, the zonal long wave reradiation cooling is unaffected [7]. The solar heat stored within the ground is transported into the lowest atmospheric regions by eddies of all scales. One can parameterize that turbulence transport by an eddy coefficient which is of the order of $\nu \simeq 15\ \mathrm{m\ s^{-1}}$ [15]. The diffusion equation governing eddy transport is [11]

$$\frac{\partial n}{\partial t} - \nu \frac{\partial^2 n}{\partial z^2} = \Delta Q_R \exp(-\omega t) \tag{1}$$

with n the enthalpy, t the time, z the height, ΔQ_R the direct short wave solar heating, and $\omega = 2\pi/27$ days $= 2.7 \times 10^{-6} \mathrm{s}^{-1}$ the frequency of the forcing function. The solution of Equation (1) is

$$n(z,t) = \{\Delta Q_R + (\Delta Q_{Ro} - \Delta Q_R) \exp\{-(1-i)z/\tilde{H}\}\} \exp(-i\omega t)/(-i\omega) \tag{2}$$

where ΔQ_{Ro} is the boundary value at the ground, and $\tilde{H} = (2\nu/w)^{1/2} \simeq 3.5$ km a positive scale height. Equation (2) describes two different processes. The first has constant amplitude ΔQ_R in the vertical The second represents an upward traveling heat wave with rapidly exponentially decaying amplitude driven by vertical diffusion from the ground.

If one parameterizes the fluctuating boundary value ΔQ_{Ro} in analogy to the mean boundary value, one obtains $\Delta Q_{Ro} \simeq 10^{-4}$ W kg^{-1} and $\Delta Q_R \simeq 10^{-5}$ W kg^{-1} if the amplitude of the forcing function is 10^{-3} of the solar constant [11]. Evidently, the vertical diffusion dominantes the short wave solar heating within the troposphere. Consequently, ΔQ_R can be neglected.

The heat input amplitude in Equation (2) represents the global mean. Its geographic distribution is determined by the zenith angle of the Sun and by the land-ocean contrast. Both effects are responsible for a standing wave pattern depending on latitude ϕ, longitude λ, and declination of the sun δ. That pattern can be displayed in terms of a Fourier series f $(\delta, \lambda, \delta)$ in λ. The coefficients depend on ϕ and δ. Kertz [13] has calculated such coefficients by implying that the temperature variation over the oceans is zero. We thus have the forcing function

$$Q = f \frac{\partial n}{\partial t} =$$

$$\Delta Q_{Ro} \exp\{(1-\varepsilon)z/(2H) - i\omega t\} \sum_{m=o} \{a^m(\phi,\delta) \cos m\lambda + b^m(\phi,\delta) \sin m\lambda\} \quad (3)$$

as the driving mechanism to generate planetary waves. Here I introduce the term

$$\varepsilon = \varepsilon_r - i\varepsilon_i \text{ sign } (\omega) \text{ with } \varepsilon_i = 2H/\tilde{H} \simeq 4; \ \varepsilon_r = 1 + \varepsilon_i \simeq 5. \quad (4)$$

H = 7.5 km is the mean pressure scale height of the troposphere and $\tilde{H}$ is from Equation (2).

I want to apply that driving force to the linear tidal theory. This requires a specification of the prevailing zonal wind. As the simplest approach I assume superrotation of the atmosphere:

$$U = U_o \cos \phi \quad (5)$$

with $U_o \simeq 10$ to 20 m s^{-1} in the troposphere and $U_o \gtrsim 20$ m s^{-1} in the stratosphere. This implies the neglect of the equatorial easterlies. However, the structure of the waves with which we are concerned is of the Rossby-Haurwitz type with pronounced amplitudes at higher latitudes and large horizontal and vertical scales, so that the real wind configuration is not expected to significantly alter the results.

3. THE GENERATION OF PLANETARY WAVES

Linear tidal theory predicts planetary waves forced by an external heat source with the following pressure and vertical wind amplitudes [23]:

$$\left.\begin{matrix} p \\ \\ w \end{matrix}\right\} = \sum \left\{\begin{matrix} p_n^m (z,\omega_D) \\ \\ w_n^m (z,\omega_D) \end{matrix}\right\} \Theta_n^m (\phi,\omega_D) \exp \{ i(m\lambda - \omega t)\} \tag{6}$$

where $\Theta_n^m (\phi,\omega_D)$ are the Hough functions depending on latitude and a Doppler-shifted frequency

$$\omega_D = \omega - mU/a \tag{7}$$

with a the Earth's radius and U_o from Equation (5). ω is the frequency of the driving force, and m and n are zonal and meridional wave numbers. With the convention $m \geq 0$, the waves propagate to the east (west) if the frequency ω is positive (negative.

The Hough functions can be determined from Laplace's tidal Equation [16]. The height structure functions p_n^m and w_n^m in Equation (6) are determined from the coupled differential equations

$$\begin{aligned} d\bar{p}_n^m/dz &= - i\bar{w}_n^m/(Hh_n^m) + (1/H - 1/h_n^m)\, \bar{p}_n^m \\ d\bar{w}_n^m/dz &= \bar{w}_n^m/h_n^m - i\,(H/h_n^m - 1/\gamma)\bar{p}_n^m + (\gamma-1)\, Q_n^m/(gH\gamma\omega_D) \end{aligned} \tag{8}$$

with $\bar{w}_n^m = w_n^m/\omega_D$, $\bar{p}_n^m = \{p_n^m + i\bar{w}_n^m\, dp_o/dz\}/p_o$, γ the ratio of the specific heats, g the Earth's acceleration and p_o the pressure of the basic state. Q_n^m is the coefficient of the heat input (Equation (3)) developed into a series of the Hough functions. Evidently, the standing wave pattern of Q in Equation (3) generates eastward and westward propagating waves. Because of the land-ocean distribution, waves of wave number m = 2 are a significant part of that wave pattern.

The equivalent depths h_n^m which are eigenvalues of the solution of Laplace's equation are shown in Figure 1 for wave number m = 2 as function of a normalized frequency ω/Ω ($\Omega = 2\pi/1$ day $= 7.29 \times 10^{-5}\ s^{-1}$ is the frequency of the Earth's rotation). The eigenvalues h_n^m are associated with two classes of waves: waves of class I or gravitational waves and waves of class II or rotational waves. The class II waves with positive h_n^m and negative frequencies are internal waves of the Rossby-Haurwitz type. Their amplitudes peak at middle and higher latitudes. Figure 2 shows the meridional structure of the Rossby-Haurwitz waves of wave number m = 2 (the normalized Hough functions Θ_n^m from Equation (6)) versus latitude. These structures are very similar to the structures of the internal class II waves. The class II waves with negative h_n^m in Figure 1 are external or evanescent waves which cannot transport wave energy vertically. The internal class I waves with periods greater than about 10 days are concentrated in the equatorial regions. Moreover, they have vertical wavelengths smaller than about

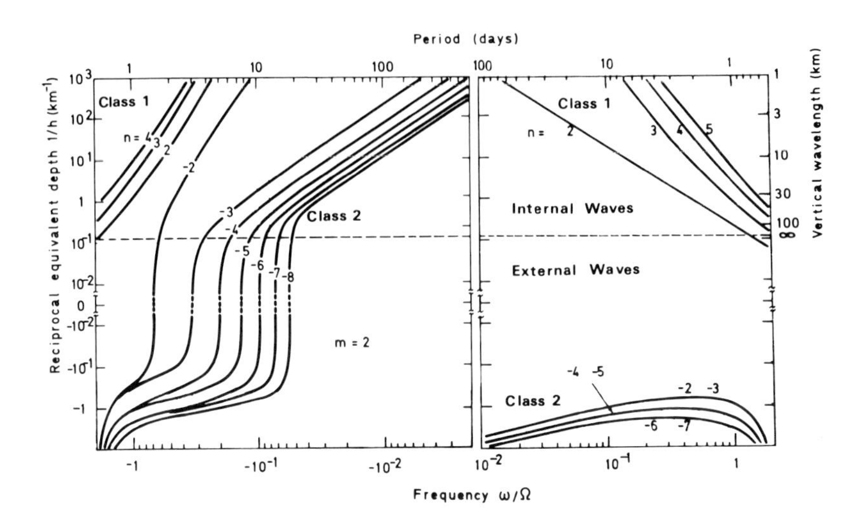

Figure 1 Reciprocal equivalent depths versus normalized frequency for planetary waves of zonal wave number m = 2. Scale on the right hand side gives vertical wavelengths. Waves with positive n: class I waves or gravitational waves; with negative n: class II waves or rotational waves. The horizontal dashed line separates the internal waves with finite vertical wavelength from the external waves with infinitely long wave-length. Note that the logarithmic ordinate includes positive and negative 1/h.

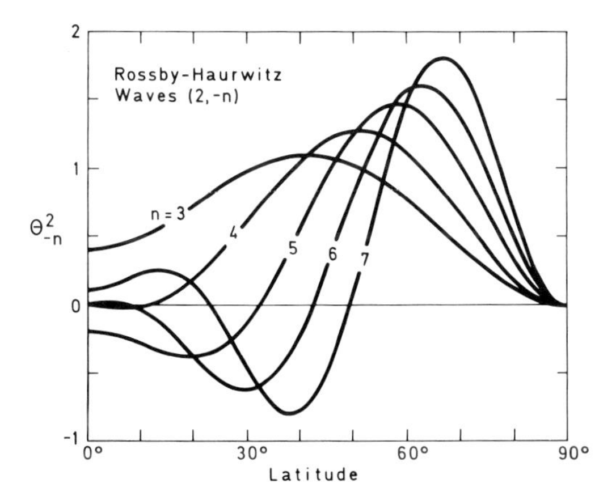

Figure 2 Normalized Hough functions of Rossby-Haurwitz waves of zonal wave number m = 2 versus latitude. Waves with odd numbers m-n are symmetric with respect to the equator, waves with even numbers are antisymmetric.

1 km and therefore are sensitive to destructive interference. Hence, extratropical wave propagation is dominated by the internal class II waves with negative Doppler-shifted frequencies.

In the case of wave number m = 2 for stationary waves (ω = 0), the Doppler-shifted frequency is negative if the prevailing wind is westerly and of amplitude $U_o < 70$ m s^{-1}. This condition is consistent with Charney and Drazin's theorem that vertical propagation of extratropical ultra-long waves can occur only in the presence of westerly winds weaker than a critical value [2].

The height structure functions of these waves can be calculated from Equation (8) by standard methods. The boundary condition at the Earth's surface is

$$w_n^m(0) = imU_o L_n^m/a \tag{9}$$

in the case of orographic forcing where L_n^m is the (m,n) component of the orographic height above sea level, or $w_n^m(0) = 0$ in the case of thermal forcing. The upper boundary condition is Sommerfeld's radiation condition at z_n. Here it is assumed that above z_n the wave propagates upward without reflection.

4. SIMPLE ANALYTIC SOLUTIONS

If one simulates the atmosphere by isothermal slabs with constant prevailing wind in each slab, Equation (8) can be solved analytically in terms of upgoing (a_i) and downgoing (b_i) characteristic waves within the i-th slab:

$$\begin{aligned} a_i &= a_{io}\exp\{(1-i\alpha_i)(z-z_i)/(2H_i)\} + Q_o G_{ai}\exp\{(1-\varepsilon)z/(2H_i)\} \\ b_i &= b_{io}\exp\{(1+i\alpha_i)(z-z_i)/(2H_i)\} + Q_o G_{bi}\exp\{(1-\varepsilon)z/(2H_i)\} \end{aligned} \tag{10}$$

with $\alpha_i = \{4(\gamma-1)H_i/(\gamma h_i)-1\}^{1/2}\,\mathrm{sign}\,(\omega)$

$$\left.\begin{matrix} G_{ai} \\ \\ G_{bi} \end{matrix}\right\} = \frac{(\gamma-1)\,(1\pm i\alpha_i-2H_i/h_i)}{i\gamma g\alpha_i\omega_{Di}\,(i\alpha_i\mp\varepsilon)} \quad .$$

The first terms on the right hand side in Equation (10) are free waves (or the solution of the homogeneous Equation (8)). The second terms are waves generated by the heat source Q (or the particular solution of Equation (8)). The characteristic waves are related to the physical wave parameters by

$$\bar{w}_i = a_i + b_i; \qquad \bar{p}_i = F_{ai}a_i + F_{bi}b_i \tag{11}$$

with

$$\left.\begin{matrix} F_{ai} \\ \\ F_{bi} \end{matrix}\right\} = \frac{2H_i/h_i - 1 \pm i\alpha_i}{2iH_i \; (H_i/h_i - 1/\gamma}$$

In Equations (10) and (11) I have dropped the indices (m,n) for convenience.

The boundary conditions are now

$$a_{1o} + b_{1o} + \overline{W}_1 Q_o = L; \quad b_{no} = 0 \tag{12}$$

where for abbreviation

$$\overline{W}_1 = G_{a1} + G_{b1}; \qquad \overline{P}_1 = F_{a1}G_{a1} + F_{b1}G_{b1} \quad .$$

Upgoing and downgoing waves exist in each slab due to partial reflection at the interfaces. The solution becomes particularly simple in the case of only one slab (isothermal atmosphere). Then

$$a_{1o} = L - \overline{W}_1 Q_o; \; b_{1o} = 0 \tag{13}$$

and

$$\overline{w} = Q_o \overline{W}_1 \; (\exp \{(1-\varepsilon) \; z/(2H)\} - \exp \{(1-i\alpha) \; z/(2H)\}) + L \exp \{(1-i\alpha) \; z/(2H)\}$$

$$\overline{p} = Q_o \; (\overline{P}_1 \exp \{(1-\varepsilon)z/(2H)\} - F_a \overline{W}_1 \exp\{ (1-i\alpha) \; z/(2H)\}) + F_a \; L \exp \{(1-i\alpha) \; z/(2H)\} \; . \tag{14}$$

These solutions show that waves generated by thermal forcing consist of two components: one component which is directly generated by the heat source and which decreases exponentially with height if Real $(\varepsilon) > 1$, and a second component which is caused by the thermally driven wave reflected at the Earth's surface from where it propagates upward as a free wave. In the case of internal waves (α real), that wave component increases exponentially upward till it reaches altitudes where it becomes dissipated. Both wave components have comparable magnitudes within the lower troposphere and therefore interfere with each other. The observed minimum in the pressure amplitude of stationary planetary waves of wave numbers one and two near 3 km altitude [5] may be such interference minimum.

On the other hand, waves generated by orographic forcing (the third terms in Equation (14))increase exponentially with height, if α is real.

If one simulates the atmosphere by two or more slabs, the upgoing waves a_i are reflected at the interfaces and give rise to downgoing components b_i. The interference between a_i and b_i within the troposphere modulates the height structure of Equation (14). Since the vertical wavelength of the internal planetary waves is of the order of the pressure scale height or larger (see Figure 1), and since the slab method gives useful results if the slab thickness is smaller than about one quarter of the wavelength, the troposphere can already be simulated by a few slabs.

5. NUMERICAL RESULTS

For the purpose of an order of magnitude estimate, it is sufficient to simulate the atmosphere by one or two isothermal slabs. In order to study a 0.1% variability of the solar heat input with a period of 27 days, I used an isothermal atmosphere of scale height H = 7.5 km. The heat input profile was taken from Equations (3) and (4) with $\Delta Q_{Ro} = Q_o = 10^{-4}$ W kg^{-1} and $\tilde{H}$ = 3.5 km. I consider waves of wave number m = 2. For prevailing wind amplitudes $9 < U_o < 80$ m s^{-1}, the Doppler-shifted frequencies of these waves are $\omega_D < 0$ and within the range of the internal class II waves in Figure 1. As a typical example, I selected the wave (2, -4).

Figure 3 shows the height structure function versus latitude of the pressure amplitude in geopotential meters ($\Delta h \simeq - H\Delta p/p_o = -Hp_{-4}^{2}/p_o$). Curves 1 and 2 belong to eastward traveling waves (ω/Ω = 1/27). The prevailing wind amplitude is U_o = 10 m s^{-1} for curve 1, and 15 m s^{-1} for curve 2. Curve 3 is for a westward traveling wave (ω/Ω = -1/27) and for U_o = 10 m/s. The corresponding vertical wavelengths of the free waves ($\lambda_z = 4\pi H/\mathrm{Real}\ (\alpha)$) are 2.3,10.5, and 32 km for curves 1, 2, and 3, respectively.

We notice that with increasing vertical wavelength, the magnitudes decrease and the interference minimum shifts to greater heights. Thus the eastward traveling waves dominate. They have magnitudes of the order of 0.5 gpm.

In order to show the influence of the heat input profile, I used for curve 4 in Figure 3 the same data as for curve 3, except for a constant heat input per unit mass ($\tilde{H} = \infty$) and the same total heat input per column ($\bar{Q}_o = Q_o\tilde{H}/(H + \tilde{H})$ = 3.2 x 10^{-5} W kg^{-1}). Evidently, as one expects, the magnitude becomes larger at greater heights.

In order to estimate the influence of the stratosphere on tropospheric ultra-long wave propagation, I applied a two-layer model with an isothermal troposphere with scale height H_1 = 7.5 km and prevailing wind U_{1o} = 10 m s^{-1}, and an isothermal stratosphere with parameters H_2 and U_{2o}. The tropopause is situated at z_{tr}. I first consider thermally forcing of the wave (2, -4) with frequency zero (or infinite period) and use $Q_o = 10^{-2}$ W kg^{-1} [4] and the heat profile of Equations (3) and (4). The parameters used in the model calculations are given in the Table in the figure caption of Figure 4. Curves 1 and 4 differing

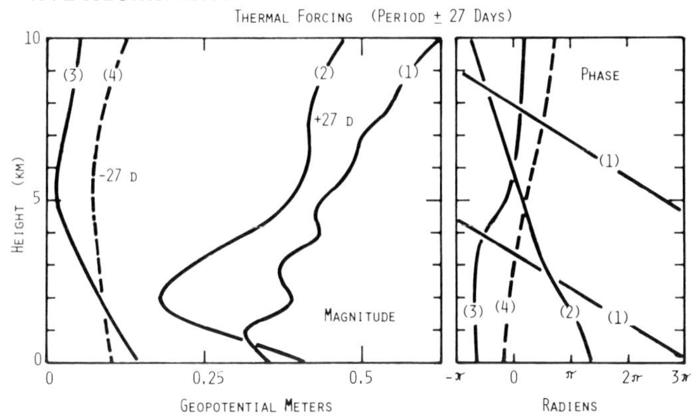

Figure 3. Thermal forcing of wave (2, -4) of period of 27 days. Magnitude (in geopotential meters) and phase (in radians) of height structure function p^2_{-4} versus altitude within an isothermal atmosphere of scale height H = 7.5 km and prevailing zonal wind U_o = 10 m s^{-1} (curves 1,3,4), and 15 m s^{-1} (curve 2). Heat input at the ground Q = 10^{-4} W kg^{-1} and scale height of heat input $\hat{H}$ = 3.5 km (curves 1,2,3), and Q_o = 3.2 x 10^{-5} W kg^{-1} and $\hat{H}$ = ∞(curve 4).

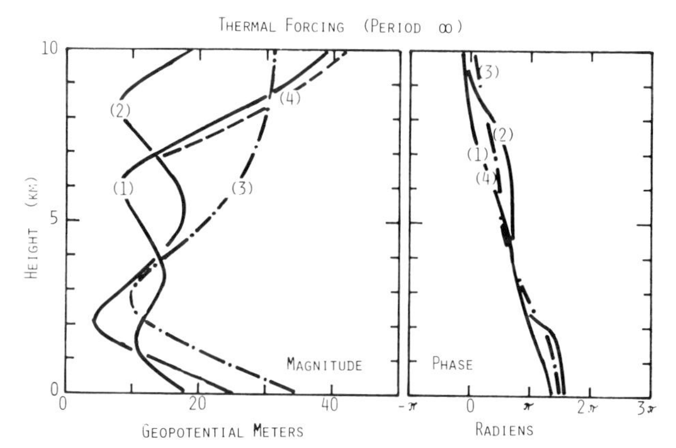

Figure 4. Thermally driven wave (2, -4) of frequency 0 (period ∞). Magnitude (in geopotential meters) and phase (in radians) of height structure function versus altitude within a two-layer atmospheric model. Heat input at the ground: Q_o = 10^{-2} W kg^{-1}. The following parameters have been used (z_{tr} in km = height of tropopause; H_1,H_2 in km = scale heights of troposphere, stratosphere; U_{10}, U_{20} in m s^{-1} = prevailing zonal winds in troposphere, stratosphere; λ_1,λ_2 in km = vertical wavelengths in troposphere, stratosphere):

curve	z_{tr}	H_1	H_2	U_{10}	U_{20}	λ_1	λ_2
1	10	7.5	6.5	10	20	16.5	32.5
2	12	7.5	6.5	10	20	16.5	32.5
3	10	7.5	7.5	10	30	16.5	16.5
4	10	7.5	6.5	10	30	16.5	54.8

by the prevailing winds in the two slabs are very similar because the reflection factor at z_{tr} has changed only from 0.6 to 0.7 in magnitude and 10° in phase. However, changing the height of the tropopause from 10 to 12 km (curve 2) gives significant differences because now the interference pattern has drastically changed.

For comparison, curve 3 is calculated for an isothermal atmosphere showing the same order of magnitude.

Orographic forcing of the (2, -4) wave is simulated in Figure 5. Here again a two slab model has been used for curves 1 and 2 with the same parameters as for curves 1 and 2 in Figure 4. Curves 1 and 2 differing by the height of the tropopause show significant differences due to the changed interference pattern between upgoing and downgoing free waves. Curve 3 calculated for an isothermal atmosphere with H = 7.5 km and U_0 = 10 m s^{-1} has greater magnitude than curves 1 and 2, indicating that the reflection at the tropopause leads to destructive interference. The magnitudes of the order of 100 gpm are larger by a factor of about three as compared with the thermally forced waves.

6. DISCUSSION AND CONCLUSION

The results in Figures 4 and 5 are in basic agreement with more sophisticated calculations of Bates [1] and Shutts [21]. (Note, however, that Shutts used a heat input at the ground of 0.1 W kg^{-1} so that his thermally driven amplitude is greater by a factor of ten as compared with the values in Figure 4.) They indicate that reflection at the tropopause of thermally and orographically forced upward propagating ultra-long waves is significant. That reflection gives rise to interference patterns within the troposphere which are sensitive mainly to changes in the height of the tropopause. Since ultra-long waves of lowest wave number are known to account for a large fraction of the total poleward transport of heat in northern mid-latitude winter [1], that mechanism may be responsible for secular changes in the climate including long term sun-weather effects.

The same mechanism predicts changes in the height of the 500 mb level of about 0.5 geopotential meter, if the solar radiation varies by 0.1% with a period of 27 days in connection with the rotation of the sun. That value is at least one order of magnitude smaller than the corresponding value reported by King *et al*. [14] and thus cannot account for that effect. Instead, Green [10] has proposed that the interference pattern of internal diurnal tidal waves generated in the lower and middle atmosphere may be changed within the troposphere if solar activity changes the stratospheric parameters on a short term time scale.

A recent spectral analysis of the height of the 500 mbar level by Schaefer [26] seems to indicate that the troposphere selects spectral bands, so that among others persistent waves with quasi-periods in the range 25 to 30 days can exist with lifetimes from months to years. These persistent waves apparently are randomly generated by internal turbulent processes independent of any trigger mechanisms from outside the atmosphere. The amplitudes of these waves are comparable with King's et al. [14] findings. If this interpretation is correct, then the "solar activity - weather - effects" involving the 27 day periodicity are merely accidental coincidences of two completely independent but persistent processes on the sun (long living active longitudes rotating with the sun) and within the earth's atmosphere (the preferred spectral band near a period of 27 days).

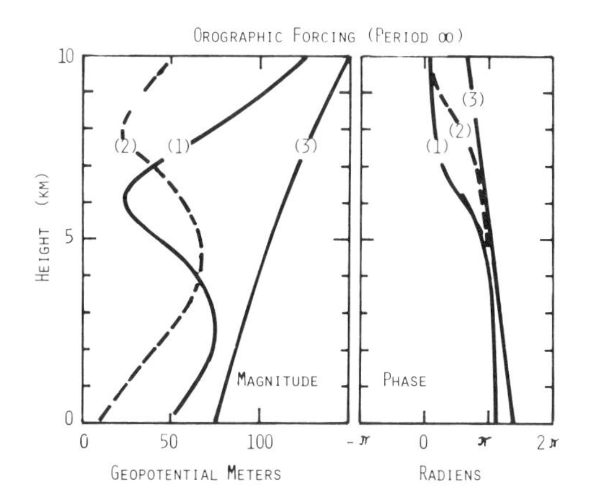

Figure 5. Orographic forcing of wave (2, -4). Magnitude (in geopotential meters) and phase (in radians) of height structure function versus altitude within two layer model with tropopause at z_{tr} = 10 km (curve 1), and 12 km (curve 2) and within isothermal atmosphere (curve 3). Amplitude of orographic forcing L^2_{-4} = 500 m [4]. Parameters of atmospheric models correspond to those of curves 1, 2, and 3 in the table in the figure caption of Figure 4.

REFERENCES

1. J. R. Bates, Quart. J. R. Met. Soc. 103, 397, 1977
2. J. G. Charney and L. P. Drazin, J. Geophys. Res. 66, 83, 1961.
3. L. B. Callis, "Solar Terrestrial Coupling Conference," Yosemite, CA 1978.
4. J. Derome and A. Wiin-Nielsen, Mon. Weath. Rev. 99, 564, 1971.
5. R. E. Dickinson, CIAP Surrey Conference Proceedings, 148, U.S. Dept. of Transportation, 1972.
6. J. A. Eddy, Science 192, 1189, 1976.
7. P. D. Falconer and W. Peyinghaus, Arch. Meteor. Geophys. Bioklim. B23, 201, 1975.
8. P. V. Foukal, P. E. Mack, and J. E. Vernazza, Astrophys. J. 215 952, 1977.
9. M. A. Geller and S. K. Avery, Joint Symposium C, IAGA/IAMAP Assembly, Seattle, WA., 1977.

10. J. S. A. Green, STP-Symposium, Innsbruck, 1978.
11. M. Hantel, Private Communication, 1978.
12. J. R. Herman and R. A. Goldberg, J. Atm. Terr. Phys. 40, 121, 1978.
13. W. Kertz, Nachr. Akad. Wissench., Göttingen, Math-Phys. Kl, 6,145,1956
14. J. W. King, A. S. Slater, A. D. Stevens, P. A. Smith, and D. M. Willis, J. Atm. Terr. Phys. 39, 1357, 1977.
15. H. Lettau, Landoldt-Bornstein, II. Band, pp. 666, Eds. J. Bartels and P. ten Bruggencate, Springer Verlag, Berlin, 1952.
16. M. S. Longuet-Higgins, Phil. Trans. Roy. Soc., London 262, 511, 1968.
17. R. Markson, Nature 273, 103, 1978.
18. J. M. Mitchell, C. W. Stockton, and D. M. Meko, Joint Symposium C, IAGA/IAMAP Assembly, Seattle, WA, 1977.
19. W. M. Neupert and V. Pizzo, J. Geophys. Res. 79, 3701, 1974.
20. H. K. Paetzold, Pure and Appl. Geophys. 106, 1308, 1973.
21. G. J. Shutts, Quart. J. R. Met. Soc. 104, 331, 1978.
22. H. Volland, Nature 269, 400, 1977.
23. H. Volland and H. G. Mayr, Rev. Geophys. Space Phys. 15, 237, 1977
24. J. M. Wilcox, J. Atm. Terr. Phys. 37, 237, 1975.
25. D. M. Willis, J. Atm. Terr. Phys. 38, 685, 1976.
26. J. Schaefer, this volume, 1979.

PLANETARY WAVES AT THE 500-MBAR LEVEL IN THE NORTHERN HEMISPHERE AND THEIR CONNECTION TO THE SOLAR ROTATION

J. Schäfer
Astronomical Institutes, University of Bonn
D5300 Bonn, W. Germany

ABSTRACT. Several methods of statistical frequency analysis and superposed-epoch analysis have been used to investigate the connection between parameters of the solar rotation and oscillations of the 500 mbar level.

1. INTRODUCTION

Recent analyses indicate that there might be an influence of the rotating Sun on certain tropospheric parameters [1,2]. According to King et al. [1], the 27.5 day periodicity in the daily sunspot number is accompanied by oscillations in the height of the 500 mbar level on the northern hemisphere. The zonal structure appears to be a superposition of planetary waves of lowest wave numbers. In this paper I want to examine the reliability of this solar-weather relationship by applying methods of superposed epoch analysis and statistical frequency analysis and by using the time series of the height of the 500 mbar level, the relative sunspot number (SSN) and the direction of the interplanetary magnetic field (IMF). The daily pressure level data at gridpoints with a distance of 5° in latitude (from 15 to 90°N) and 10° in longitude extend from 1949 to 1975 and were provided by the German weather service (Deutscher Wetterdienst). The series of IMF dates back to 1926 [3].

2. POWER SPECTRA

To obtain an impression of the 27 day variability of SSN and IMF, Figure 1 shows a plot of the daily deviation from a running 27 day mean of SSN and IMF. Each row contains four solar rotations (108 days) and is displaced by 27 days against the preceding. By repeating a single day four times, prominence is given to the periodic character in the 27-day period region. One notices "waves" which remain constant for a certain time, then decay and after a more or less disturbed area revive with a new phase and a different frequency. However, there is no coherence between SSN and IMF over the whole time period (as proven by coherency spectra), and therefore the 500 mbar level height cannot be

B.M. McCormac and T.A. Seliga (Eds.): Solar-Terrestrial Influences on Weather and Climate. 275-281.

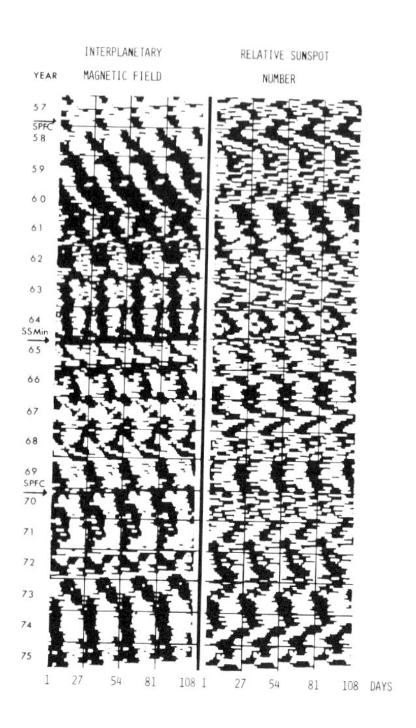

Figure 1. Daily deviations from the medium value of sunspot numbers (SSN) and interplanetary field direction (IMF). Black areas indicate days with a sunspot number larger than a running 27 day mean and days with an "away"(from the sun) polarity of IMF, resp. (see text). SPFC: Solar polar field change, SSMin: Sunspot minimum.

correlated to both of them at the frequency considered. As expected, the power spectra of SSN and IMF in Figure 2 show large power amplitudes around the solar rotation frequency f_o. In addition, IMF shows a remarkably high peak at 14 days, indicating that a four-sector structure sometimes dominates.

However, the same analysis of the 500 mbar level at single gridpoints does not give significant amplitudes near these periods. The spectral estimates are based on samples of 9861 days using a Parzen lag window of length 500 d and frequency spacing of 0.001 d^{-1}. Therefore the spectral bandwidth is 0.0037 d^{-1}, and the 95% confidence limit for the spectral density C_{xx} is (0.73,1.3) C_{xx}[4]. Figure 2 shows typical spectra at two gridpoints at midlatitudes, which exhibit a marked red spectrum with a distinct peak at the 1 yr period (not on the picture) but no additional power around the frequency f_o. On the other hand, if one plots the power at f_o against the 36 gridpoints at a fixed latitude, one obtains a zonal dependency with a dominant $m = 2$ component at midlatitudes [2]. In attempting to determine whether this behavior is unique for f_o, the power values at each frequency of the 36 spectra have been Fourier-analyzed. Figure 3 shows the first three modes for the latitude circle

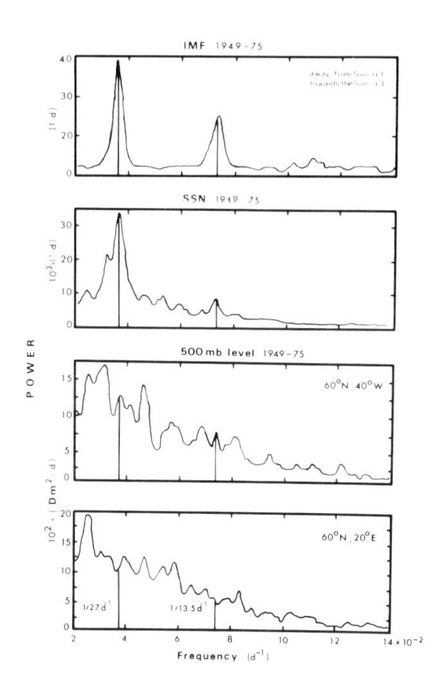

Figure 2. Power spectra of IMF, SSN and the 500 mbar level height at two gridpoints.

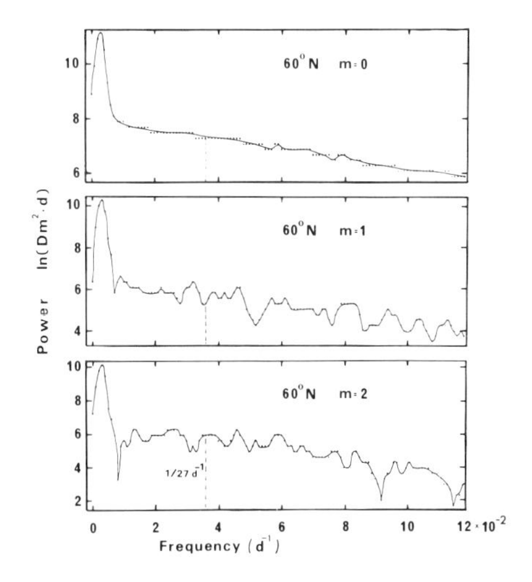

Figure 3. Mode separation of the 36 spectra at 60°N.

60°N. One notices that these mode-separated spectra resemble those of single grid points and that no specific characteristics near f_o can be detected. The same is true for lower and higher latitudes.

3. SUPERPOSED-EPOCH ANALYSES

One could suggest that the solar influence on the atmosphere is not large enough to overcome the noise in the spectra but nevertheless triggers a 27 day variation at the 500 mbar level. Such dependency should be detectable by applying the superposed-epoch analysis (Buys-Ballot-pattern). This analysis was applied to the time series of the pressure height values (9861 days) using equally spaced key dates as well as key dates calculated from the SSN and IMF series by fitting a cosine cycle to each 27 day interval. Indeed some of the analyses reveal an amazing periodic variation, as can be seen in Figure 4, but in spite of that the amplitudes are not significant. This can be tested by several methods. The first one is to shake the pattern, i.e., to shift each row by any given number of days (say up to 40) to the right or left, and to compare the variations [5]. The significance levels are given by the F-distribution of Fisher [6]. Another method is to use several sets of key dates which are equally spaced by a different interval for each set and to compare the amplitudes. This has been done using intervals of 20 d, 40 d, 26.90 d, 26.91 d,..., 27.80 d. Finally a method appropriate for the SSN and IMF key dates is to turn around the cycles, so that the key date within the last cycle now becomes the key date within the first cycle and so on, or to shift the cycles, so that the last half of the key dates is interchanged with the first half. After this manipulation a coherence between SSN/IMF and the pressure level data, if it existed before, can no longer be expected. All these methods have been employed to the pressure level data at midlatitudes over the whole time period. Winter and summer seasons as well as the mean pressure level height in the auroral zone and at the pole were also analyzed. The results of these investigations can be summarized as follows:

1) With all the mentioned sets of key dates one obtains nearly always a periodic variation, regardless of the exact nature of the key dates (see Figure 4a,b,c). As was already pointed out, the periodic variation must be trivial if similar results using SSN and IMF key dates occur.

2) The analyses of the shaken patterns, as well as those using the turned and shifted SSN and IMF key dates, show amplitudes and variations which are of the same order as the correct ones (see Figure 4c).

3) The significance test (F-test), which was always employed, gives generally no convincing probabilities for the periodic variations.

In particular it is clear that, if one shifts the key dates by half a period (e.g., by taking at first day 8 and then day 22 of the solar rotation, as done by King et al.[2]), one obtains a variation which has the opposite deviation from the original. I therefore conclude: A 27 day variation of the height of the 500 mbar level at one grid point and a coherence with the SSN and IMF data cannot be confirmed to be significant. Concerning the "resolving power" of this method, the amplitude of such a variation must exceed 4 Dm (geopotential decameters) in order to be considered significant. It should be mentioned that the

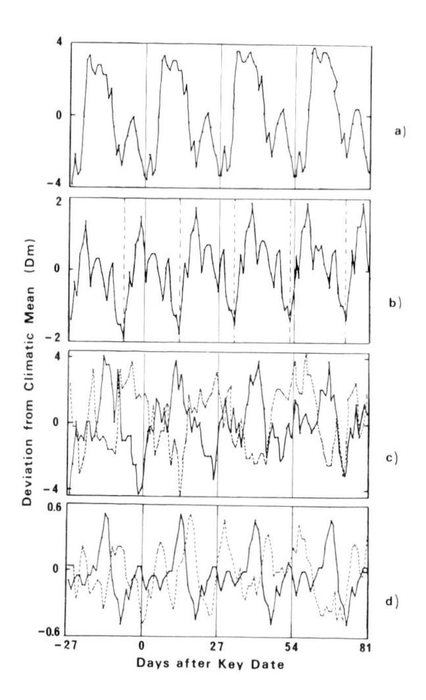

Figure 4. Superposed-epoch analysis: (a) the 500 mbar level at 60°N, 60°W using key dates separated by 27 days; (b) the 500 mbar level at 60°N, 60°W using key dates separated by 20 days; (c) the 500 mbar level at 60°N, 20°W using key dates calculated from SSN (bold line), and using a shifted set of SSN key dates (see text, dashed line); and (d) the sine term of the m = 1 mode at 55°N using key dates calculated from IMF (bold line), and using the same key dates but a shaked pattern (see text, dashed line).

same negative results come out if one uses harmonic dials. In a further attempt, the pressure level data for each latitudinal zone were first separated into modes by a Fourier analysis for each day, thus isolating the planetary scale (lowest wave numbers) signals from the highly noisy synoptic scale (large wave numbers) signals. In this way, new time series with daily values for the sine and cosine term (or the real and imaginary part) for each mode were created. Similar to the grid point spectra, the spectra of the sine and cosine series do not show additional power around f_0. The superposed-epoch analysis has been employed again to the time series for m = 0, 1, 2, at all latitudes. The SSN and IMF key dates were used and also dates equally spaced by 20 and 27.5 days. The general outcome of this analysis is that the periodicity using equally spaced key dates looks even better than that by using the SSN and IMF key dates, regardless of the length of the interval, and that the significance levels (F test) are always of the same order. Apart from that, the shifted and turned series were also analyzed. As expected, one obtains the same periodic behavior with comparable amplitudes. Figure 4d shows the superposed-epoch analysis of the sine term (m = 1, 55°N) using IMF key dates, as well as an analysis applying a shifted pattern [5]. Obviously the variances are of the same order and the F test is negative. The elimination of the noise has increased the resolving power. The amplitude of a m = 1 wave (Figure 4d) generated by the solar rotation has to exceed at least about 0.6 Dm in order to become distinguished from the noise. Similar values hold for m = 0,2. In conclusion, the results of the superposed-epoch analyses are not significant in spite of sometimes amazingly clear periodicities, and no influence of SSN, IMF or the constant solar rotation on the 500 mbar level has been found!

4. SQUARED COHERENCY ESTIMATES

In order to verify this result, the squared coherency estimates between the time series of SSN and IMF and the series of the mode coefficients for m = 0, 1, 2 were calculated, and its significances were tested using formulas given by Jenkins and Watts [4]. At first sight, these analyses again seem to prove a connection between the solar rotation and planetary waves, because about one third of the 60 coherency spectra (sine and cosine terms at 15 latitudes against SSN and IMF) for m = 1 reach or surpass the 95% confidence limit at f_o. For m = 0 and m = 2 this fraction is about 20%. In order to test these results once again, the time series of the mode coefficients were shifted in the way described above, so that the first half of the SSN/IMF series faced now the last half of the coefficient series, and the coherence should vanish because of the large time lag. It is no longer so astonishing, that in this case the fraction of spectra with "significant" peaks at f_o is just the same and the peaks are just as high as in the original spectra. To explain this characteristic, one can show that the 27 day period (especially of the m = 1 wave) has a rather large persistence tendency, similar to SSN and IMF (Figure 1), so that high correlations may occur. The probability for a uniform distribution of phase differences over the whole unit circle reduces to the same degree as the number of time intervals with a persistent 27 day period (large variance of the distribution of phase differences). This leads to a preference of a certain phase and to a large coherency. If one therefore examines only a short time period, e.g., some years or winter seasons, there will be most probably a large coherency. This, however, is no proof for a physical connection between solar rotation and planetary waves. It should also be mentioned that a separation of each mode into west- or eastward traveling waves and into standing waves (see e.g. Pratt [7]) also does not reveal any peculiar behavior at a period around 27 days. An amplitude of at least about 0.4 Dm is required for significant detection in this case.

5. CONCLUSION

Several methods of statistical frequency analysis and superposed-epoch analysis have been used in an attempt to establish a connection between parameters of the solar rotation and oscillations of the 500 mbar level. In spite of the long time series of 9861 days it was not possible to detect any significant effect near the solar rotation period of 27 days, whatever method was applied. If there are planetary waves generated by the solar rotation, their amplitudes at this height must be smaller than about 0.4 Dm. This result corresponds to that of Volland [8] who predicts changes of the 500 mbar height of about 0.1 Dm, which is below the detection level.

REFERENCES

1. J. M. Wilcox, L. Svalgaard and P. H. Scherrer, J. Atm. Sci., 33, 1113, 1976.
2. J. W. King, A. J. Slater, A. D. Stevens, P. A. Smith, and D. M. Willis, J. Atm. Terr. Phys., 39, 1357, 1977.
3. L. Svalgaard, Interplanetary magnetic sector structure, Geophysical papers R-29, Danish Meteorological Institute, Charlottenlund, 1972.
4. G. M. Jenkins and D. G. Watts, Spectral analysis and its applications, Holden-Day, San Francisco, 1968.
5. J. Taubenheim, Statistische Auswertung geophysikalischer und meteorologischer Daten, Akademische Verlagsgesellschaft Geest & Portig, Leipzig, 1969.
6. R. A. Fisher, Statistical methods for research workers, Oliver and Boyd, Edinburgh and London, 1941.
7. R. W. Pratt, J. Atm. Sci., 33, 1060, 1976.
8. H. Volland, this volume, 1979.

THE ROLE OF GRAVITY WAVES IN SOLAR-TERRESTRIAL ATMOSPHERE COUPLING AND SEVERE STORM DETECTION

R.J. Hung
The University of Alabama in Huntsville, AL 35807 USA
R. E. Smith
NASA/Marshall Space Flight Center, AL 35812 USA

ABSTRACT. The role that gravity waves play in the coupling of the solar wind and the ionosphere, the coupling of high and low latitude ionospheres, and the coupling of the ionosphere and the neutral atmosphere are discussed. Examples are given of the detection of tornado touchdowns from locating the sources of the gravity waves.

1. INTRODUCTION

Gravity waves are the product of a number of impulsive events, such as Joule heating, perturbations of the auroral electrojet during geomagnetic substorms, reversals in the dynamo/equatorial electrojet system, and also severe thunderstorms, tornadoes and hurricanes in the troposphere. A thorough understanding of the sources and sinks of the gravity waves can greatly advance our knowledge of the energetics of the solar wind-ionosphere-atmosphere coupling processes.

The excitation of large scale gravity waves is dependent upon the intensity of geomagnetic activity. Titheridge [1] has shown that a $K_p > 5$ is a necessary condition; and Testud, [2] has proposed from his observations of the F-region neutral wind that large scale gravity waves occur when $\Sigma K_p > 16$. Here K_p is the 3-hour geomagnetic index and Σ is its sum for a day.

Medium scale gravity waves may be excited by various sources including both severe storms and geomagnetic disturbances. Recently, experimental observations, based on CW Doppler sounding of the upper atmosphere, have shown that there are ionospheric disturbances associated with severe weather [3-6].

2. COUPLING OF SOLAR-TERRESTRIAL ATMOSPHERE

Roberts and Olson [7], and Wilcox [8] have studied the developments of 300 mb low pressure trough systems in the North Pacific and North

B.M. McCormac and T.A. Seliga (Eds.): Solar-Terrestrial Influences on Weather and Climate. 283-287.

American region. They find that troughs formed in the Gulf of Alaska 2 to 4 days after the reversals of the polarity of the solar magnetic field, or the passing of the interplanetary magnetic field (IMF) sector boundary. These correlations of the IMF reversals and changes in the upper atmosphere, and the energy and momentum transfers from the solar wind to the Earth's atmosphere can be studied in detail by studying the sinks and sources of gravity waves.

2.1 Solar magnetic field-ionosphere coupling

Based on the OGO-6 and AE-C ion composition data, Taylor _et al_. [9] showed the apparent correlation between enhanced ion drift velocity and IMF reversals observed at the auroral zone/polar cap boundary. Separate studies of the OGO-6 data and AE-E data show wave-like fluctuations in the H^+, O^+, NO^+ concentrations. Analyses of the sinks and sources of gravity waves will improve our understanding of the process coupling the solar wind perturbations to anomalous events in the auroral and equatorial zones. Results of the study of Taylor _et al_. [10] of the morphology of a variation in the concentration of H^+ in the equatorial anomaly associated with variable equatorial electric fields and neutral winds suggest that the systematic analysis of the origins of gravity waves associated with longitudinal/temporal variations in the ion composition can improve our understanding of the external and internal processes affecting the magnetosphere-ionosphere-atmosphere system.

2.2 High latitude-low latitude ionosphere coupling

It is well-recognized that geomagnetic substorms and auroral electrojet perturbations play a significant role in the magnetosphere-ionosphere-atmosphere coupling. The auroral and equatorial electrojets also generate gravity waves.

Identification of the sources of large scale gravity waves has been attempted using a reverse ray tracing technique on waves observed in mid-latitude at F-region heights [4]. The results of more than ten case studies show that the computed locations of the wave sources were in the auroral zone at auroral altitudes. Figure 1 shows the computed ray path for one such event observed during the 2200-2400 UT time period on 3 April 1974 by the HF-CW Doppler sounder Array Huntsville, AL. The $\sum K_p$ was 36 and the three-hourly K_p index was 6 when these waves which took 3 h and 20 min to travel from the computed source to the receivers were excited. The results show that studies of the sources and sinks of gravity waves can help to understand processes that couple the high and low latitude ionospheres.

2.3 Ionosphere-atmosphere coupling

Gossard and Sweezy [11] have suggested that thunderstorms or fronts could excite gravity waves in the atmosphere. Hung _et al_. [5] and Hung and Smith [6] have detected gravity waves at F2 region heights associated with severe storms using an HF-CW Doppler sounder. Results of studies of more than 20 ray tracing cases show that wave sources were located near the locations where tornadoes touched down more than 1 h later. These results illustrate that gravity waves also play a major role in the coupling between the ionosphere and troposphere during time periods with severe storm activity.

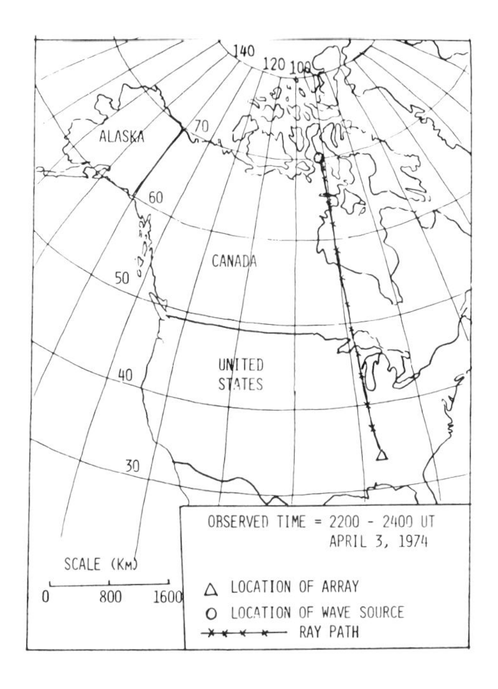

Figure 1. Computed ray path of large scale gravity waves associated with auroral activity during 2200-2400 UT, April 3, 1974 and the location of computed wave source.

Results of a case study of the gravity waves associated with three isolated tornadic storms on January 13, 1976 are presented. Figure 2 shows the ray paths of three gravity waves observed during the 1830-1930, 1930-2030, and 1830-2030 UT time periods. An isolated tornado touched down at 1915 UT. The computed wave traveling time from the source to the receivers was 55 min; therefore, the gravity waves were excited more than 1 h prior to the touchdown. Two other isolated tornadoes touched down at 2014 and 2050 UT. Figure 3 shows that the tornado touchdown locations started in the southwest section of Indiana, moved to central Indiana, and finally to central southern Indiana in excellent agreement with the computed locations of the wave sources observed during the 1830-2200 UT time period and the movement of a squall line during the 1735-2035 UT time period.

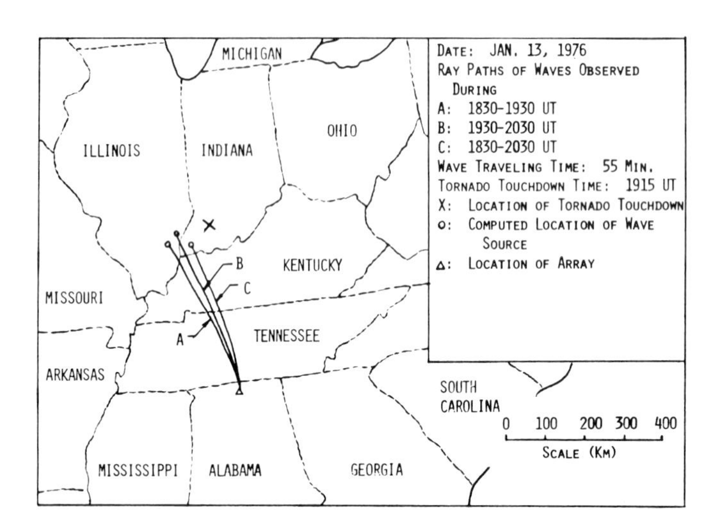

Figure 2. Computed ray path and location of wave sources for gravity waves observed at 1830-2030 UT, January 13, 1976, and the location of the tornado touchdown.

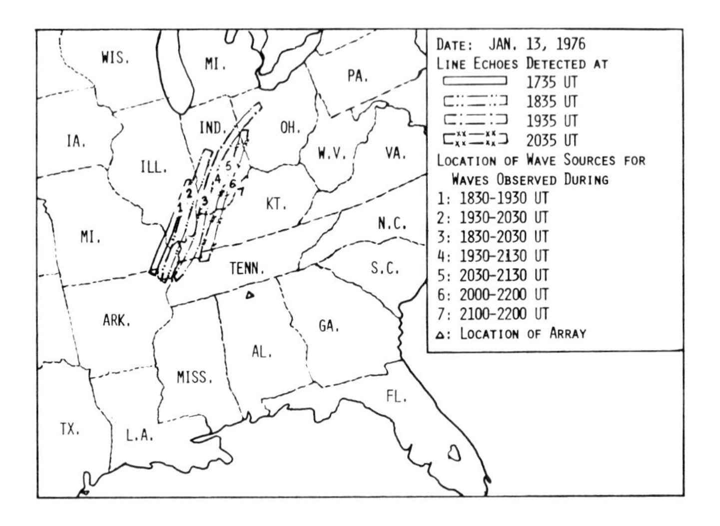

Figure 3. Comparison of the movement of computed locations of wave sources and the movement of squall line during 1735-2035 UT, Jan. 13, 1976, when the observed gravity waves were being excited.

3. CONCLUSIONS

Evidence of the coupling between the solar wind and the terrestrial atmosphere is contained in the apparent correlations between enhanced ion drift velocity and IMF reversals. It has also been shown that gravity waves play an important role in the transport of momentum and energy between the solar wind and the terrestrial atmosphere. Studies also have shown that gravity waves couple the high and low latitude ionospheres.

As for the correlation between gravity waves and atmospheric dynamics, it is believed that gravity waves can be excited by penetrative convection. Our present study shows that gravity waves associated with tornadoes were excited at the squall line verifying the suggestion of Gossard and Sweezy [11]. It seems reasonable to assume that gravity wave studies can improve our understanding of the ionosphere-atmosphere coupling process as well as our understanding of severe storm generation mechanisms.

A technique for using simultaneous observations from in situ measurements on orbiting spacecraft, balloons, and ground-based instruments coupled with reverse ray tracing computations to establish the sources and sinks of gravity waves may be a most powerful tool for studying the dynamical coupling between the solar and terrestrial atmospheres.

REFERENCES

1. J.E. Titheridge, J. Geophys. Res., 76, 6955, 1971.
2. J. Testud, J. Atmos. Terr. Phys., 32, 1793, 1970.
3. R.E. Smith and R.J. Hung, J. Appl. Meteor., 14, 1611, 1975
4. R.J. Hung, T. Phan and R.E. Smith, J. Atmos. Terr. Phys., 40, June 1978 (in press).
5. R.J. Hung, T. Phan and R.E. Smith, AIAA Journal, 16, 763, 1978.
6. R.J. Hung and R.E. Smith, J. Appl. Meteor., 17, 3, 1978.
7. W.O. Roberts and R.H. Olson, J. Atmos. Sci., a3, 135, 1973.
8. J.M. Wilcox, Solar Activity and the Weather, in Possible Relationships Between Solar Activity and Meteorological Phenomena, NASA SP-366, 1975.
9. H.A. Taylor, H.C. Briton, J.M. Grebowsky and B. Blackwell, Geophys. Res. Lett., 1978, (in press).
10. H.A. Taylor, H.G. Mayr, S.L. Hsieh, and G.R. Cordier, Rev. Geophys. Space Phys., 16, 267, 1978.
11. E. Gossard and W.B. Sweezy, J. Atmos. Sci., 31, 1540, 1974.

INFLUENCES OF SOLAR ACTIVITY ON THE EXCHANGE INTENSITY BETWEEN STRATOSPHERE AND TROPOSPHERE

Reinhold Reiter
Institute for Atmospheric Environmental Research of the Fraunhofer Society, D-8100 Garmisch-Partenkirchen
West Germany

ABSTRACT. Recordings of Be7 and O_3 in the air from a mountain observatory (3 km altitude, Zugspitze) constitute the basis for a study of the question whether the frequency of intrusions of stratospheric air into the troposphere is influenced by solar events. An influx of stratospheric air passing the measuring station is indicated by a noticeable increase in the concentration of these stratospheric constituents. Since 1973 we have been able to identify a 40% increase in the frequency of stratospheric intrusions 2 to 3 days after solar flares and sector structure boundary passages. Recent studies aimed at a physical interpretation revealed a close correlation between stratospheric intrusions triggered by solar action and the Forbush effect. Recently, it has become possible to show that the intrusions are also coupled with an increase in the O_3 concentration in the stratosphere, mainly in its lower layers.

In previous publications [1,2] we have been able to show on the basis of several years' measurements that the frequency of stratospheric intrusions significantly increases after solar events. The present paper provides the most recent results based not only on the extended data base but also on new, specific investigations such as the profile of stratospheric O_3 as a possible physical link. Space limitations do not permit discussion herein of the relevant literature.

Figure 1 illustrates simply and schematically the experimental part of our studies. In the lower stratosphere, i.e., in a layer of about 15 to 25 km, high energy protons generate radionuclides, e.g., Be7, P32, P33, S35, by spallation out of gas atoms. The production rate in the troposphere is much lower and can be neglected. The radioactive aerosols are injected from the stratosphere (residence time there between 30 and some 100 days) into the troposphere (residence time there only a few days) by intrusion processes (Reiter [3]) mainly in regions of a tropopause folding in the vicinity of a Jet stream.

In the troposphere the radionuclides are transported on isentropic surfaces, depending on the large-scale meteorologic situation, to our

B.M. McCormac and T.A. Seliga (Eds.): Solar-Terrestrial Influences on Weather and Climate. 289-296.

measuring station at 3 km altitude where they have been recorded since Fall 1969. The station is mostly above the mixing layer in free advection. Sharp peaks of e.g. Be7 indicate that the station is reached by a stratospheric air mass. The O_3 is locally recorded at the same station. Besides, after solar events we recently conducted O_3 soundings with a rawinsonde supplemented by measurements of the total O_3 by means of an O_3 spectrometer. It must be emphasized at this point that the variations in the concentration of radionuclides at the station Zugspitze (amplitude about one order of magnitude) cannot be explained by a variation in the production rate but rather by the influx from the stratosphere on the one hand, and the tropospheric sinks for aerosols (mainly washout) on the other.

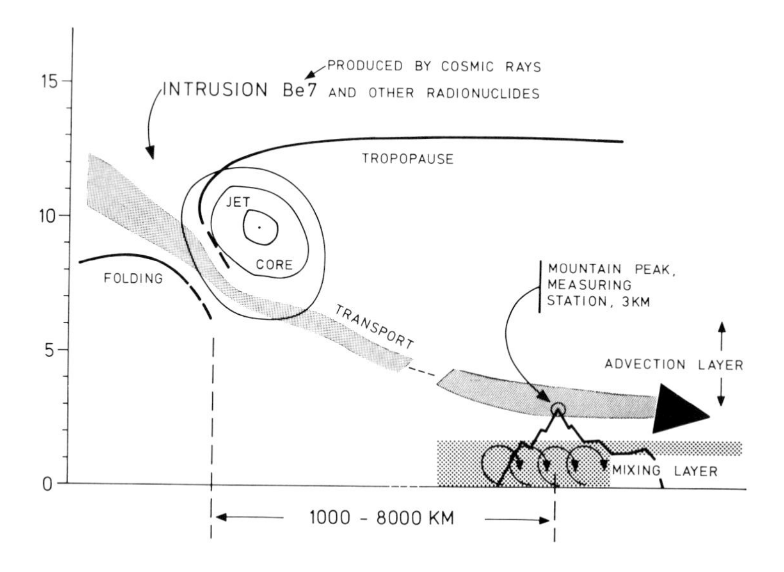

Figure 1. Schematic representation of our measuring principle for determining stratospheric intrusions through radionuclides which have been continuously recorded since Nov. 69 at the station Zugspitze at 3 km altitude which is above the mixing layer.

It has been possible through augmented data records covering 8 yr to confirm our earlier findings that the frequency of stratospheric intrusions (indicated by a marked rise of the Be7 concentration) increases significantly after (+/-) or during (-/+) passages of sector boundaries of the interplanetary magnetic field although the amplitudes were found to be somewhat smaller for the latter passages and during decreasing solar activity. In particular it was shown that an effect after passages +/- is much weaker than after passages -/+. In the latter case we observe <u>on the day</u> of the passage a doubling of the frequency of stratospheric intrusions in winter and spring. In contrast thereto, we note in summer and fall a minimum on the day of the passage but a maximum (doubling) of the intrusions on the 5th day after the passage. This will be specified later in another paper.

In full agreement with our earlier findings, Figure 2 shows the behavior of stratospheric intrusions after solar H_α-flares. We correspondingly find in both the total data records (8 yr, 216 key days) and in different phases of the solar activity an increase by about 40 to 50% beginning with the day of the flare until the second day thereafter. The standard deviation (vertical bars) confirms the significance of this correlation. Figure 3 now shows that a variation of the O_3 in phase with that of the Be7 can be observed at the station Zugspitze (amplitude about 30%) even during the phase of decreasing solar activity. (Since it is only 2 yr that recordings of O_3 are available the number of cases is comparatively small.) Obviously then, after solar events O_3 is also entrained in the stratospheric air flow to the measuring station along with the Be7. Also striking is the fact that the variation of the atmospheric total O_3 shows a completely different pattern (Figure 3, bottom line): We obtain a significant maximum of the total O_3 (amplitude 5 to 10%) on the day before the flare (during decreasing solar activity: 2 days before the flare). That means, that a process which affects the stratosphere sets in as the active region (M-region) approaches the central meridian of the Sun. We made a step towards an understanding of the correlations in that we considered also the neutron density (caused by the galactic cosmic rays) as it is recorded on the Earth surface (using data of the station Sulfur Mountains).

In connection with active regions and solar flares the known Forbush effect is observed, i.e., a reduction of the galactic ray intensity in the lower stratosphere and in the troposphere which is linked up with the arrival in the vicinity of the Earth of magnetic field "blobs." These blobs often appear to form at the leading edge of high speed streams in the solar wind which originate either in flare or M-regions [4]. The galactic cosmic rays are incidentally the widely dominant source for the production of ions and excited N atoms below 20 km [4,5] which is not even exceeded by the rare strong emissions of solar protons [4].

We selected some key days from the super neutron monitor recordings which indicate a marked Forbush effect. Figure 4 shows that these key days must be looked upon as significant with regard to solar activity (solar radio flux) and the neutron density as a measure of the galactic components of cosmic rays, as well as with regard to the effects in the Magnetic field of the Earth (Ap) and the radio propagation as reflecting the perturbations in the ionosphere. When using these Forbush key days in the superposed epoch analysis of the Be7 concentration (Figure 5) it is noted that the Be7 increases from the 3rd day before the Forbush key day onwards and maximizes significantly 1 day before the key day (amplitude about 40%). During the period of decreasing solar activity, however, the number of Forbush key days is still too small to permit a definitive statement. Figure 6 shows in summary the situation when we include neutron density, total O_3, and the Be7, and use 50 H_α- solar flares as key days during the period of decreasing solar activity. We note the following: Simultaneous in time with the decrease of the neutron density (1) the total O_3 in the atmosphere rises by about 8%(2.).

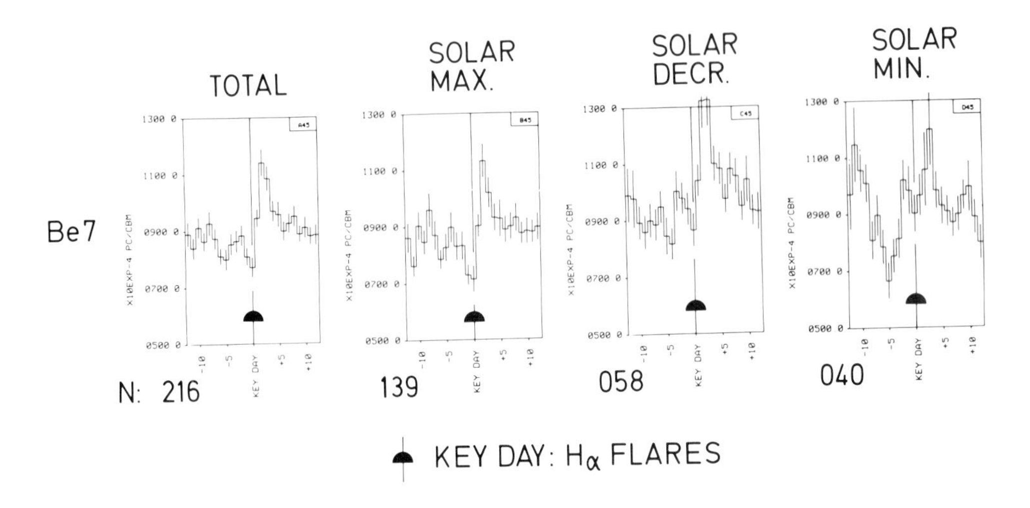

Figure 2. Concentration of the Be7 as a measure of the frequency of stratospheric intrusions prior to and after solar flares in H_{α} (max. 2 - 3 days after flare).

Figures 2 - 6 give the results of our superimposed epoch analysis, the kind of the respective key day being indicated at the bottom of the diagram. The epoch analysis in principle covers the period from 12 days prior to until 12 days after the key day. The analyzed parameter is given on the left margin of the figures, the vertical bars denote the standard deviation. N means the number of key days being used in each individual analysis. The items on the upper margin of the figures stand for:

TOTAL: Total period from Nov. 1969 through Dec. 1977, i.e., 8 yr; SOLAR MAX: Nov. 69 - 1972 inclusive; SOLAR DECR: 1973 - 1975 incl.; SOLAR MIN: 1975 - 1977 incl.; hence this classification refers to the respective phase of solar activity.

Two days later the increase of the Be7 sets in, reaching its maximum 2 to 3 days later (3.) after the flare (+40%).

Even though at the present moment this time sequence will surely not allow us to infer a causal link, we nevertheless may at least suspect that an increase in O_3 concentration is brought about by a reduction of the production rate of excited N atoms and nitrous gases during the beginning Forbush effect. According to numerical models available so far we may, of course, assume that gas reactions in the lower stratosphere are extremely slow but it is not quite sure whether all possible reactions can be described by the present one-dimensional chemical models.

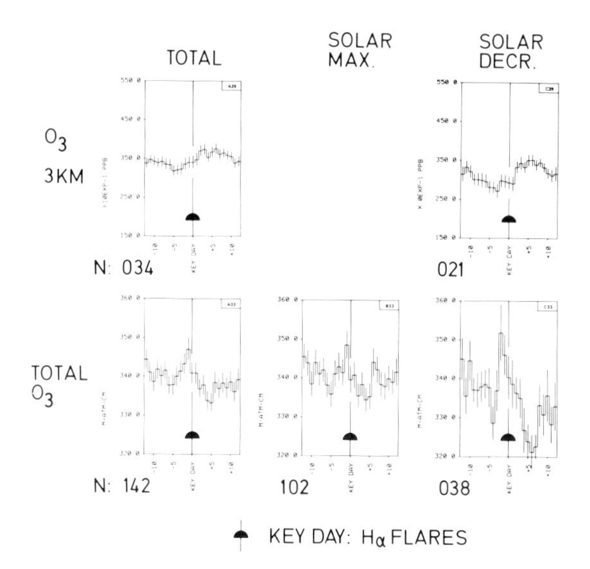

Figure 3. Analogous to Figure 2 but for O_3 at the recording station for Be7 (3 km alt.) and for the total atmospheric O_3 (the latter with max. 1 day before the flare).

Now, what does a link between O_3 increases in the stratosphere and triggering of intrusions possibly look like? The Table below gives progressively from day to day the numbers for two stratospheric intrusions. All interesting solar and geophysical data are entered here. From 13 April on we performed daily O_3 rawinsonde flights. From those we derived both the total O_3 (in reasonable agreement with the spectrometric total O_3 measurements) and the amount of O_3 which was present in the space between the tropopause and a height of 150 mb above the tropopause. We note that after the flare the amount of O_3 above the tropopause increases even stronger than the total O_3. A rise of temperature might be expected as a result of the increase of the stratospheric O_3 and indeed, after O_3 enhancement the temperature at tropopause level rises considerably, its height lowers, and Be7, fallout, and then O_3 at the station Zugspitze in 3 km altitude increase strongly. That means, the injection of stratospheric air occurred.

We will now continue our careful studies and analyze in some detail the nature of processes following violent solar flares in time isolation. This concerns in particular the determination of the fine structure of the O_3 profile and vertical or horizontal O_3 transport, the total O_3 (spectrometrically), and the lapse rates of temperature, wind, and humidity. Only on the basis of a sufficient number of analyses of that kind we hope to make some progress in our search for the physical link which obviously connects solar event and stratospheric intrusion. It cannot be excluded that Ney [6] was right in considering the Forbush effect as a possible link.

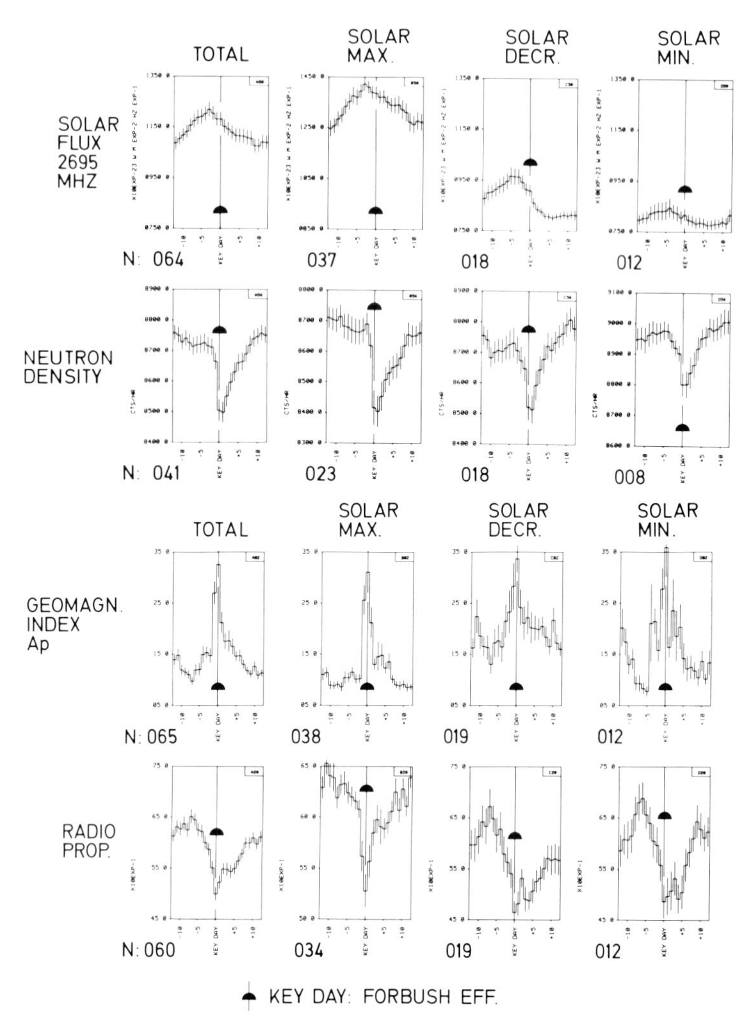

Figure 4. Variation of different solar and geophysical parameters around key days with Forbush effect.

REFERENCES

1. R. Reiter, J. Atm. Terr. Phys., 38, 503, 1976.
2. R. Reiter and M. Littfass, Arch. Met. Geoph. Biokl. Ser. A., 26, 127, 1977.
3. E. R. Reiter, U. S. Atomic Energy Commission, 1972.
4. R. E. Dickinson, Bull. Amer. Met. Soc. 56, 1240, 1975.
5. M. G. Heaps, B R L Report No. 1938, 1976.
6. E. P. Ney, Nature 183, 451, 1959.

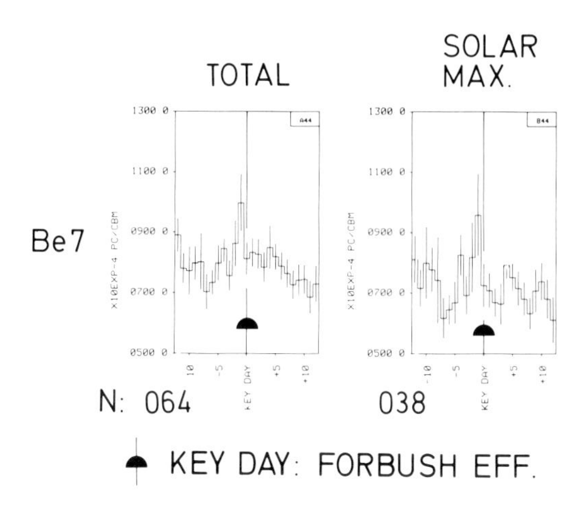

Figure 5. Be7, indicating frequency of stratospheric intrusions, around key days with Forbush effect (max. 1 day before the flare).

APRIL 1978

DAY:	9	10	11	12	13	14	15	16	17	18	19	20
H_α FLARES			2B	1B	1B		1N			2B		
SUNSPOT NO.	126	111	109	107	93	75	61	85	99	100	107	115
SOLAR FLUX 10.7 cm	156	155	180	155	145	138	140	137	132	133	145	139
RADIO PROP. INDEX	12.9	12.3	10.9	13.4	13.2	11.8	13.5	13.5	13.9	12.9	13.8	12.2
NEUTRON DENSITY		65585	63817	63433	64682	64840	61928	60593	59130	58653	58021	59539
GEOMAGN. IND. Ap	6	10	64	29	24	51	19	8	7	19	38	23
TOTAL O_3 SPECTROM m Atm cm: AROSA	–	391	–	414	417	413		–	433	385	–	389
TOTAL O_3 SPECTROM m Atm cm: GARMISCH	–	–	–		–	425		–	461	425	431	442
TOTAL O_3 SPECTROM m Atm cm: HOHENPEISS	385	388	–	426	422	421		422	425	399	395	402
RADIOSO.: INTEGR. TOTAL O_3 SONDE GAP, nb	–	–	–		435	429	508	–	385	425	375	426
RADIOSO.: O_3 FROM TROPOP. TO +150 mb	–	–	–		125	119	146	–	84	117	119	105
TROPOPAUSE TEMP. °C	-64	-63	-59	-55	-56	-52	-59	-61	-49	-58	-57	-54
INTERDIURNAL TEMP. CHANGE		+1	+4	+4	-1	+4	-7	-2	+12	-9	+1	+3
TROPOPAUSE HEIGHT KM	11.3	11.0	9.5	9.1	9.5	9.0	9.1	9.5	7.5	9.8	10.2	9.2
ZUGSPITZE 3 KM: O_3 ppb	49.0	48.8	41.7	41.3	44.3	44.2	37.2	41.2	39.9	43.4	52.2	45.6
ZUGSPITZE 3 KM: Be 7 10^{-2} pcie m^{-3}	12.1	4.9	5.2	16.5	15.7	7.7	3.6	4.5	8.5	18.9	10.9	6.7
ZUGSPITZE 3 KM: FALLOUT 10^{-2} pcie m^{-3}	3.7	1.3	1.3	2.7	2.2	1.7	1.1	0.8	1.2	2.2	1.6	1.0
ZUGSPITZE 3 KM: REL. HUMIDITY %	81	98	96	87	95	80	95	99	75	40	20	75
TIME LAG				⊢ 1-2 DAYS ⊣				⊢ 4 DAYS ⊣				

Table: Daily means of significant solar, geophysical, and meteorological parameters during two intrusions; one due to the flare on 11 April 1978, the other after the flare of 13 April. Note above all the variations of O_3, temperature, and height of the tropopause after the decrease of neutron density and before the arrival of the stratospheric material (O_3, Be7, fallout) at 3 km altitude.

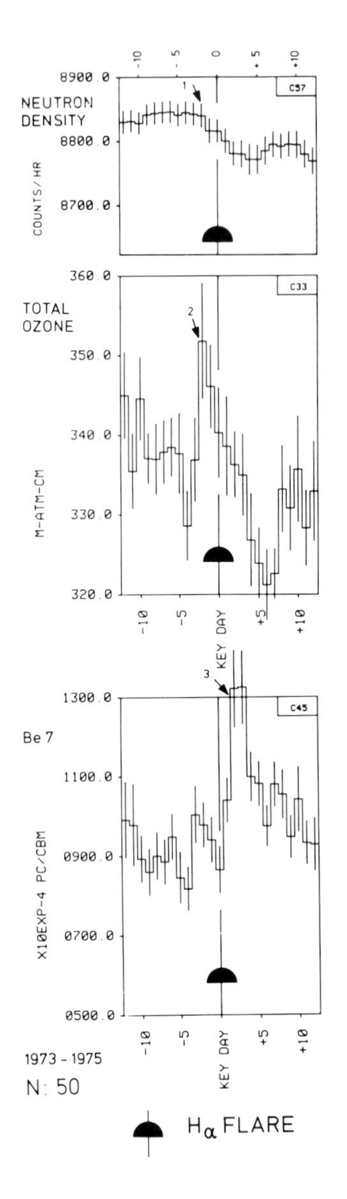

Figure 6. Time sequence of the reaction of neutron density, total O_3 and stratospheric intrusions around H_α- flare key days in the period of decreasing solar activity.

SOLAR INFLUENCE ON NORTH ATLANTIC MEAN SEA LEVEL PRESSURE

Dr. P. M. Kelly
Climatic Research Unit
School of Environmental Sciences
University of East Anglia
Norwich NR4 7TJ
England

ABSTRACT

A principal component analysis of a 101 year set of winter mean sea level pressure anomalies for the North Atlantic-European sector has revealed evidence of an 11 yr cycle. This cycle is apparently related to the long-term modulation of solar flare activity during the sunspot cycle.

The principal components represent independent atmospheric circulation anomaly patterns. The variation of the strength of these patterns with time is given by a series of coefficients.

The variance spectrum of the coefficients of the second principal component (PC2) contains statistically significant peaks at 11, 5 and 3-1/2 years, indicating the presence of a non-sinusoidal 11 yr fluctuation. The main feature of the spatial pattern of PC2 is an area of intensified cyclonic activity to the west of the British Isles. This pattern, identified in <u>seasonal</u> data, is consistent with the pattern of <u>24 hr</u> 500 mb height change following solar flare events reported elsewhere in this volume. The cyclonic intensification is located beneath a region of divergence ahead of an anomalous upper air trough. It is proposed that the 11 yr fluctuation arises as the frequency of solar flares, and the atmospheric responses on the day-to-day time scale, changes with the sunspot cycle.

This compatibility between the short term solar-weather and long term solar-climate relationships suggests an approach which may avoid a problem frequently encountered in investigations of the latter. Although the data length may be sufficient to generate a hypothesis, there is rarely additional data available to test that hypothesis rigorously. The approach would be to base hypotheses not on analysis of the limited climatic data, but on the results of short term solar-weather investigations, i.e., to deduce their implications, or integrated effects, on climatic time scales.

B.M. McCormac and T.A. Seliga (Eds.): Solar-Terrestrial Influences on Weather and Climate. 297-298.

The phase locking between the sunspot cycle and the 11 yr fluctuation in PC2 was reasonable during the period 1920 to date, but broke down about the turn of the century. Although no firm conclusions can be drawn from the limited data available, three possible explanations felt worthy of further investigation are advanced. First, sunspot numbers may not be a good indicator of the component of the solar output affecting the atmospheric circulation. Second, the volcanic dust loading of the stratosphere was minimal during the period 1920 to 1960. This may have resulted in a stronger solar control on the atmospheric circulation during that period; at other times, the volcanic control may mask the solar effects. Third, the amplitude of the sunspot cycle was far greater after 1920 than it was before.

A version of this paper has appeared in Nature (Vol. 269, No. 5626, pp. 320-322, September 22, 1977). This research was supported by the Natural Environment Research Council, U.K.

SOME ASPECTS OF STRATOSPHERIC CHEMICAL RESPONSE TO SOLAR PARTICLE PRECIPITATIONS: Potential Roles of $N_2(A^3\Sigma)$ and Ion-Chemistry

Sheo S. Prasad
Jet Propulsion Laboratory, California Institute of Technology
Pasadena, CA 91103 USA

ABSTRACT. Large amounts of long lived $N_2(A^3\Sigma)$ are created by the energy degradation of precipitating solar particles. Laboratory data suggest that in the stratosphere $N_2(A^3\Sigma)$ are efficiently converted into N_2O. Through reactions with $O(^1D)$, N_2O may gradually release NO and thereby influence the long term aspects of stratospheric chemical response. During the daytime, negative ions may transform an active NO_X into an inactive HNO_3. At night both negative and positive ion chemistry generate HO_X. Omission of ionic chemistry results in considerable underestimation of O_3 depletion during the initial phases of solar particle events, and thereby introduces significant error in the estimation of the nature of the prompt response. For further progress more refined model calculations as well as additional laboratory verification of some of the key assumptions are needed.

1. INTRODUCTION

From time to time various suggestions, as to how solar activity could affect the atmosphere, have been made. These suggestions include: (i) involvement of atmospheric electricity, (ii) direct and indirect roles of ions produced by cosmic rays and bremsstrahlung X-rays in nucleating clouds in the vicinity of the tropopause, and (iii) influence of possible changes in stratospheric-mesospheric chemical composition, heat balance and dynamics in response to variability in solar corpuscular and e.m. radiation [1-5]. Obviously electrical characteristics of the atmosphere can change almost instantly and their effects upon thunderstorm activity can be prompt. But the connecting links between enhanced thunderstorm activity and key elements of weather/climate, and the corresponding time constants are still in the realm of vague conjectures. Experimental and theoretical studies are needed to clarify the role of ionization in the triggering of extensive cirrus cloud formation. There are several quantitative or semi-quantitative studies which indicate that changes in stratospheric-mesospheric wind field, and chemical and thermal structure can affect terrestrial weather/climate on time scales ranging from a few days to eons [6-8]. It is, therefore, essential that all those physical processes that may sig-

B.M. McCormac and T.A. Seliga (Eds.): Solar-Terrestrial Influences on Weather and Climate. 299-304.

nificantly control the nature of the stratospheric-mesospheric response to variable solar activity be identified and studied as completely as possible within the present state of the art. This paper presents preliminary results of our efforts in that direction, and concentrates on solar particle events.

2. POTENTIALS OF EXCITED $N_2(A^3\Sigma)$.

Since N_2 is the major constituent of the atmosphere, significant amounts of excited triplet states of N_2 are created in the course of the energy degradation of the precipitating solar particles. Higher triplet states $C^3\Pi_u$ and $B^3\Pi_g$ cascade to the lowest $A^3\Sigma$ state by the emission of the 1st and 2nd positive band system. Combination of the rate of direct excitation and the contribution from cascading leads to a production rate of about 0.35 $N_2(A^3\Sigma)$ per ion pair. This state is metastable with a long radiative lifetime of about 1.27 to 2.5 s. In the stratosphere-lower mesosphere $A^3\Sigma$ is, therefore, expected to undergo numerous collisions. Since N_2 is an inefficient quencher and the concentration of O is very low, collisions of $A^3\Sigma$ with O_2 only need to be considered in the stratosphere. Conversion of a large amount ($\sim$ 6 eV) of electronic energy into the translational degree of freedom being difficult, the following processes appear preferable over mere physical quenching:

$$N_2(A^3\Sigma) + O_2 \rightarrow N_2(X^1\Sigma) + O_2(B^3\Sigma) \quad (1a)$$
$$\rightarrow N_2O + O(^3P, ^1D) \quad (1b)$$
$$\rightarrow N_2(X^1\Sigma) + 2O(^3P) \quad (1c)$$
$$\rightarrow 2NO \quad (1d)$$

Process (1d) requires considerable rearrangement and may be neglected in comparison with the others. There are no direct determinations for the probabilities of either (1a) or (1c). On the other hand, bombardment of an N_2-O_2 mixture by 14 MeV neutrons in laboratory experiments [9] produced only two products $N^{13}N$ and $N^{13}N$ O. Accordingly, we have assumed that Equation (1b) is the most probable channel in the reactions of $N_2(A^3\Sigma)$ with O_2 which occur with a rate coefficient of 8 x 10^{-12} cm^3 s^{-1} [10]. With this assumption, almost all of the $N_2(A^3\Sigma)$ produced directly or indirectly becomes N_2O in the stratosphere-lower mesosphere.

3. IMPLICATIONS OF ION-CHEMISTRY

Generation of HO_X by positive ion chemistry is known to play a decisive role in the mesospheric O_3 depression in response to proton precipitation [11]. The points to be made here are: (a) even though HO_X does not directly affect O_3 balance in the stratosphere, it may still do so indirectly; e.g., through the release of active CℓO, and (b) implications of negative ion chemistry may also be quite significant. To clarify this new point (b), a simplified version of negative ion chemistry which omits the hydration part is
version of negative ion chemistry which omits the hydration part is

given in Figure 1. Obviously, O_4^-, CO_2^-, CO_3^- consume odd nitrogen (NO, NO_2) before they become terminal NO_3^- ions. In the stratosphere almost all the negative charge is carreid by hydrated negative ions, and the ionization is removed by ion-ion neutralization. For hydrated ions, neutralization by simple electron transfer becomes improbable not only on the basis of energetics [12-14] but also due to unfavorable Franck-Condon factors [15]. For the stratospheric-lower mesospheric negative ions, therefore, neutralization would most probably occur through coalescence with positively charged hydrated ions. Consequently, the formation of HNO_3 in the neutralization of $NO_3^- \cdot (H_2O)_n$ and of HO_X in the neutralizations of $CO_3^- \cdot (H_2O)_m$ become quite possible. In this possible scenario, during daytime the negative ion chemistry takes an active NO_X out of circulation and transforms it into the form of an inactive HNO_3. During nighttime, when the terminal ions are O_4^-, CO_3^-, or CO_4^-, however, there is no effective difference between positive and negative ion chemistry; both generate HO_X.

4. MODEL CALCULATION

In order to test the potentialities of the two classes of physical processes explained in Sections 2 and 3, some calculations were performed using a one-dimensional photochemical-dynamical model atmosphere developed by Turco and Whitten [16]. An ionization profile calculated by Reagan [17] for the August 1972 SPE at its peak was switched on at midday and held constant in magnitude for 6 h. Thereafter it was allowed to decay with a time constant such that the ionization rate decreased to one tenth of its peak value in 24 h. The calculations for the positive and negative ion and electron densities were done approximately using effective recombination coefficients and lumped parameters pertinent to solar proton events [18].

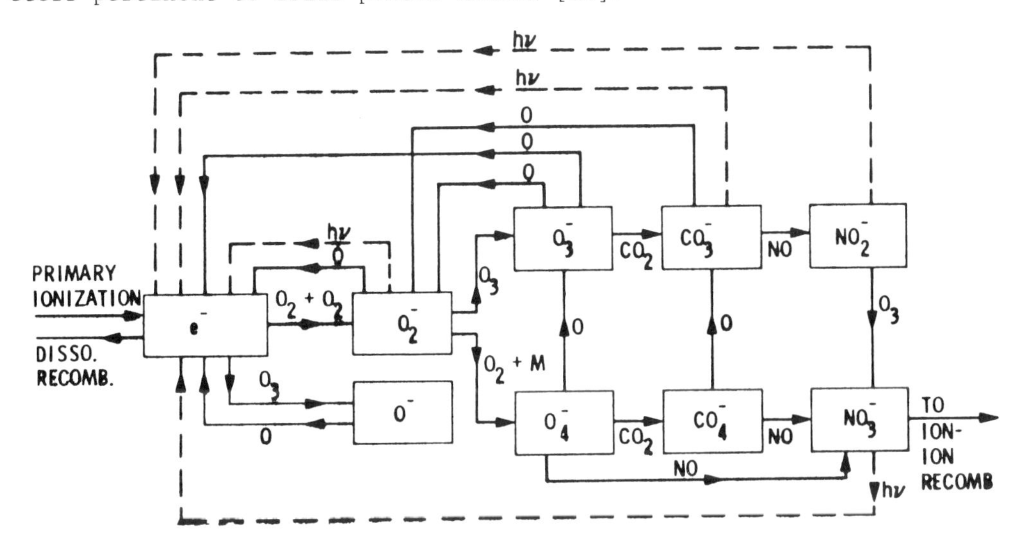

Figure 1. Simplified schematic representation of negative ion chemistry.

5. RESULTS AND DISCUSSIONS

Table I shows the results pertaining to N_2O formation by particle precipitation. For the assumed ionization profile there is clear enhancement of the ambient N_2O concentration at all altitudes above 40 km. The percent enhancement increased with altitude, and was ∿60% in the stratopause region. The enhanced concentration persisted undiminished for several days after the precipitation was effectively reduced to zero. This was a consequence of the very long lifetime of N_2O in the atmospheric region of interest. To realize the significance of the increases in N_2O, it is most instructive to note that this quantity was about 33% of the total increase in odd nitrogen (NO + NO_X) at each altitude. It is possible that after the precipitation has stopped, reactions of $O^1(D)$ with the excess N_2O may continue to compensate for the removal of the initial NO_X deposition. Presently, theoretical calculations [19-20] predict that by day 19 the initial stratospheric O_3 depletion created by the August 1972 event should have recovered almost completely. The experimental observations [19], however, show continued depletion long after day 19. Hopefully, the gradual release of NO from the excess N_2O may possibly account for the observational data. Unfortunately, on these time scales one-dimensional calculations become devoid of physical relevance because of the neglect of lateral transport. It is, therefore, suggested that the possibility hypothesized in the present paper be tested by two-dimensional model calculations.

TABLE I. Enhancement of N_2O concentration under the influence of a model particle precipitation event.

Alt. (km)	Reference N_2O conc.	N_2O concentration (cm^{-3}) after the onset of the model event 6 h.	24 h.	72 h.
40	9.0 (9)*	9.2 (9)	9.3 (9)	9.3 (9)
44	3.7 (9)	3.9 (9)	4.0 (9)	4.0 (9)
48	1.6 (9)	1.8 (9)	1.9 (9)	1.9 (9)
52	7.2 (8)	8.8 (8)	9.8 (8)	9.8 (8)
56	3.4 (8)	4.7 (8)	5.5 (8)	5.5 (8)
60	1.7 (8)	2.7 (8)	3.3 (8)	3.3 (8)
64	8.5 (7)	1.6 (8)	2.1 (8)	2.1 (8)
68	4.4 (7)	1.0 (8)	1.4 (8)	1.4 (8)

*$A\ (\pm B) \equiv A \times 10^{\pm B}$

In the stratosphere HNO_3 has a long lifetime of several (∿6) hours. Due to this, for daytime ionization events, a dramatic HNO_3 enhancement should be a prominent feature of the prompt response of the stratosphere. This is clearly seen from the results presented in Table II for our model precipitation event which started and remained at its peak during the daytime hours between 12.00 and 18.00.

TABLE II. Comparison of predicted HNO_3 concentration with and without the contributions from negative ion chemistry

Alt.	HNO_3 concentration (cm^{-3}) after the onset of a model particle precipitation			
	6 h		24 h	
	without	with	without	with
44	4.51 (6)*	9.1 (8)	6.8 (6)	2.7 (7)
50	1.0 (5)	2.15 (8)	3.2 (5)	3.0 (7)
60	1.4 (3)	4.0 (7)	5.6 (4)	5.1 (6)

*$A(\pm B) \equiv A \times 10^{\pm B}$

Omission of the effects of ion chemistry may also produce significant underestimation of O_3 depletion in the initial phases of the particle precipitation event. The stratosphere, however, relaxes quickly from the effect of ion chemistry. Thus, over a period of 24 h (by which time the intensity of precipitation has decreased to one-tenth of its peak value) the underestimations of O_3 depletion due to neglect of ion chemistry are only 11.5%, 8.7%, and 30% at 44, 50, and 60 km against 15%, 40%, and 73% underestimation over a time period of 6 h when the precipitation was at its peak. Over a three day period the inclusion or neglect of the ion chemistry did not make much difference. Clearly, the ion chemistry must be properly taken into account in describing the prompt response of the stratosphere to solar particle events. Since the stratospheric dynamics responds to major solar flares on a 24 h [21] time scale, our conclusion regarding the importance of ion-chemistry is relevant in the context of sun-weather/climate relationships. For this reason more accurate theoretical models for the effects of ion chemistry are currently under study by this author. In the meantime, it is also advisable that our assumptions regarding N_2O formation from $N_2(A^3\Sigma)$ and HNO_3 and HO_x formation from negative ion chemistry be subjected to refined laboratory tests, even though these assumptions were based on currently available convincing laboratory data.

ACKNOWLEDGMENTS. This paper presents the results of one phase of research carried out at the Jet Propulsion Laboratory, California Institute of Technology, under Contract No. NAS 7-100 sponsored by the National Aeronautics and Space Administration.

REFERENCES

1. R. Markson, Nature, 273, 103, 1978
2. J. R. Herman and R. Goldberg, J. Atmos. Terr, Phys., 40, 121, 1978.
3. W. O. Roberts and R. H. Olson, Rev. Geophys. Space Phys., 11, 731, 1973.
4. R. E. Dickinson, Bull. Am. Met. Soc., 56, 1240, 1975.
5. W. R. Bandeen and S. P. Maran (Ed.), Possible Relationships between

Solar Activity and Meteorological Phenomena, NASA-SP-366, 1975, National Aeronautics and Space Administration, Washington, D.C., USA, 1975.

6. J. Chamberlain, J. Atmos. Sci., 34, 737, 1977.
7. G. R. Reid, I. S. A. Isaksen, R. E. Holzer, and P. J. Crutzen, Nature, 259, 177, 1976.
8. J. R. Bates, Quart. J. R. Met. Soc., 103, 397, 1977.
9. D. J. Malcombe-Lawis, Nature, 247, 540, 1974.
10. J. W. Dreyer, D. Perner and C. R. Roy, J. Chem. Phys., 61, 3164, 1973.
11. W. Swider and T. K. Kaneshea, Planet. Space Sci., 21, 1969, 1973.
12. R.A. Bennett, D. L. Huestis, J. T. Moseley, J. Mukherjee, R. E. Olson, S. W. Benson, J. R. Peterson, and F. T. Smith, AFCRL-TR-74-0417, Air Force Cambridge Research Laboratory, Hanscom, MA, 1974.
13. D. L. Huestis, J. T. Moseley, D. Mukherjee, J. R. Peterson, F. T. Smith and H. D. Zeman, AFCRL-TR-75-0606, Air Force Cambridge Research Laboratory, Hanscom, MA, 1975.
14. F. C. Fehsenfeld; Private communication.
15. F. T. Smith, Private communicated cited by D. S. Smith, M. J. Church and T. M. Miller, J. Chem. Phys., 68, 1224, 1978.
16. R. P. Turco and R. C. Whitten, NASA-TP-1002, National Aeronautics and Space Administration, Washington, D.C., 1977.
17. J. B. Reagan, Ionization Processes in Dynamical and Chemical Coupling, Ed. by B. Grandal and J. A. Holtet, D. Reidel Publishing Company, Dordrecht, Holland.
18. J. C. Ulwick, Effective Recombination Coefficients and Lumped Parameters in the D Region During Solar Particle Events in Proceedings of COSPAR Symposium on Solar Particle Events of Nov. 1969, AFCRL Report 72-0474, Air Force Cambridge Res. Lab., Hanscom, MA, 1972.
19. D. F. Heath, A. J. Krueger and P. J. Crutzen, Science, 197, 866, 1977.
20. W. J. Boruki, D. S. Colburn, R. C. Whitten, L. A. Capone, and M. Covert, EOS, Trans. Am. Geophys. Union, 59, 284, 1978.
21. C. S. Zerefos, Planet. Space Sci., 23, 1035, 1975.

SPECTRAL ANALYSIS OF LONG-INTERVAL TEMPERATURE DATA

C. G. Maclennan and L. J. Lanzerotti

Bell Laboratories
Murray Hill, NJ 07974 U.S.A.

ABSTRACT. Searches for periodicities in climatological data, and their associations with periodicities in solar-terrestrial phenomena, are common procedures in the investigation of possible solar influences on weather and the mechanisms involved. The dynamic spectral analysis technique, applied to sufficiently long data sets, can provide evidence for short-lived (in comparison to the entire data set) periodicities that would be "washed out" in an analysis of an entire data set. We have performed such analyses, using a Fast Fourier Transform algorithm and the Thomson data window applied in the time domain, to long interval (∿200 yr) monthly temperature data acquired at Baltimore, Philadelphia, and New Haven, U.S.A., and in central England. We find that at times broad enhancements in the power can occur, centered around 11 yr. The central England data show less evidence for any 11 yr variations. A strong ∿2 yr periodicity ("quasi-biennial oscillation") in temperature was observed between ∿1871 and ∿1891 at Baltimore and Philadelphia, with a less noticeable periodicity seen at New Haven. No strong enhancement at this period was observed at this time in England. These dynamic spectral results demonstrate the power of the technique and the problems inherent in studying regional climatic conditions and their associations with solar phenomena.

1. INTRODUCTION

Several studies have demonstrated, via time-series analyses, that long-time periods of temperature data do not show evidence for any strong periodicities related to the solar cycle [1 - 3]. However, there is not unanimity in this regard [e.g., 4 - 5]. There are not many sets of long time intervals of reliable climatic data that can be reasonably used to study periodicities. Such long-interval observations exist largely by historical "accident " and, as such, may or may not be representative of the climatic conditions over a broad geographical region. In this paper we compare temperature periodicities determined from data obtained at four locations covering overlapping time intervals. To examine the data for short-term periodicities we use a dynamic

B.M. McCormac and T.A. Seliga (Eds.): Solar-Terrestrial Influences on Weather and Climate. 305-310.

spectral analysis technique. We show that there are significant differences in the temperature fluctuations observed between stations in eastern North America and central England. Further, we show that there are also differences in the North American data for stations separated by only 200 to 300 km.

We have performed power spectral analyses and digital dynamic power spectra analyses on the time series of temperature data from three locations in North America. The North American data (monthly averages) are from Baltimore (39°18'N, 76°38'W geographic; 1817-1977), Philadelphia (40°0'N, 75°10'W geographic; 1790-1977), and New Haven (41°18'N, 72°55'W geographic; 1781-1977). These data were obtained through the National Climatic Center, Asheville, N.C. Analysis was also made of monthly mean temperature records of central England (1659-1973) as constructed by Manley [6]. Power spectral analyses of these data by the maximum entropy method were reported by Mason [1]. The English centigrade records were converted to Fahrenheit for purposes of comparison with the North American data.

The spectral analysis technique used is that previously employed by us for geomagnetic studies [e.g., 7-9]. After first applying the Thomson window to the data in the time domain [10] the analyses were performed using a fast Fourier transform algorithm. Dynamic spectra were calculated in the same manner, but taking 256 or 1024 point segments of data and sliding 2 yr after each analysis.

2. RESULTS

Shown in Figure 1 are power spectra of the English data for four consecutive time intervals between 1659 and 1959. There are strong yearly and twice yearly (6 month) peaks in all of the spectra. (The two-sigma deviation at each spectral point is equal to $\sim\pm$ 3 times the value at the given point.) A strong peak at 2 yr is seen in the 80 yr data set from 1739-1819; an $\sim$1.6 yr peak is observed in the 60 yr data set from 1899-1959. No significant peak is found in the range $\sim$10 to 12 yr in either the spectra of Figure 1 or in the spectra of the 300 yr data set taken as a whole (not shown here). This is consistent with the results of Mason [1].

The spectra vary considerably from interval to interval in Figure 1. Thus, a spectra of the entire data series would not reveal short-term periodicities in the local climatic conditions. Shown in Figure 2 in the left-hand panel is a dynamic spectrum of the central England data showing frequencies from about 0.25 to 2.5 cycles yr^{-1} for the spectra centered at the years 1671-1961. The 1 yr period is clearly evident, as is a 6 mo period. This last periodicity recurs throughout the 300 yr interval, although at times (e.g., $\sim$1821 to $\sim$1851) it essentially disappears. No strong, continuous enhancement is seen at an $\sim$2 yr period.

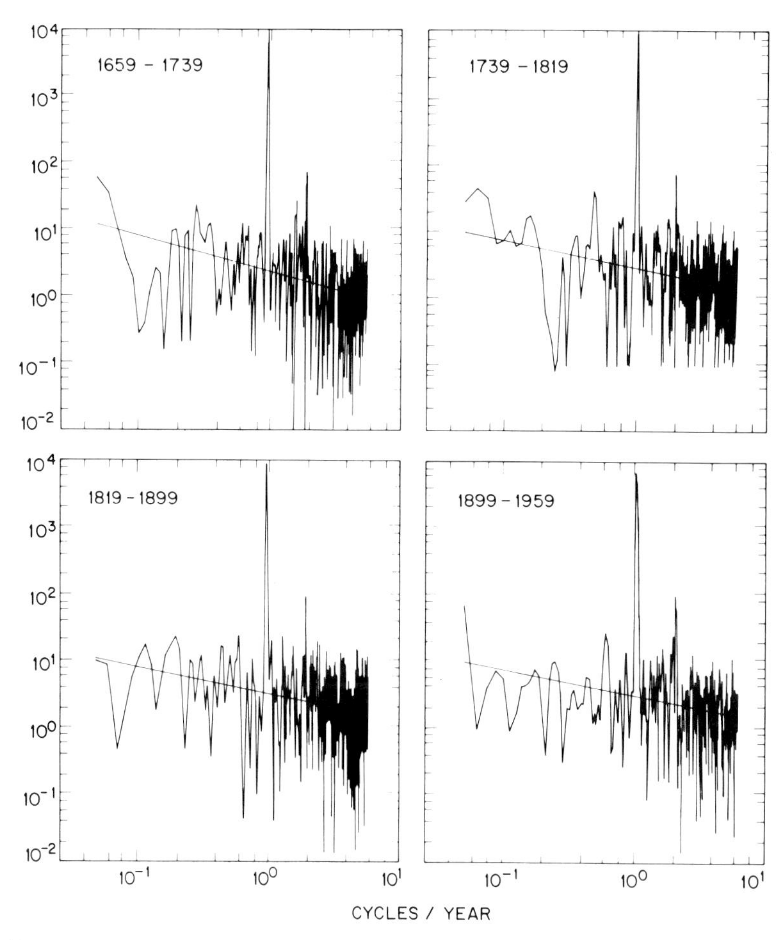

Figure 1. Power spectra of central England monthly temperature data for four time intervals from 1659 to 1959.

The ∿ 2 yr period reported by Mason [1] in a spectral analysis of the entire central England data set was evident but was certainly not the strongest peak in the spectrum. The right-hand panel of Figure 2 shows the corresponding spectrum of the New Haven data beginning at ∿ 1790. Note that the 6-mo periodicity is clearly absent here, although a 2-yr periodicity is somewhat stronger.

Plotted in Figure 3 are dynamic spectra of the data from three North American stations in the interval 1820 to 1940. The most striking feature in these spectra is the enhancement in the periodicity with period ∿ 2 yr observed during the interval between ∿ 1870 and 1900. This enhancement was observed most strongly, and to start earliest, at the most southerly station, Baltimore. It began later at Philadelphia, and ended about the same time that it did at Baltimore. The enhancement observed at the most northern station, New Haven, was not as strong. The dynamic spectra of the central England data at this time show no significant enhancements.

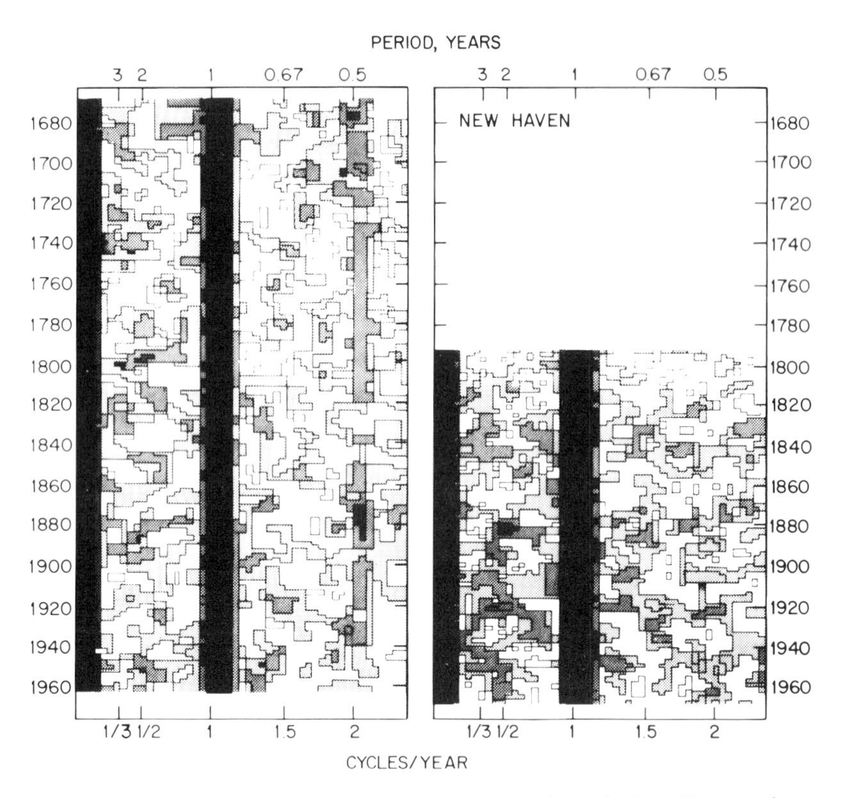

Figure 2. Dynamic spectra of the central England and New Haven data sets for the period band ∿0.25 to 4 yr. Two consecutive intensity levels differ by ∿10 dB. Note the enhancements at 1 yr and 6 mo periods in the central England data, but the absence of a similar 6 mo enhancement in the New Haven data.

There is some evidence for enhancements in the spectra with periods between 11 and 22 yr in the data of Figure 3. The enhancements occur at different time intervals at the three stations. For example, an enhancement at a period of ∿15 yr was observed at New Haven between ∿1860 and 1890. An enhancement at a period of ∿10 yr was seen at Philadelphia at this time and no enhancement (rather a depression) was observed in the Baltimore data.

3. DISCUSSION

The major finding from these analyses is that there are short-lived time variabilities in the temperature records as recorded by individual stations. These variabilities may not be the same at measurement sites located within only a few hundred kilometers of one another. These short-lived periodicities can be "washed out" when a spectrum is made of an extended time series of data.

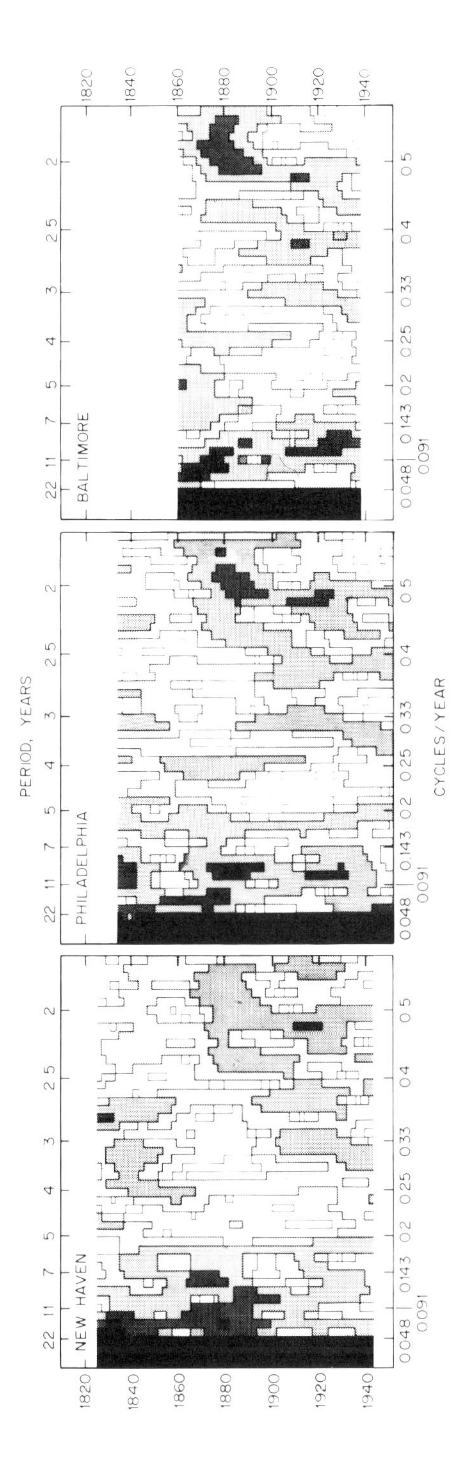

Figure 3. Dynamic spectra of temperature data from three North American stations using data from 1781-1977.

The results also show that for these particular stations there are no significant relationships between the periodicities observed on the two continents. In particular, an approximately 6 mo periodicity clearly evident in the central England records for most of the 300 yr interval was not observed on the east coast of North America.

The approximately 2 yr periodicities observed at times in the temperature spectra can be called "quasi-biennial oscillations" (e.g., [11], [12], [1]). The spectral results demonstrate clearly that the 2 yr periodicities are sporadic in occurrence and can be very strong at times at one station (e.g., Baltimore) while being almost absent at another within a few hundred km (e.g., New Haven; see Figure 3). Landsberg and Kaylor [12] have shown that the quasi-biennial oscillation appears stronger if only winter data are spectral analyzed. Our dynamic spectral results demonstrate that definite temporal and spatial variations also exist in the oscillation; these variations must be considered in developing theoretical mechanisms.

Sugiura and Poros [13] noted a time variability in the quasi-biennial variation of geomagnetic activity. The relationship, if any, of such a 2 yr variation to global climatic occurrences is not obvious from the data presented here.

ACKNOWLEDGMENT

We thank Dr. G. P. Gregori for helpful discussions on this paper and for pointing out the 6 mo periodicity in the English data from his superposed epoch study. We also thank R. Quayle and H. Diaz of the NOAA National Climatic Center, Asheville, NC, for supplying the North American data.

REFERENCES

1. B. J. Mason, Quart. J. R. Met. Soc., 102, 473, 1976.
2. A. S. Monin and I. L. Vulis, Tellus, 23, 337, 1971.
3. S. J. Mock and W. D. Hibler, III, Nature, 261, 484, 1976.
4. R. G. Currie, J. Geophys. Res., 36, 5657, 1974.
5. J. W. King, Astronaut. Aeronaut., 13 (4), 10, 1975.
6. Gordon Manley, Quart. J. R. Met. Soc., 100, 389, 1974.
7. L. J. Lanzerotti and M. F. Robbins, J. Geophys. Res., 78, 3816, 1973
8. H. Fukunishi and L. J. Lanzerotti, J. Geophys. Res., 79, 142, 1974.
9. C. G. Maclennan, L. J. Lanzerotti, and H. Fukunishi, Spatial and temporal variations of ULF disturbances, in Magnetospheric Physics (ed. by B. M. McCormac), D. Reidel, Dordrecht-Holland, 1974.
10. D. J. Thomson, M. F. Robbins, C. G. Maclennan, and L. J. Lanzerotti, Phys. Earth. Planet. Int., 12, 217, 1976.
11. H. E. Landsberg, Beiträge zur Physik der Atmosphäre, 35, 184, 1962.
12. H. E. Landsberg and R. E. Kaylor, J. Interdisciplinary Cycle Res., 7, 237, 1976.
13. M. Sugiura and D. J. Poros, J. Geophys. Res., 82, 5621, 1977.

PRELIMINARY REPORT OF CHANGES IN TROPOPAUSE HEIGHT COMPARED TO VARIATIONS IN SOLAR BEHAVIOR

John P. Basart* and Douglas N. Yarger+
Iowa State University
Ames, IA 50011, U.S.A.

ABSTRACT. A study has been made relating changes in the tropopause height to magnetic sector boundary crossings. Superposed epoch analyses did not clearly indicate a relationship. Modeling the fluctuations with a damped oscillation triggered by boundary crossings did give features varying in time which closely approximate the data.

In the near future, investigators at Iowa State University and the University of Wisconsin will undertake a series of experiments with a forward scatter radar to investigate correlations between tropopause behavior and solar activity on time scales varying in length from a few minutes to a few days. Preparatory to these experiments we have been studying tropopause behavior recorded at several Canadian stations and searching for correlations with solar activity on time scales ranging from several days to a year. The solar variable chosen for most of the study has been the magnetic sector boundary crossing.

Superposed epoch analyses with magnetic sector boundary crossings as key days were completed on daily time series of tropopause heights for stations Alert, Resolute, Baker Lake, Trout Lake, and Maniwaki [1] for 1968 and for the same five stations plus Moosonee for 1969. None of these analyses, carried out over a time range of $\pm$ 30 days about key days, gave a result greater than 95% confidence for a correlation. The analyses were done separately on plus-to-minus crossings, on minus-to-plus crossings, and on both types of crossings combined. All crossings for 1968 and 1969 as given by Svalgaard [2] were used. An analysis was carried out on the change of height for all five stations combined from one day to the next using the results of the superposed epoch analyses.

A more intensive study of tropopause behavior was made by inspecting daily changes in the raw data. A time series of daily height measurements shows that a peak in tropopause height could be identified with a particular boundary crossing. However, a variation in time existed

*Department of Electrical Engineering and +Department of Earth Sciences

B.M. McCormac and T.A. Seliga (Eds.): Solar-Terrestrial Influences on Weather and Climate. 311-315.

as to the exact day a peak occurred compared to the exact day a crossing occurred. This was evidence that a superposed epoch analysis would give a negative result because of the smearing out effect. Further evidence for a one-to-one relationship between data peaks and boundary crossings was the similarity between a Fourier analysis of the tropopause heights and the average time interval between crossings in 1969. The maximum coefficient for the Fourier series, other than the mean, was about 14 days for most stations and the average crossing interval was 14 days.

To find a possible cause-and-effect relationship between an individual change in tropopause height and an individual boundary crossing, an empirical model for the tropopause height variations was constructed. The atmosphere was considered to be a lossy resonant system with the magnetic sector boundary crossings representing the input, and the tropopause height variations the output. The impulse response of this system is a damped sinusoidal wave. The mathematical function for the simplest model considered is

$$h(t) = \sum_{i=1}^{N} u(t - T_{di}) A e^{-\xi\omega_r (t - T_{di})} \cos[\omega_n (t - T_{di}) + \phi]$$

where t = time as days of the year
T_{di} = day of magnetic sector boundary crossing
ω_n = natural resonant angular frequency
$\omega_r = \omega_n \sqrt{1 - \xi^2}$ damped frequency
ξ = damping ratio
ϕ = phase
A = amplitude of wave
N = number of boundary crossings
$u(t)$ = unit step function
$h(t)$ = tropopause height variations

If indeed the crossing of the magnetic sector boundary leads to a change in tropopause height, the equivalent input function to the atmospheric system could have any arbitrary wave shape. In this case, harmonic functions of the fundamental frequency ω_n must be added to the model. However, a single sinusoidal wave was sufficient in many attempts to fit the model to the data. This indicates that many times the boundary crossing could be considered as an impulse function. The effect of the model, then, is to start a new cosine wave at each boundary crossing. Each new wave is superposed with several previously triggered waves and later fades away as determined by the damping ratio. Various effective amplitudes and periods can be simulated by the superposition. If the tropopause height changes do not occur on the day of a boundary crossing, the parameter T_{di} can be incremented to produce an advance or a delay of the triggered wave.

Data from the Canadian stations exhibit four basic types of fluctuations: (1) a sinusoidal type variation, (2) a single narrow peak, (3) a double or triple peak, and (4) a broad maximum that may last for up to 20 days. Clearly the first type is the one most easily satisfied by the basic model.

Type 3 frequently has peaks separated in time by harmonic values of the type 1 variations so triggered harmonics are necessary to model it. Type 4 requires a longer wave (than type 1) to be present and often can be a subharmonic of type one. The single narrow peak (type 2) which is not spaced correctly in time to flow with a sinusoidal variation can be difficult to fit. Modeling of these four types of variation has been attempted with the fewest parameters possible. A successful example of type 1 modeling will be discussed.

The lower trace in Figure 1 illustrates a 100 day segment of the tropopause height over Maniwaki in 1969. The data have been de-trended by removing the mean and Fourier sine and cosine components with a period of 365 days. The result was smoothed twice with a 3 point moving average. Variations in the tropopause height follow a sinusoidal behavior but the time interval between the zero crossing fluctuates to such an extent that a sine wave does not fit them. Above the data in Figure 1 is a plot of a very simple model. The parameters are $A = 1.5$ km, $\phi = 0$, $\xi = 0.122$, while the waves are triggered on the day of the boundary crossings. In this example, as was found with many others, the model fits best when the cosine starts on a positive peak for minus-to-plus crossings, and on a negative peak for plus-to-minus crossings. With one exception near day 118, the model closely agrees with the data. The zero crossings of the model shift to the left or right as do those of the data, and the model amplitude grows near the peak on day 139 as does that of the data. This example is the best model obtained with the least parameters. Other sections of the data can be fit by introducing time delays, phase shifts, and harmonics.

A model that depends upon magnetic sector boundary crossings must, of course, be flexible enough to fit the data for more than one meteorological site. Comparisons between data from Maniwaki, Moosonee, and Trout Lake showed that most strong features on one data set show up on another data set with amplitude differences and perhaps phase shifts. A composite curve of the data from the three sites was used to determine which features should be modeled and which features were statistical fluctuations. Features in common among the data sets stood out clearly from the statistical variations that averaged to small amplitudes. Data from Maniwaki, Moosonee, and Trout Lake were most suitable for modeling because of their large fluctuations in tropopause height. The smaller fluctuations in the higher latitude data from Baker Lake, Resolute, and Alert were harder to separate into common features and into statistical variations. An interesting feature of the model is that it has the capability of producing large and small amplitudes as a function of time according to whether the supposed waves are triggered in or out of phase as the time interval between boundary crossings changes.

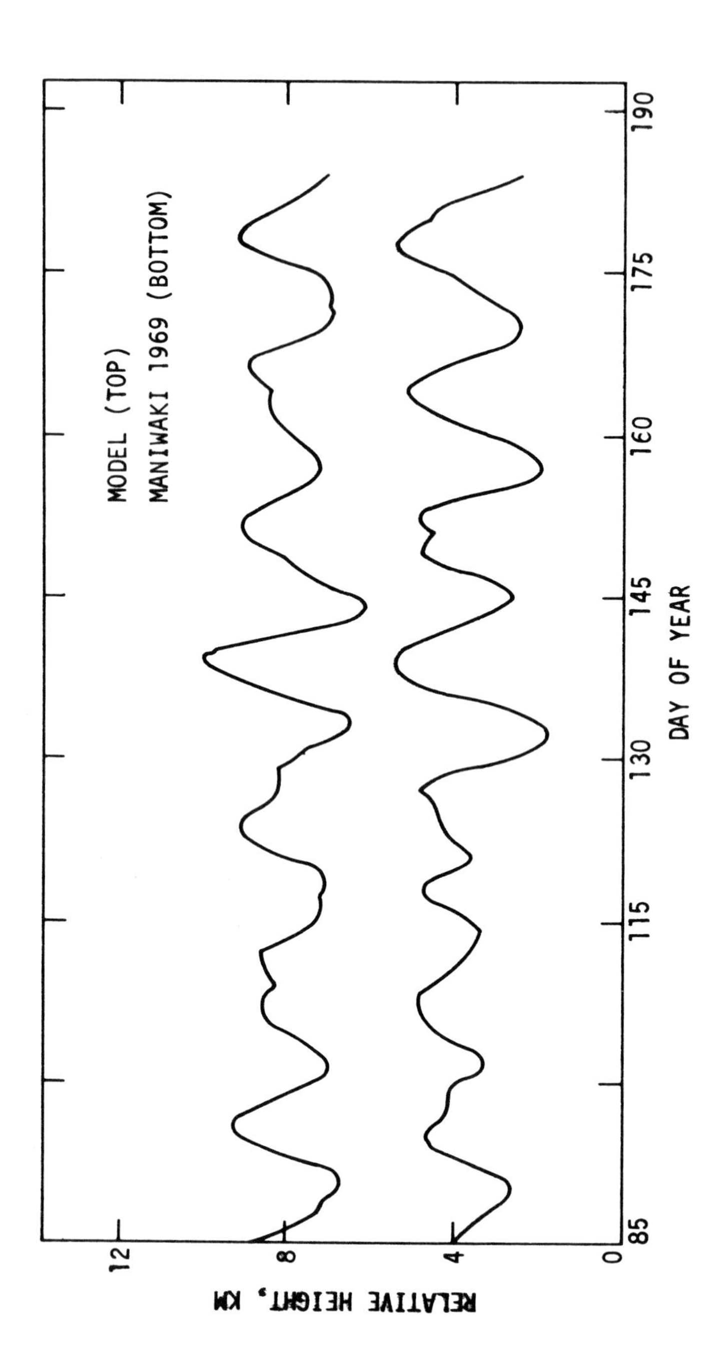

Figure 1. Lower trace: Height changes of tropopause in km over Maniwaki in 1969. Lower trace: Model based on triggering a damped cosine wave on each magnetic sector boundary crossing.

Another piece of evidence favors a deterministic relationship between boundary crossings and tropopause height fluctuations. Time differences between adjacent model and data maxima were plotted against days for a year. The model was the same as that used for Figure 1. The plot had a pronounced seasonal variation changing from a positive difference to a negative difference over the period of a year. A relationship for this change as a function of time was put into the model. Overall agreement between model and data was improved but no segments had agreement as good as that shown in Figure 1. Thus, the seasonal trend, while not accounting for all variations between model and data, did strengthen the hypothesis of a relationship between boundary crossings and tropopause heights.

Various comparisons between magnetic sector boundary crossings and variations in tropopause heights, and modeling relating the two, have not conclusively shown that events relating to boundary crossings cause oscillations in the height of the tropopause, but they do show that a model can reproduce many of the data features.

ACKNOLWEDGMENTS

We thank Curtis Stephany and Edward Hall for assistance in the data analysis. John Basart thanks the Engineering Research Institute at Iowa State University for support during this project.

REFERENCES

1. Monthly Bulletin, Canadian Upper Air Data, Department of Transport, Meteorological Section, Toronto, Canada, 1968, 1969.
2. Solar-Terrestrial Physics and Meteorology: A Working Document, issued by Special Committee for Solar-Terrestrial Physics, c/o National Academy of Sciences, 1975.

THE STRATOSPHERIC MECHANISM OF SOLAR AND ANTHROPOGENIC INFLUENCES ON CLIMATE

K. Ya. Kondratyev
Department of Radiation Studies
Main Geophysical Observatory
Leningrad, U.S.S.R., and

G. A. Nikolsky
Department of Atmospheric Physics, Institute of Physics,
Leningrad University, Leningrad, U.S.S.R.

ABSTRACT. Analysis of the results of balloon soundings in 1962-70 and the calculations of the shortwave solar radiation attenuation made it possible to reveal the reason for abnormal attenuation of the total flux of direct solar radiation in the middle and upper stratosphere in 1962-66 and in the winter periods of 1967-68-69 and 1970-71. It has been discovered that the influence of the 1961-62 nuclear tests was connected with a considerable increase of NO_x concentrations in the stratosphere of high and moderate latitudes of the Northern Hemisphere (NH) which increased absorption of the solar flux above 30 km altitudes. Since the decrease in the attenuation of S was followed by a global increase in the ozone content, NO_2 should be considered one of the main climate-forming factors of the stratosphere and, in the end, the troposphere.

1. ANTHROPOGENIC IMPACT ON THE STRATOSPHERE IN 1961-68

Detailed analysis of the data from balloon soundings carried out at Leningrad University during 1961-70 has confirmed the supposition that the abnormal attenuation of direct solar radiation (DSR) at heights above 26 km has been caused by the presence of nuclear explosion products in the upper stratosphere [1]. A correlation was marked between the dates of nuclear tests and the periods of the DSR attenuation in layers above 30 km. The DSR integral flux in some cases decreased by the extra-atmospheric value (S_0) by 6 to 8%. Besides, it has been noted [1] that in the upper atmosphere during the period of 1968-70 there was a systematic increase in the attenuation of the DSR flux and by 1971 amounted to about 1.5%. We attributed this to the influence of atmospheric aerosol.

In recent years, a number of papers have appeared where calculations have been made of the quantities of NO_x which are generated during ex-

B.M. McCormac and T.A. Seliga (Eds.): Solar-Terrestrial Influences on Weather and Climate. 317-322.

plosions of hydrogen bombs of different yields. Apart from this, it has been established that with yields exceeding 1 megaton (MT) of trinitrotoluene (TNT) equivalent, the products of explosions near the Earth's surface are injected directly into the stratosphere. With powers of the order of 30 to 50 MT, the fire ball stabilizes at heights of 30 to 40 km [2]. According to approximate estimations, the number of NO_x molecules on the average amounts to 10^{32} per 1 MT. The maximum calculation gives 1.5 x 10^{32} NO_x molecules. Hampson mentioned [3] that with explosions in the upper atmosphere, the number of molecules of NO_2 generated may increase by two orders of magnitude. According to the available data [4], the cumulative yield of explosions during the period of 1961-62 was about 340 MT. As a result, about 1.3 to 1.7 MT of NO_x, or 5 x 10^{34} molecules, were injected into the upper stratosphere at heights 20 to 25 km.

Our calculations have shown that the content of NO_2 in a vertical column of unit cross-section penetrating the semiglobal ring-shaped cloud produced by nuclear tests in the moderate and high latitudes of the Northern hemisphere (NH) must amount to 1.6 x 10^{17} molecules cm^{-2} to be able to cause additional attenuation of the DSR integral flux of 2.6%. The above nuclear detonation data give the actual content of about 3.5 x 10^{16} molecules. If we take into consideration the fact that similar tests were performed in previous years (1954-58) as well, and the increased efficiency of explosions in the stratosphere (from the point of view of the amount of NO_x produced) in comparison with those near the Earth's surface, the difference between the estimates listed above will be reduced to a minimum.

The realistic character of the estimations of the considered quantities of NO_2 molecules in the vertical atmospheric column at 50°N in the second half of 1962 has been confirmed by the simultaneously obtained data from balloon filter measurements of DSR spectral fluxes (in the 305 to 370 nm range). Thus, for the ascent of July 7, 1962 (ascent No. 9 in Figure 1) it has been observed that the number of NO_2 molecules above the 26 km layer is 3.4 x 10^{17} molecules cm^{-2}. Therefore, it may be concluded with confidence that the effects of nuclear tests in 1961-62 which resulted in substantial values of DSR attenuation (2.6% and more) are of global scale and considerable duration (1962-68).

Figure 1 shows the rate of decrease of additional attenuation of DSR in the upper stratosphere of moderate latitudes which is 0.4% annual (up to 1968). The continued deficit of solar radiation in the troposphere and on the Earth's surface in the moderate and high latitudes of the NH in 1962-63 affected synoptic processes and weather conditions [5]. Such consequences of anthropogenic influences result in the disturbance of normal circulation conditions in the stratosphere and troposphere, i.e., in climate perturbations.

The decrease of DSR absorption with time should be regarded as due not only to the natural sink of NO_2 in the NH troposphere, but also (on

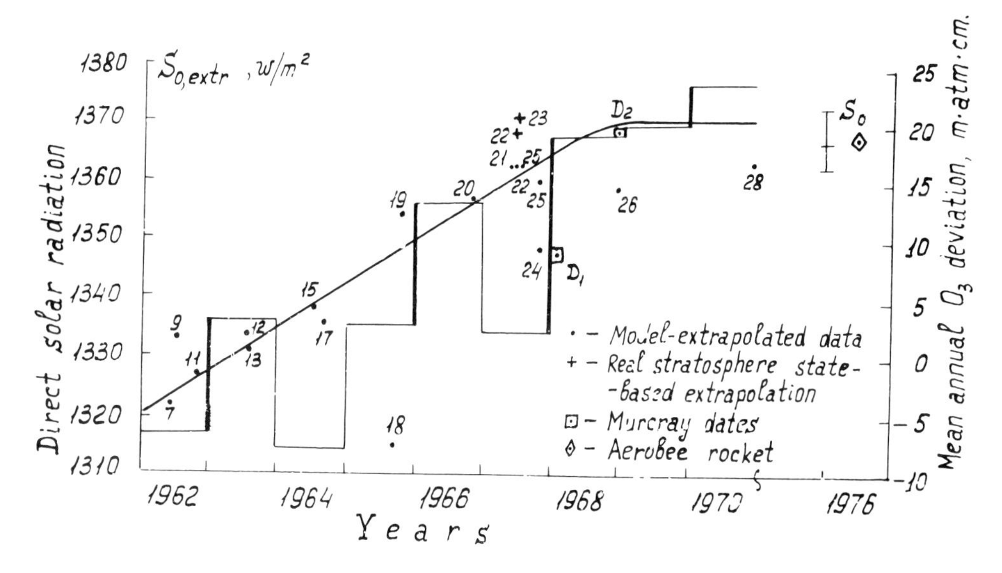

Figure 1. Chronological variations of the direct solar radiation flux measured at heights of about 30 km and extrapolated to outside the limits of the atmosphere by taking account of the attenuating properties of the upper layer which are characteristic for the period of stratospheric purification (1967-68). Numbers put near points correspond to the numbers of balloon ascents. D_1 and D_2 designate the Denver University ascents; . - the model-extrapolated data; + - extrapolation based on the real state of the stratosphere; .the value of S_0 according to the data of measurements from the "Aerobee" rocket. The broken line shows deviations of annual mean global values of the total O_3 content from the value averaged over a prolonged period. Solid vertical lines mark the dates of the strongest unexpected stratospheric warmings.

the basis of the data on the propagation of nuclear explosion products from the NH high latitudes to the equator and then to the Southern Hemisphere (SH) [4] to the NO_2 transfer in the upper stratosphere to the SH where intensive abnormal attenuation of DSR took place during the period of 1963-66. Comparison of experimental data and calculations makes it possible to state that the superposition of nuclear and volcanic effects took place in the SH. Rocket spectral measurements (λ = 3800 Å) at heights 20 to 40 km at Woomera (30°S, 1965) discovered the presence of an attenuating component having at 28 km an optical thickness equal to the Rayleigh one, and at 22 km — a 2.5 times higher optical thickness than the latter [6]. Such attenuation cannot be fully attributed to the influence of the aerosol component. Estimations of the DSR absorption for λ = 3800 Å which corresponds to the maximum of absorption of NO_2, show that above 22 km an atmospheric column must contain 7 x 10^{16} molecules cm^{-1} of NO_2. This surplus amount of NO_2 will surely provide for absorption equal to 1% of S_0 in the vertical column above 22 km.

2. THE SOLAR CONSTANT AND ITS VARIATIONS

Figure 1 presents data on DSR fluxes measured at heights 26 to 33 km and extrapolated to outside the limits of the atmosphere on the basis of model representations of the attenuating properties of the upper layers of the air. For comparison, the data of rocket measurements of the solar constant in 1976 [8] are given. The rocket measurements of S_o were performed with four absolute cavity radiometers of different construction. The value averaged over the readings of three well-correlated radiometers is 1367 ± 6 W m^{-2}. According to data from automatic interplanetary stations "Mariner-6 and -7" and "Nimbus-6" satellite, variations of S_o do not exceed 0.15%. It should also be mentioned here that in connection with the discovery of the rings of the planet Uranus, a strong arugment in favor of the variability of S_o — the changing brightness of the planet — apparently becomes superfluous.

The conversion of our balloon data [7] to outside the limits of the atmosphere was made in accordance with the model representation of the transparency of upper stratospheric layers for 1968. Therefore, all the model-extrapolated data reveal manifestations of the variability of the upper stratosphere transparency, as compared to the conditions of 1968, and they should be regarded as data on the "meteorological solar constant" ($S_{o,M}$) referenced to the absoltue value of the astronomical solar constant $S_{o,astr.}$ = 1371 W m^{-2} determined from the balloon data of 1967-68 and not being subject to any notable variability in dependence on solar activity. However, the "meteorological solar constant" is characteristic of such a dependence which, on the whole, coincides with that suggested earlier for $S_{o,astr.}$ [1]. The notion of $S_{o,M}$ which, according to the representations assumed, must characterize the "sub-O_3" value of $S_{o,astr.}$ determined by the DSR flux at the level of the tropopause, should be specified more accurately. It would be expedient to consider $S_{o,M}$ as a characteristic of the influence of surplus stratospheric NO_x (located above 26 km) on $S_{o,astr.}$.

3. SOLAR INFLUENCES ON THE STRATOSPHRE

Analysis of the results of the winter ascents in 1966-70 has made it possible to ascertain the fact that the deficit of solar radiation in winter periods (up to 1.2%) is the consequence of not so much the anthropogenic influences, as the solar ones and seasonal effects in the absorption of DSR. During those years, solar activity reached its maximum, and there were frequent and intense energetic solar proton events which affected significantly the processes of ionization and dissociation of the N_2, O_2, O_3, and other components of the upper and middle stratosphere. According to [9] an intense solar cosmic rays (SCR) event leads to the generation in the upper stratosphere at high latitudes (ψ > 60°) of the same amount of NO as is produced by a nuclear explosion with a yield of 50 MT or more. There is also a possible decrease of the O_3 concentration at the height of 45 km by 25%, and an increase in the concentration of NO by 2 to 3 times, which will result in the growth of the amount of NO molecules in the vertical column of 6 x 10^{15} cm^{-2}. This increase is approximately 4 times as large as the contribution due to galactic cosmic rays (GCR) in the same region of latitudes.

An important feature of the influence of GCR is their great penetrability. Therefore, the main absorption of GCR takes place at heights 9 to 16 km where the local decrease of the concentration of water vapor and O_3, as well as the local maximum of the concentration of NO_2, are located [10]. The impact of SCR and GCR on stratospheric composition differs both in time (different periods of solar activity cycles), and in height: the maximum of the absorption of SCR is located at heights 30 to 40 km where the maximum of the NO_2 mixing ratio is located.

The combined influence of SCR and GCR apparently determines the dependence of $S_{o,M}$ on solar activity, as is shown in Figure 2. The dependence has been obtained after the exclusion of the trend due to anthropogenic effects of 1961-67. In comparison with the initial curve for S_o[1], the location of the maximum is somewhat biased, and it is now in the region of smaller Wolf numbers (R_z = 40 to 70), and the amplitude of variations does not exceed 1.2%.

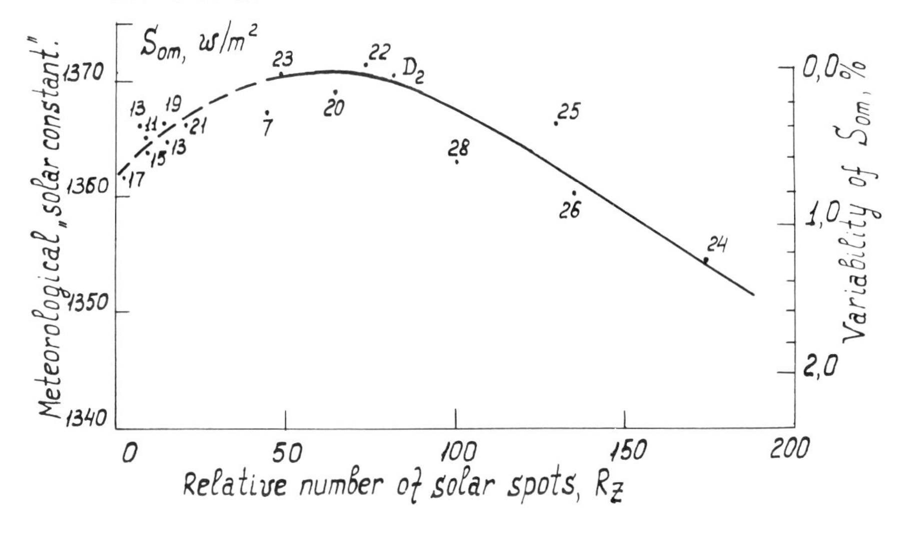

Figure 2. Variation of the "meteorological solar constant" (MSC) in dependence on solar activity. MSC indicates the degree of influence of cosmic rays on the absorption characteristics of the upper half of the stratosphere. The dotted part of the curve corresponds to the region of great errors in the variation of the dependence.

The decrease of $S_{o,M}$ for the range of Wolf numbers exceeding 100 is connected with the effect of energetic solar protons on the chemical composition of the middle and upper stratosphere which mainly is observed as an increase of NO_x at heights 35 to 45 km in the moderate and especially high latitudes. The fall of $S_{o,M}$ in the transition to the area of small values of R_Z is due to the influence of GCR whose intensity has its maximum with the minimum of solar activity [11]. It should be noted that the influence of GCR on the "balloon" $S_{o,M}$ must be less notable than that of SCR because the main GCR absorption takes place in the lower stratosphere.

Thus, the effect of solar activity on the processes in the atmosphere is mainly felt in changes in its chemical composition (the ionization and dissociation of N_2 and O_2, the formation of NO and then NO_2) with subsequent active participation of NO_2 in the transformation of solar radiation (the region of 350 to 550 nm) into heat. The "modulation" of stratospheric composition and, accordingly, of the radiative flux divergence due to SCR and GCR is an important mechanism of solar-tropospheric relations where the stratosphere plays the part of the control element.

REFERENCES

1. K. Ya. Kondratyev, G. A. Nikolsky, Izvestiya of USSR Academy of Sciences, Physics of Atmosphere and Ocean, (VI), 3, 1970.
2. H. Johnston, J. Birks, G. Whitten, J. Geophys. Res., 78, 27, 1973.
3. J. Hampson, Surface Insolation and Climatic Potential Effect of Atmospheric Contamination, Preprint, Univ. of Western Ontario, London, Canada, 1977.
4. M. A. Ruderman and J. W. Chamberlain, Planet. Space Sci., 23, 247, 1975.
5. J. K. Angell and J. Korshover, Monthly Weather Rev., 105, 375, 1977.
6. A. W. Brewer, A. W. Wilson, Quart. J. Roy. Met. Soc., 92, 267, 1966.
7. K. Ya. Kondratyev, G. A. Nikolsky, D. G. Murcray, J. J. Kosters, and P. R. Gast, The Solar Constant from Data of Balloon Investigation in the USSR and the USA. Space Research XI, Akademie-Verlag, Berlin, 1971.
8. Ch. H. Duncan, R. G. Harrison, J. R. Hickey, J. M. Kendall, Sr., M. P. Thekaekara, and R. G. Wilson, Appl. Optics, 16, 2690, 1977.
9. P. J. Crutzen, J. S. A. Isaksen, and G. C. Reed, Solar Proton Events: Stratospheric Sources of Nitric Oxide, Science, 189, 457, 1975.
10. M. Ackerman, J. Atm. Sci., 32, 1649, 1975.
11. M. Nicolet, On the Production of Nitric Oxide by Cosmic Rays in the Mesosphere and Stratosphere, Proc. Fourth Conf. on CIAP (Feb. 1975), 292, 1976.

HIGH LATITUDES ANTICYCLOGENESIS AND ENHANCED ATMOSPHERIC LATENT HEAT TRANSFER PROCESSES

J. H. Shirley
Life Support Systems Group, Ltd.
2432 N. W. Johnson
Portland, OR 97210 U.S.A.

ABSTRACT. Several investigators have demonstrated a positive correlation between enhanced solar activity and the development of strong anticyclones in high latitudes ($\pm$ 70°). Enhanced cyclonic vorticity is shown in latitudes 40° to 60°. These effects are demonstrated for both hemispheres during local winter.

This paper briefly summarizes the changes in surface weather conditions which occur along with this enhancement of pressure systems. A latitudinally differentiated CO_2/latent heat transfer/cloud/sea surface temperatures feedback system is described; and a potential synergism between these as yet poorly understood phenomena is linked with the strongly anomalous winter conditions of 1976-77 and 1977-78.

1. SOLAR ACTIVITY INDUCED CHANGES IN THE TROPOSPHERIC CIRCULATION

The work of Schuurmans [1], Roberts [2], Wilcox [3], King [4], Svalgaard [5], and others has demonstrated that pronounced changes in the height of atmospheric pressure surfaces during local winter appear to be associated with solar events such as the passage of the solar magnetic sector boundary, the central meridian passage of solar active regions and large plages, and the occurrence of solar flares. These effects are strongest during periods of intense solar activity. The greatest changes in pressure surfaces occur in preferred longitudinal zones; increased cyclonic vorticity over the eastern North Pacific and over the North Atlantic; enhanced anticyclogenesis over northeastern Siberia, the American midwest, etc.

This enhancement of pressure differentials in latitudes 40° to 70° increases the degree of meridional energy transport accomplished by the atmospheric circulation; an increase in thunderstorm activity and in the incidence of blocking anticyclones is also seen. King, referring to a 90 decimeter change in the height of pressure surfaces over the North Artlantic, has characterized this as a "powering up" of the atmospheric system.

B.M. McCormac and T.A. Seliga (Eds.): Solar-Terrestrial Influences on Weather and Climate. 323-327.

A cascading series of surface effects follows from the initial change in the height of pressure surfaces; these include:

a. Increased frontal turbulence and enhanced wind velocities in middle latitudes; this results in an increase in surface wind and wave activity, which in turn:

b. Increases the rate of latent heat transfer from high latitudes ocean surfaces in areas of enhanced cyclonic vorticity, resulting in:

c. Enhanced cloud formation, which may:

d. Increase total precipitation (depending upon a number of other variables); a portion of this may modify surface albedo in sensitive areas, with a consequent negative impact on surface air temperatures in temperature latitudes.

Changes in the amplitude and frequency of solar disturbance may thus influence the degree of winter storminess in temperate zones; for example, unusual solar disturbance in late Fall could result in early snowfall, by aiding the development of early winter storms. The same activity in March could extend winter conditions well into Spring.

2. LATENT HEAT TRANSFER AND ANTROPOGENIC CARBON DIOXIDE

Latent heat transfer accounts for a substantial percentage (∿20%) of the total poleward energy transport of the atmospheric system. The global evaporation maxima are centered over subtropical ocean surfaces at about 25° N and S latitude; vapor is transported both equatorward and poleward from these maxima. The maxima correspond to the zones of anticyclonic subsidence of the Hadley circulation, in which clear skies permit maximum insolation of the ocean surface.

Carbon dioxide is a trace gas of atmosphere which helps maintain warm surface temperatures by capturing and re-radiating IR radiation from the planetary surface. The concentration of atmospheric CO_2 has been increased by ∿13% over the past 100 yr due to the combustion of fossil fuels and harvesting of forests, with the largest part of this increase coming within the past 25 yr.

Most of the researchers engaged in modeling the atmospheric system postulate an increase in globally averaged surface temperatures due to increased CO_2, with the strongest increase coming in high latitudes of the northern hemisphere. This trend has not materialized; apparently other factors (increased aerosols, perhaps stronger solar activity) are exerting a stronger influence on the system.

The behaviors of water within the atmospheric system are not adequately treated by any of the various models. Möller (1963) [6] pointed out that changes in the radiation budget due to a 10% increase in CO_2 may be completely compensated for by small increases in water vapor content (3%) and cloudiness (1%), without any variation in surface temperatures. If the atmospheric system in fact responds to the increased CO_2 IR retention by "powering up" the hydrologic cycle, then an enhancement of the greenhouse effect <u>and</u> and increase in atmospheric albedo are indicated. If these opposing feedbacks cancel exactly, then globally averaged temperatures may remain unchanged.

This simple model obscures a critical real world time lag between cloud formation and negative feedback to surface temperatures at the subtropical evaporation maxima.

As noted above, water vapor is exported from the evaporation maxima, moving poleward into the temperate zone (where the largest concentration of anthropogenic aerosols is found), and equatorward; subsequent cloud formation in these regions does not reduce insolation or evaporation at the evaporation maxima. The negative feedback to these zones comes in the form of reduced ocean surface temperatures, a consequence of the reduction in insolation of the ocean surface at higher latitudes. This feedback is subject to a time lag introduced by the circulation parameters and large thermal inertia (energy storage) of the oceans; the mixing of cooled waters from the North Pacific, for example, with warm equatorial waters via currents and eddies results in a delayed reduction in sea surface temperatures in the Kuroshiwa region which is <u>not</u> commensurate with the initial increase in total evaporation. (Note: The ocean currents flow in response to a number of conditions which are, in energetic terms, of far larger magnitude than the simple insolation of the ocean surface; the effect postulated above is thought to operate within the context provided by these larger energy flows.)

Thus it appears that continually increasing concentrations of atmospheric CO_2 may result in a small increase in atmospheric albedo without generating commensurate limiting feedback, since the system is in effect drawing upon the vast energy storage of the oceans to dampen and reduce the negative feedback. This finding suggests that the long term consequence of increasing atmospheric CO_2 may in fact be a reduction in surface temperatures in the temperate zone; especially in the case of an atmosphere carrying a burden of $\sim 10^{12}$ kg of anthropogenic aerosols as well.

Observed trends of the past 20 to 30 yr, which include:

a. A strong reduction in high latitudes ocean surface temperatures;

b. A less pronounced drop in average surface tmperatures in middle latitudes; and

c. An increase in atmospheric humidity in some regions of the temperate zones

are consistent with the operation of the system as described above.

However, the operation of this mechanism is not in itself adequate to account for the observed trends. A number of other factors are present (increased aerosols, stronger solar disturbance) which may have a more significant effect in determining the overall trend. We can only state that, in the context of the atmospheric system as a whole, increased atmospheric CO_2 may be enhancing meriodional latent heat transfer, with a consequent increase in atmospheric albedo, by drawing upon energy stored within the oceans.

Unfortunately, it is apparent that this CO_2-induced modification of planetary scale energy flows probably cannot be adequately measured, modeled, or resolved at the present time.

3. THE POTENTIAL SYNERGISM

The most recent sunspot minimum (1976) was one of unusual solar disturbance. The present solar cycle (#21) has already shown an abrupt increase in the number of sunspots, and may be one of the most active in this century, according to Sargent [7], Kane [8], Gribbin [9], and others. As noted above, strong solar activity, by increasing pressure differentials and enhancing vorticity in latitudes 40° to 70°, may result in enhanced winter storminess in the temperature zones.

We are entering this period with a strongly increased level of atmospheric CO_2, even relative to the late 1950's (the most recent period of strong solar activity). This increase is, in the historic sense, probably highly anomalous.

It is the hypothesis of this author that:

a. The "unverifiable" model of CO_2-enhanced latent heat transfer outlined above is an accurate representation of system energy flows on a planetary scale;

b. That the anomalous winter conditions of the past two years are a "normal" result of this complex of conditions; and that

c. If the perturbations of Earth's atmospheric system due to the unusual events of this solar cycle continue to incrase, that we may expect a succession of adverse winters as the present solar cycle peaks and declines, each contributing to create a strong cooling trend for the middle latitudes of the noerthern hemisphere over the near term future.

REFERENCES

1. C. J. E. Schuurmans, this volume, 1979.
2, W. O. Roberts, this volume, 1979.
3. J. M. Wilcox, this volume, 1979.
4. J. W. King, this volume, 1979.

5. L. Svalgaard, in D. E. Page (ed.), Correlated Interplanetary and Magnetospheric Observations, D. Reidel Pub. Co., Boston, MA, 1974.
6. F. Möller, J. Geophys. Res. 68 13, 1963.
7. H. H. Sargent, Paper Am. Geophys. Union, San Francisco, CA, 1977.
8. R. P. Kane, Nature, 274, 139, 13, 1978.
9. J. R. Gribbin and S. Plagemann, The Jupiter Effect, Vintage Books, Books 76,1974.

THE POTENTIAL OF THE CHATANIKA RADAR IN INVESTIGATING POSSIBLE SOLAR/ ATMOSPHERE COUPLING

Norman J. F. Chang

SRI International
233 Ravenswood Ave.,
Menlo Park, CA 94025, U.S.A.

ABSTRACT. The Chatanika radar, located near Fairbanks, Alaska, has been used for incoherent-scatter studies of the ionosphere since 1971. In this paper we describe some of the results obtained from ionospheric measurements. Recent tropospheric and stratospheric measurements and experiments to determine possible atmospheric and ionospheric interaction are also described.

1. INTRODUCTION

The Chatanika radar is a coherent radar with the following typical parameters: operating frequency of 1290 MHz with peak power around 3 to 4 MW; parabolic dish antenna 27 m in diameter with a 0.6° beamwidth, fully steerable with hemispheric coverage; system noise temperature of 110°K. For more details see Leadabrand *et al.* [1].

Because of the radar's high-latitude location, ionospheric electric field measurements made when it rotates beneath the aurora oval generally can be considered measurements of the magnetospheric electric fields. These measurements are important because the magnetospheric electric fields redistribute and structure F-region ionization, drive the electrojet current system, and are responsible for Joule heating of both the ionosphere and the neutral atmosphere.

During the past 2 yr the operational range of the radar has been extended into the lower atmosphere as an outgrowth of the work done by Balsley and Farley [2] that established the usefulness of the Chatanika radar for tropospheric/stratospheric wind measurements. Since then the technique has been refined and further developed. The radar currently can obtain profiles of wind speed and direction from about 4 to 25 cm with a range resolution of 750 m and a time resolution of about 6 min.

With the Chatanika radar currently developed to the point where significant measurements can be made in the troposphere/stratosphere, ionosphere, and magnetosphere, it appears that coordinated measurements made

B.M. McCormac and T.A. Seliga (Eds.): Solar-Terrestrial Influences on Weather and Climate. 329-334.

over the whole radar height range can contribute significantly to studies of solar/atmosphere coupling. In the next section we give an example of an energy deposition event due to Joule heating, to illustrate the kinds of energy-input measurements possible, and present examples of wind measurements and measurements of the turbulence structure constant, C_n^2, made in the troposphere/stratosphere. We conclude by suggesting measurements that should prove valuable in studies of tropospheric/stratospheric coupling and of solar/atmosphere interaction.

2. CAPABILITY OF THE CHATANIKA RADAR

2.1 Ionospheric/Magnetospheric Measurements

During a large precipitation event the ambient electron densities around altitudes of 90 to 100 km are greatly enhanced through ionizing collisions between the atmospheric constituents and the incoming auroral electrons. Profiles of electron density obtained from measurements of the backscatter signal strength allow determination of the ionization rate, the precipitating electron energy distribution, and the mean energy input [3].

The amount of energy deposited by a Joule heating can be found from the frequency spectrum and strength of the backscatter returns. The first moment of the spectrum allows derivation of the electric field perpendicular to the magnetic field and of the neutral wind vector. These parameters, combined with the E-region conductivities derived from the electron density profile, allow the energy input to be found (Baron and Chang [4]; Wickwar *et al.* [5]).

Particle precipitation and Joule heating are important because the heat generated by them influences the dynamics of the thermosphere and has a far-reaching effect through the winds and waves generated to redistribute the incident energy. Because these winds and waves can perturb and modulate the mid-latitude thermosphere and perhaps even the underlying atmosphere, Joule heating and particle precipitation events are good candidates for case studies of cause-and-effect relationships between the upper and lower atmosphere.

Figure 1 shows an example of a Joule heating event. Associated with this event were a marked rise in the F-region ion temperature and a 150°K increase in the exospheric temperature (not shown). This example illustrates the amount and temporal variation of energy available; the maximum heat input from solar radiation above 90 km at lower latitude is less than 5 erg $cm^{-2}s^{-1}$. After integrating the energy input for the duration of this event, we find that roughly 2 or 3 x 10^5 erg cm^{-2} were deposited in the ionosphere above the radar.

Precipitation events have energy inputs comparable to the Joule heating example shown in Figure 1 and typically last an hour or two. In terms of energy deposition 2 or 3 x 10^5 erg cm^{-2} appears to be representative.

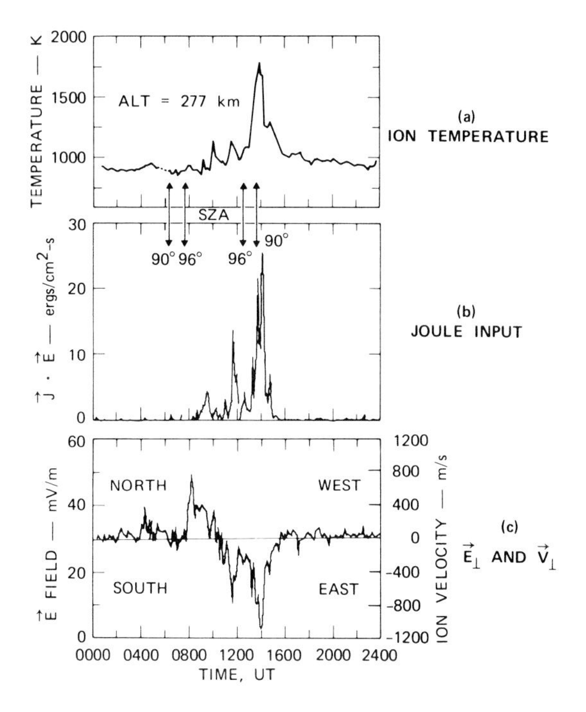

Figure 1. F-region ion temperature, height-integrated Joule input, and north-south electric field at Chatanika on 13 August 1975. There is a major heating event centered at 1400 UT shortly after local sunrise. The maximum energy input rate was approximately 25 erg $cm^{-2}s^{-1}$. SZA is the solar zenith angle [6].

2.2 Tropospheric/Stratospheric Measurements

Below about 60 km, the number of free electrons decreases to a point where incoherent scatter from them is no longer detectable. At about 30 to 35 km, however, sensitive radars with large power-aperture products [7] can obtain backscatter signals from clear air. Unlike incoherent scatter, which is due to scatter from density fluctuations caused by the thermal motion of the free electrons, these backscatter signals are due to scatter from refractive index variations caused by turbulence. Since the refractive index irregularities have been shown to be advected by the ambient wind, the spectra of the backscatter signals provide a means of extracting information on the wind field as well as on the degree of turbulence [2, 8, 9].

Two different modes have been used for wind measurements. In one mode, the antenna is kept at a constant elevation and rotated at a constant rate through 360°. The display obtained at a particular altitude when the mean Doppler estimate (the first moment of the backscatter spectrum) is plotted as a function of azimuth results in a velocity-azimuth display (VAD). Assuming that the winds are uniform over the time (12 min) and space (circle 19 km in diameter at 9.4 km altitude) probed by the radar, the Doppler estimates should exhibit a sinusoidal variation with azimuth. On this basis, a sine wave can be fitted to the data points using a least-square criterion. The amplitude, phase, and bias of the sine wave indicate the speed, direction, and vertical motion, respectively, of the wind field.

An example of a wind profile generated by an azimuth-scan is shown in Figure 2a. In this Figure, the radar-derived speeds and directions are compared with rawinsonde measurements made from Poker Flat, located about 3 km north of the radar. The excellent agreement between the two measurements leaves little doubt of the radar as a wind measuring tool.

A measurement not possible with rawinsonde, but important for studies of tropospheric/stratospheric coupling, is that of vertical motions. An example of vertical velocities derived from VADs is shown in Figure 2b. A more direct measurement of vertical motions can be made with the radar pointed vertically. Measurements of this type made during April 4, 1978 show peak velocities in agreement with those of Figure 2b and also show wavelike oscillations in the velocities at 4 km with periods of the order of a few minutes. Vertical velocities of this magnitude are very significant and would have a major effect on energy exchange between the various levels of the atmosphere.

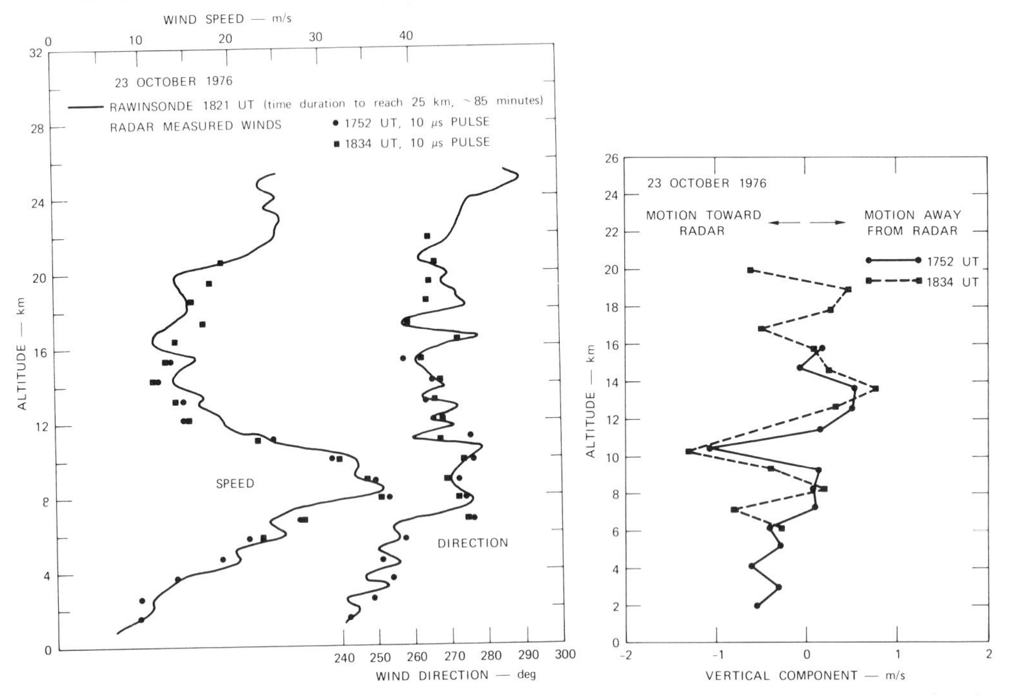

Figure 2. (a) Comparison of horizontal wind speed and direction derived from radar measurements (points) and rawinsonde (solid lines); and (b) Vertical motions obtained from VAD's for October 23, 1976.

A second mode used recently to obtain wind profiles is a "position" mode, in which the antenna is moved to fixed positions in a predetermined sequence. Since the velocity vector is determined with three independent measurements, this mode allows wind profiles to be constructed in 6 min or less. In addition to having better time resolution than in the azimuth-scan mode, the position mode is well suited to extended operation. Recent measurements in April 1978 used the position mode to continuously monitor the wind field over a 48 h period. Thus, the radar can monitor the wind field with a temporal resolution and duration not possible with other techniques.

The backscatter signal power is directly related to C_n^2, the refractivity structure constant. This parameter provides information on atmospheric structure in two ways: (1) the temporal and spatial variations of C_n^2 trace the gross structure, evolution, and motion of turbulence in the atmosphere, and (2) the variability of C_n^2 provides quantitative information of the fine-scale variability of temperature and water vapor and on the small-scale (comparable to the radar wavelength) velocity field.

Although the position mode would probably be preferred over the VAD mode in extended measurements of the winds aloft, the latter VAD mode is better suited for a detailed study of the wind field and the turbulence structure constant. For example: (1) the departure of the data points from a sine wave provide information on the kinematic properties of the wind field (horizontal, stretching, and shearing deformation), and (2) variations of the received power as a function of azimuth provide information on the spatial scale and spatial variation of C_n^2. Thus, choice of a particular mode depends on the application.

III. CONCLUSION

We have shown that the Chatanika radar can measure a number of parameters important to the dynamics and/or energy balance of the ionosphere and the troposphere/stratosphere. Since these measurements can be made on a temporal and spatial scale not possible with other sensors, the radar can be an important tool in studies of solar/weather relationships.

An experimental probem to contribute to our understanding of the coupling and interaction among the various regions of the atmosphere might consist of:

(1) Regular measurement of the wind vector (especially vertical motions) and of the turbulence structure constant between 4 to 25 km.

(2) Coordinated measurement of the ionosphere and lower atmosphere during specific events, e.g., active aurora periods, solar flare, polar cap absorption, stratospheric warming.

The first type of measurement should provide information on stratospheric/tropospheric coupling and general circulation in the atmosphere. As a data base is accumulated, it may be possible to relate the occurrence of specific solar-related events (e.g., sector boundary passage, solar flare) to variations in wind speed, wind direction, or degree of turbulence. In addition, the data can reveal the presence of vertically or horizontally propagating waves that may, in turn, be linked with a specific source.

The second type of measurement will allow a study to be made of possible cause-and-effect relationships between a well-defined energy deposition event in the ionosphere and the lower atmosphere.

These suggestions illustrate the types of measurements that are possible. More detailed experiments can be defined when specific theories or coupling mechanisms are proposed.

ACKNOWLEDGMENTS

The work reported here has been supported by the National Science Foundation under Grant ATM77-16213 and by the Internal Research and Development Program at SRI International.

REFERENCES

1. R. L. Leadabrand, M. J. Baron, J. Petriceks, and H. F. Bates, Radio Sci., 7, 1972.
2. B. B. Balsley and D. T. Farley, Geophys. Res. Letts., 3, 525, 1977.
3. R. R. Vondrak and M. J. Baron, Radio Sci., 11, 939, 1976.
4. M. J. Baron and N. J. Chang, "Icecap '73a--Chatanika radar results,' Tech. Rep. 4, SRI Proj. 3118, Stanford Research Institute, Menlo Park, CA, 1975.
5. V. B. Wickwar, M. J. Baron, and R. D. Sears, J. Geophys. Res., 80, 4634, 1975.
6. J. D. Kelly, Private communications, 1978.
7. B. B. Balsley and J. L. Green, "Coherent radar systems for probing the troposphere, stratosphere, and mesosphere," in Preprint Vol. Fourth Symposium on Meteorological Observations and Instrumentation, April 10-14, 1978, Denver, CO, 1978.
8. B. B. Balsley, N. Cianos, D. T. Farley, and M. J. Baron, J. of Applied Meteor., 16, 1235, 1977.
9. H. Ottersten, "Radar observations of the turbulent structure in shear zones in the clear atmosphere," in Preprint of 14th Radar Meteorology Conf., Tucson, AZ, Amer. Meteorol. Soc., Boston, MA, pp. 111-116, 1970.

LIST OF PARTICIPANTS

Louis K. Acheson, Jr.
Hughes Aircraft Company
17721 Marcello Place
Encino, CA 91316

Susan Avery
Dept. of Electrical Engineering
University of Illinois
Urbana, IL 61801

Fred L. Bartman
Dept. of Atmospheric & Oceanic Sci.
University of Michigan
Ann Arbor, MI 48109

John P. Basart
Iowa State University
Coover Hall
Ames, IA 50011

Michael A. Bilello
U.S. Army Cold Regions Research
and Engineering Lab
Snow and Ice Branch
Hanover, NH 03755

Joseph H. Binsack
Laboratory for Space Experiments
Center for Space Research
Room N51-330
Massachusetts Institute of
Technology
Cambridge, MA 02139

Edward G. Bowen
5010 Maxwell Avenue
West River, MD 20881

Dale D. Branch
National Weather Service
129 S.E. Philip
Des Moines, IA 50315

Glenn W. Brier
Dept. of Atmospheric Science
Colorado State University
Fort Collins, CO 80523

Sreedevi K. Bringi
Atmospheric Sciences Program
The Ohio State University
469/2015 Neil Avenue
Columbus, OH 43210

Madeleine Briskin
Department of Geology
University of Cincinnati
Old Tech Building
Cincinnati, OH 45221

Charles M. Brown
Naval Research Laboratory
Code 7146
Washington, D.C. 20375

Jerry A. Carlson
Professional Farmers of America
219 Parkade
Cedar Falls, IA 50613

Charles P. Cato
Atmospheric Sciences Program
The Ohio State University
469/2015 Neil Avenue
Columbus, OH 43210

Norman J. F. Chang
Radio Physics Lab
SRI International
333 Ravenswood Avenue
Menlo Park, CA 94025

Robert S. Davis
DeNardo & McFarland Weather
Services, Inc.
Allegheny County Airport
West Mifflin, PA 15122

Henry Diaz
National Climatic Center
Federal Building
Asheville, NC 28805

John A. Eddy
National Center for Atmospheric
Research
High Altitude Observatory
Box 3000
Boulder, CO 80307

Hugh W. Ellsaesser
Lawrence Livermore Laboratory
P.B. Box 808
Livermore, CA 94550

Michael R. Foster
Department of Aeronautical and
Astronautical Engineering
The Ohio State University
2036 Neil Avenue Mall
Columbus, OH 43210

Sigmund Fritz
Meteorology Program
Space Sciences Building
University of Maryland
College Park, MD 20742

Stuart Gathman
Naval Research Laboratory
Code 8327
Washington, D.C. 20375

Richard A. Goldberg
Laboratory for Atmospheric Sciences
NASA/Goddard Space Flight Center
Greenbelt, MD 20771

Pat Haines
University of Dayton Research Inst.
300 College Park
Dayton, OH 45469

Richard D. Hake, Jr.
SRI International
333 Ravenswood Avenue
Menlo Park, CA 94025

Itamar Halevy
Department of Physics
University of Toronto
Toronto, Ontario
Canada M5S1A1

Gerald P. Hannes
Department of Geography
California State University/
Fullerton
800 N. State College Blvd.
Fullerton, CA 92634

Karen Harrower
Department of Geology
University of Kansas
Lawrence, KS 66045

Richard E. Hartle
Planetary Aeronomy Branch,
Code 621
NASA/Goddard Space Flight Center
Greenbelt, MD 20771

Alan D. Hecht
Climate Dynamics Program
National Science Foundation
Washington, D.C. 20550

Randall J. Heydinger
2309 North High St.
Columbus, OH 43210

Robert Hill
Box 5484
Montecito, CA 93108

Edward W. Hones, Jr.
University of California
Los Alamos Scientific Laboratory
Los Alamos, NM 87545

Frederick B. House
Department of Physics &
Atmospheric Sci.
Drexel University
Philadelphia, PA 19104

Douglas Hoyt
National Oceanic & Atmospheric Admin.
Environmental Research Lab
R329
Boulder, CO 80303

J. Hughes
Office of Naval Research
Dept. of the Navy
Arlington, VA 22217

R. J. Hung
University of Alabama in Huntsville
P. O. Box 1247
Huntsville, AL 35807

James C. Jafolla
Dept. of Physics & Atmospheric Sci.
Drexel University
Philadelphia, PA 19104

Benton Jamison
State University of New York/Albany
E5114-1400 Washington Avenue
Albany, NY 12222

Gracen Joiner
Atmospheric & Ionospheric Sciences Prog.
Office of Naval Research-Code 465
Room 315, 800 N. Quincy St.
Arlington, VA 22217

Ramesh K. Kakar
NASA Headquarters
Code ST-5
Washington, D.C. 20546

John Keller
University of Dayton Research Inst.
300 College Park
Dayton, OH 45469

Philip M. Kelly
Climatic Research Unit
School of Environmental Sciences
University of EAst Anglia
Norwich, NR4 8TJ, Norfolk
England

Earl C. Kindle
Geophysical Sciences
Old Dominion University
Norfolk, VA 23508

J. W. King
Appleton Laboratory
Ditton Park
Slough SL39JX, Berkshire
United Kingdom

R. A. Koehler
York University
4700 Keele St., Petrie Bldg.
Downsview, Ontario
Canada M3J 1P3

Herbert W. Kroehl
National Oceanic & Atmospheric Admin.
Environmental Data Service - D64
Boulder, CO 80302

Andre Lachapelle
Atmospheric Environment Service
CCAI, 4905 Dufferin St.
Downsview, Ontario
Canada M3H 5T4

Miguel F. Larsen
Cornell University
120 Phillips Hall
Ithaca, NY 14853

James Lax
Atmospheric Sciences Program
The Ohio State University
469/2015 Neil Avenue
Columbus, OH 43210

Mae D. Lethbridge
Pennsylvania State University
506 Walker Building
University Park, PA 16802

Curt A. Levis
Dept. of Electrical Engineering
The Ohio State University
1320 Kinnear Road
Columbus, OH 43212

W. C. Livingston
Kitt Peak National Observatory
P. O. Box 26732
Tucson, AZ 85726

Charles Lundquist
NASA/Marshall Space Flight Center
Huntsville, AL 35812

Jeffrey T. Lutz
Office of the Geographer
INR/RGE, Room 8742
U.S. Department of State
Washington, D.C. 20520

John W. MacArthur
Marlboro College
Box 15
Marlboro, VT 05344

Carol G. Maclennan
Bell Telephone Laboratories
Room 1E-436
Murray Hill, NJ 07974

K. Kurian Mani
Georgia Institute of Technology
School of Aerospace Engineering
Atlanta, GA 30332

Ralph Markson
Dept. of Aeronautics & Astronautics
Massachusetts Institute of Tech.
Building W-91
Cambridge, MA 02139

Nelson C. Maynard
NASA/Goddard Space Flight Center
Laboratory for Planetary Atmospheres
Code 625
Greenbelt, MD 20771

Billy M. McCormac
Lockheed Palo Alto Research Laboratory
Dept. 52-10, Bldg. 202
3251 Hanover St.
Palo Alto, CA 94304

David G. McFarland
DeNardo & McFarland Weather
Services, Inc.
Allegheny County Airport
West Mifflin, PA 15122

Michael Mendillo
Astronomy Department
Boston University
Boston, MA 02215

John H. Mercer
Institute of Polar Studies
The Ohio State University
125 South Oval
Columbus, OH 43210

J. Murray Mitchell, Jr.
NOAA Environmental Data Service
Room 625, Grammax Building
8060 13th St.
Silver Spring, MD 20910

Walter E. Mitchell, Jr.
Department of Astronomy
The Ohio State University
174 W. 18th Avenue
Columbus, OH 43210

James Monahan
Dept. of Aeronautical & Astro-
nautical Engineering
The Ohio State University
2036 Neil Avenue
Columbus, OH 43210

Hans T. Mörth
School of Environmental Sciences
University of East Anglia
Norwich, NR4 7TJ
United Kingdom

M. S. Muir
Department of Physics
University of Natal
King George V Avenue
Durban, Natal 4001
South Africa

Sam I. Nakagama
Kidder, Peabody & Co, Inc.
10 Hanover Square
New York, NY 10005

Jerome Namias
Scripps Inst. of Oceanography
University of CAlifornia/San Diego
La Jolla, CA 92093

Gregory D. Nastrom
Control Data Corporation
P. O. Box 1249
Minneapolis, MN 55440

James J. Nasuti
Department of Geography
University of Texas at Austin
Austin, TX 78712

James C. Neill
Illinois State Water Survey
P. O. Box 232
Urbana, IL 61801

Marcel Nicolet
External Geophysics
Brussels University
30, Avenue Den Doorn
1180 Brussels, Belgium

Roger Olson
Aspen Institute for Humanistic Studies
1919 Fourteenth St., #811
Boulder, CO 80302

Bruce C. Parker
Department of Biology
Virginia Polytechnic Institute
and State University
Blacksburg, VA 24061

Barrie Pittock
Laboratory of Tree-Ring Research
University of Arizona
Tucson, AZ 85721

Sheo S. Prasad
Jet Propulsion Laboratory
4800 Oak Grove Drive
Pasadena, CA 91103

John T. Prohaska
Solar Climatic Research Institute
8 Mechanic Street
Marblehead, MA 01945

Boyd E. Quate
Boyd E. Quatc and Associates
P. O. Box 7065
Holland Station
Suffolk, VA 23437

S. Ramakrishna
Dept. of Aeronautical Engineering
Indian Institute of Science
Banglore 560012, India

Reinhold Reiter
Inst. for Atmospheric Environmental Research
Kreuzeckbahnstrasse 19
D-8100 Garmisch-Partenkirchen
West Germany

Walter Orr Roberts
Aspen Institute for Humanistic Studies
1919 Fourteenth St., #811
Boulder, CO 80302

Raymond G. Roble
National Center for Atmospheric Research
P. O. Box 3000
Boulder, CO 80307

Guillermo Rodriguez Rodriguez
Licenciado en Ciencias Fisicas
Geofisica y Astrofisica
Las Tricias, Isla de la Palma
Canary Islands, Spain

Juan G. Roederer
Geophysical Institute
University of Alaska
Fairbanks, AK 99701

Donald Rote
Energy-Environment Sys. Division
Argonne National Laboratory
EES-12, 9700 Cass Avenue
Argonne, IL 60439

Ann W. Rudolph
Battelle Columbus Laboratories
505 King Avenue
Columbus, OH 43201

Bill C. Ryan
United States Dept. of Agric.
Forest Service
2291 Quartz Place
Riverside, CA 92507

H. H. Sargent, III
Space Environment Resources Cntr.
Boulder, CO 80303

Philip H. Scherrer
Institute for Plasma Research
Stanford University
Via Crespi
Stanford, CA 94305

Michael Schlesinger
Dept. of Atmospheric Science and
Climate Research Inst.
Oregon State University
Corvallis, OR 97331

C.J.E. Schuurmans
Royal Netherlands Meteorological
Inst.
Wilhelminalaan 10, Postbus 201,
3730 AE De Bilt
The Netherlands

Robert Seals
NASA Headquarters
Code ST-5
400 Maryland Avenue
Washington, D.C. 20546

C. F. Sechrist, Jr.
Aeronomy Laboratory
Electrical Engineering Dept.
University of Illinois
Urbana, IL 61801

Thomas A. Seliga
Atmospheric Sciences Program
The Ohio State University
469/2015 Neil Avenue
Columbus, OH 43210

Glenn Shaw
Geophysical Institute
University of Alaska
Fairbanks, AK 99701

Neil R. Sheeley, Jr.
Naval Research Laboratory
Code-7172
Washington, D.C. 20375

James H. Shirley
Life Support Systems Group, Ltd.
2432 N.W. Johnson St.
Portland, OR 97210

Jag J. Singh
NASA/Langley Research Center
Mail Stop 235
Hampton, VA 23665

Rama N. Singh
Institute of Technology
Baranas Hindu University
Varanasi 220005
India

H. Prescott Sleeper
Kentron International
2003 Byrd Springs Road
Huntsville, AL 35802

Alphonsa Smith
NASA/Langley Research Center
Mail Stop 234
Hampton, VA 23665

Jesse B. Smith, Jr.
NOAA/Marshall Space Flt. Center
ES52
Huntsville, AL 35812

Nelson W. Spencer
NASA/Goddard Space Flt. Center
Code 620
Greenbelt, MD 20771

William F. Spires
Northern Illinois University
221 Eldon Avenue
Columbus, OH 43204

Richard S. Stolarski
NASA/Goddard Space Flt Center
Mail Code 624
Greenbelt, MD 20771

Thomas Sullivan
Lawrence Livermore Laboratory
L-262
Livermore, CA 94550

H. Suzuka
Canada Climatic Center
Atmospheric Environment Service
4905 Dufferin
Downsview, Ontario M3W 5T4
Canada

J. Terasmae
Department of Geological Sciences
Brock University
St. Catharines, Ontario L2S 3A1
Canada

Ellen Thompson
Institute of Polar Studies
The Ohio State University
125 S. Oval Mall
Columbus, OH 43210

Lonnie G. Thompson
Institute of Polar Studies
The Ohio State University
125 S. Oval Mall
Columbus, OH 43210

Jack Villmow
Geography
Northern Illinois University
DeKalb, IL 60115

Hans Volland
Radioastronomical Institute
University of Bonn
Auf dem Huegel 71
5300 Bonn
West Germany

Garvin L. Von Eschen
Department of AEronautical and Astronautical Engineering
The Ohio State University
2036 Neil Avenue
Columbus, OH 43210

Paul Waite
Iowa Department of Agriculture
Room 10, Municipal Airport Term.
Des Moines, IA 50321

M. Walt
Lockheed Palo Alto Research Lab
3251 Hanover Street
Palo Alto, CA 94304

Eric K. Walton
Electroscience Laboratory
The Ohio State University
1320 Kinnear Road
Columbus, OH 43210

Andrew J. Weinheimer
Space Physics Department
Rice University
Houston, TX 77001

John M. Wilcox
Institute for Plasma Research
Via Crespi
Stanford University
Stanford, CA 94305

Stanley Woronko
Atmospheric Environment Service
4905 Dufferin Street
Downsview, Ontario M3H 5T4
Canada

S. T. Wu
University of Alabama
P. O. Box 1247
Huntsville, AL 35807

Douglas N. Yarger
Iowa State University
313 Curtiss Hall
Ames, IA 50011

Edward J. Zeller
Department of Geology
University of Kansas
Lawrence, KS 66045

PRESS:

Robert M. Boyce
OSU COMMUNICATION SERVICES

Robert C. Cowen
CHRISTIAN SCIENCE MONITOR

Susan Kitterman
OSU LANTERN

Joe McKnight
ASSOCIATED PRESS

Susan West
SCIENCE NEWS

INDEX